Steam Turbines

ABOUT THE AUTHORS

Heinz P. Bloch (West Des Moines, Iowa) is a consulting engineer. Before retiring from Exxon in 1986 after over two decades of service, Mr. Bloch's professional career included long-term assignments as Exxon Chemical's Regional Machinery Specialist for the United States. He has also held machinery-oriented staff and line positions with Exxon affiliates in the United States, Italy, Spain, England, The Netherlands, and Japan. He has conducted over 500 public and in-plant courses in the United States and at international locations.

Dr. Murari P. Singh (Bethlehem, Pennsylvania) is Consulting Engineer/Probabilistic Lifing Leader of GE Oil & Gas for all products in the Chief Engineers' Office. Dr. Singh has been involved in the design, development, and analysis of industrial turbomachinery for more than 30 years with Turbodyne Corporation, Dresser Industries, Dresser-Rand Company, and, most recently, Safe Technical Solutions Inc., where he served as Director of Engineering Technology. Dr. Singh has extensive knowledge and experience with fatigue and fracture mechanics, stress and vibration of structures, reliability, life analysis, and probabilistic analysis. His practical application experience is with a variety of rotating equipment including warm gas and FCC expanders, steam turbines, and centrifugal compressors. He developed the widely used SAFE diagram for reliability evaluation of turbine blades. Dr. Singh has authored more than 35 technical papers on topics relating to turbomachinery.

Steam Turbines

Design, Applications, and Rerating

Heinz P. Bloch
Murari P. Singh

Second Edition

McGraw Hill

New York Chicago San Francisco Lisbon London
Madrid Mexico City Milan New Delhi San Juan
Seoul Singapore Sydney Toronto

The McGraw-Hill Companies

Cataloging-in-Publication Data is on file with the Library of Congress.

McGraw-Hill books are available at special quantity discounts to use as premiums and sales promotions, or for use in corporate training programs. To contact a special sales representative, please visit the Contact Us page at www.mhprofessional.com.

Steam Turbines, Second Edition

The first edition was published as *A Practical Guide to Steam Turbine Technology* (McGraw-Hill, 1996).

1 2 3 4 5 6 7 8 9 0 DOC/DOC 0 1 4 3 2 1 0 9 8

ISBN 978-0-07-150821-6
MHID 0-07-150821-X

Sponsoring Editor
Stephen S. Chapman

Acquisitions Coordinator
Alexis Richard

Editorial Supervisor
David E. Fogarty

Project Manager
Virginia Carroll

Copy Editor
Virginia Carroll

Proofreaders
Roberta Burkert, Linda Enterline, Carol Heisey, Sue Miller, Stewart Smith

Indexer
Murari P. Singh

Production Supervisor
Pamela A. Pelton

Composition
North Market Street Graphics

Art Director, Cover
Jeff Weeks

To my father. He would be pleased.

H.P. Bloch

To my parents. They would be pleased.

M.P. Singh

Contents

Preface xiii
Acknowledgments xvii

Chapter 1. Introduction **1**

1.1 Why Mechanical Drive Steam Turbines Are Applied 1
1.2 Overview of Steam Turbine Fundamentals 2
1.2.1 Steam turbine staging can vary 5
1.2.2 Modern impulse design 5
1.2.3 Single-valve vs. multivalve construction 5
1.2.4 Steam balance considerations 9
1.3 Overview of Steam Turbine Types and Controls 9
1.3.1 Straight noncondensing 14
1.3.2 Automatic extraction noncondensing 15
1.3.3 Automatic extraction condensing 15
1.3.4 Basic steam control considerations 18
1.3.5 Automatic extraction condensing controls 21
1.3.6 Geared and direct-drive types 21
1.3.7 Modular design concepts 23

Chapter 2. Turbine Casing and Major Stationary Components **29**

2.1 Casing Design 29
2.2 Steam Admission Sections 33
2.3 Steam Turbine Diaphragms and Labyrinth Packing 36

Chapter 3. Bearings for Mechanical Drive Turbines **51**

3.1 Journal Bearings for Industrial Turbomachinery 51
3.1.1 Fixed-geometry journal bearing stability 52
3.1.2 Tilting-pad journal bearings 56
3.1.3 Advanced tilting-pad journal bearings 61
3.1.4 Lubrication-starved tilting-pad bearings 65
3.2 Key Design Parameters 68
3.3 Thrust Bearings for Turbomachinery 69
3.4 Active Magnetic Bearings 75

Chapter 4. Rotors for Impulse Turbines 81

4.1 Long-Term Operating Experience 81
4.2 Pitch Diameter and Speed 82
4.3 Steam Temperature 83
4.4 Built-Up Construction 84
4.5 Solid Construction 89
4.6 Shaft Ends 90
4.7 Turbine Rotor Balance Methods 91
4.7.1 At-speed rotor balancing 92
4.8 Balance Tolerance 94

Chapter 5. Rotors for Reaction Turbines 95

5.1 Solid Rotors 95
5.2 Materials for Solid Rotors 99
5.3 Welded Rotor Design 100
5.4 Welded Rotor Materials 105

Chapter 6. Turbine Blade Design Overview 109

6.1 Blade Materials 111
6.2 Blade Root Attachments 111
6.3 Types of Airfoils and Blading Capabilities 113
6.4 Guide Blades for Reaction Turbines 114
6.5 Low-Pressure Final Stage Blading 120

Chapter 7. Turbine Auxiliaries 125

7.1 Lube Systems 125
7.2 Barring or Turning Gears 128
7.3 Trip-Throttle or Main Stop Valves 129
7.4 Overspeed Trip Devices 132
7.5 Gland Seal Systems 135
7.6 Lube Oil Purifiers 135

Chapter 8. Governors and Control Systems 137

8.1 General 137
8.2 Governor System Terminology 140
8.2.1 Speed regulation 140
8.2.2 Speed variation 141
8.2.3 Dead band 141
8.2.4 Stability 141
8.2.5 Speed rise 141
8.3 NEMA Classifications 143
8.4 Valves 144
8.4.1 Single-valve turbines 144
8.4.2 Multivalve turbines 145
8.5 PG Governors 145
8.6 Electronic Governors 148
8.7 Governor Systems 150
8.7.1 General 150
8.7.2 Extraction control 150

Chapter 9. Couplings and Coupling Considerations 157

9.1 Power Transmission 157
9.2 Shaft Alignment 160
9.3 Maintenance 162
9.4 Influence on the Critical Speeds 162
9.5 Differential Expansions 162
9.6 Axial Thrusts 163
9.7 Limits of Application 163

Chapter 10. Rotor Dynamics Technology 165

10.1 Rotor Model 165
10.2 Dynamic Stiffness 166
10.3 Effects of Damping on Critical Speed Prediction 169
10.4 Bearing-Related Developments 170
10.5 Refinements 172
10.6 Bearing Support Considerations 173
10.7 Foundations 174
10.8 Impedance 174
10.9 Partial Arc Forces 178
10.10 Design Procedure 179
10.11 Rotor Response 180
10.12 Instability Mechanisms 180
10.13 Subsynchronous Vibration 180
10.14 Service Examples 183
10.15 Labyrinth and Cover Seal Forces 185
10.16 Rotor Stability Criteria 187
10.17 Experimental Verification 187

Chapter 11. Campbell, Goodman, and SAFE Diagrams for Steam Turbine Blades 189

11.1 Goodman Diagram 189
11.2 Goodman-Soderberg Diagram 190
11.3 Campbell Diagram 191
11.3.1 Exciting frequencies 195
11.4 SAFE Diagram—Evaluation Tool for Packeted Bladed Disk Assembly 197
11.4.1 Definition of resonance 198
11.4.2 Mode shape 198
11.4.3 Fluctuating forces 200
11.5 SAFE Diagram for Bladed Disk Assembly 203
11.6 Mode Shapes of a Packeted Bladed Disk 209
11.7 Interference Diagram Beyond *N*/2 Limit 211
11.8 Explaining Published Data by the Use of Dresser-Rand's SAFE Diagram 214
11.9 Summary 217

Chapter 12. Reaction vs. Impulse Type Steam Turbines 219

12.1 Introduction 219
12.2 Impulse and Reaction Turbines Compared 220

12.3 Efficiency 220
12.4 Design 223
12.4.1 Rotor 223
12.4.2 Blading 224
12.5 Erosion 230
12.6 Axial Thrust 232
12.7 Maintenance 233
12.8 Design Features of Modern Reaction Turbines 233
12.9 Deposit Formation and Turbine Water Washing 235

Chapter 13. Transmission Elements for High-Speed Turbomachinery 243

13.1 Spur Gear Units 243
13.2 Epicyclic Gears 245
13.3 Clutches 246
13.4 Hydroviscous Drives 253
13.5 Hydrodynamic Converters and Geared Variable-Speed Turbo Couplings 257
13.5.1 Function of the multistage variable-speed drive 261
13.5.2 Design and operating details 261
13.5.3 Working oil and lube oil circuits 264
13.5.4 Lubricating system 264
13.5.5 Lubricant oil containment on gear and variable-speed units 265

Chapter 14. Shortcut Graphical Methods of Turbine Selection 267

14.1 Mollier Chart Instructions 267
14.2 Estimating Steam Rates 271
14.3 Quick Reference Information to Estimate Steam Rates of Multivalve, Multistage Steam Turbines 303

Chapter 15. Elliott Shortcut Selection Method for Multivalve, Multistage Steam Turbines 309

15.1 Approximate Steam Rates 309
15.2 Stage Performance Determination 313
15.3 Extraction Turbine Performance 320

Chapter 16. Rerates, Upgrades, and Modifications 329

16.1 Performance and Efficiency Upgrade 331
16.1.1 Brush seals and labyrinth seals 332
16.1.2 Wavy face dry seals 336
16.1.3 Buckets 348
16.2 Reliability Upgrade 352
16.2.1 Electronic controls 352
16.2.2 Monitoring systems 356
16.3 Life Extension 356
16.4 Modification and Reapplication 358
16.4.1 Casing 359
16.4.2 Flange sizing 360
16.4.3 Nozzle ring capacity 362
16.4.4 Steam path analysis 362

16.4.5 Rotor blade loading 363
16.4.6 Thrust bearing loading 363
16.4.7 Governor valve capacity 364
16.4.8 Rotor 364
16.4.9 Shaft end reliability assessment 364
16.4.10 Speed range changes 366
16.4.11 Auxiliary equipment review 366
16.4.12 Oil mist lubrication for general-purpose steam turbines 367
16.4.13 Problem solving 376
16.5 Summary 376

Appendix A. Glossary 377
Appendix B. Units of Measurement 385
Bibliography and List of Contributors 399

Index 407

Preface

In order to efficiently and reliably drive compressors and other fluid movers, virtually every industry depends on steam turbine drivers. The various types of fluid movers often require variable input speeds, and steam turbines are capable of providing these without too much difficulty.

Situations may arise using applications during which a process plant needs large quantities of heat. The modern mechanical drive steam turbine proves capable of adding to plant efficiency by allowing the motive steam to first expand through a series of blades and then be used in the process of heating elsewhere in the plant, or as utility steam for heating buildings on-site or in the community.

The economy and feasibility of these and a multitude of related applications depend on the reliability of steam turbines. There is also a strong dependency on the capability of the selected models and geometries of steam turbines to handle a given steam condition at the desired throughput or output capacity.

Similar considerations will prompt the engineer to survey the field of available drivers for process and/or utility duty. We note that in most large, complex petrochemical plants, particularly plants where steam is either generated or consumed by the process, mechanical drive steam turbines have been the prime mover of choice. These large variable-speed units are a critical component in continuous-flow chemical processes and, in most cases, are placed in service without backup capability. This kind of application demands the highest reliability and availability performance. These two requirements form the cornerstone of the development programs under way at the design and manufacturing facilities of the world's leading equipment producers.

More than ever before, petrochemical and other industries are facing intense global competition, which in turn has created a need for lower-cost equipment. Global competition has also created a demand for the modification of existing steam turbines to gain efficiency, that is, an increase in power output per ton of steam consumed (although purists will find this to be the layman's definition). Making this equipment without compromising quality, efficiency, and reliability is not easy, and only the industrial world's best manufacturers measure up to the task. It is equally important that a contemplative, informed, and discerning equipment purchaser or equipment user can be expected to spot the right combination of two desirable and seemingly contradictory requirements: low cost and high quality.

The starting point of machinery selection is machinery know-how. From know-how, one can progress to type selection, such as condensing versus extraction versus backpressure turbine, or reaction steam turbine versus impulse steam turbine. Type selection in turn leads to component selection, such as fixed land thrust bearing versus tilting-pad thrust bearing. These could be exceedingly important considerations, since both type selection and component selection have a lasting impact on the maintainability, serviceability, availability, and reliability of steam turbines. As a result of these considerations, the ultimate effect will be plant profitability or possibly even plant survival.

This second edition text is intended to provide the kind of guidance that will enable the reader to make intelligent choices. We have added Chapter 16 on the upgrading of steam turbines, completely revised the chapter on bearings, and added new information on bearing protector seals, brush seals, oil mist lubrication, and wavy face mechanical seals that promise to replace carbon ring seals in small steam turbines. While the text cannot claim to be all-encompassing and complete in every detail, it was the coauthors' hope to make the material both readable and relevant. We believe we have succeeded in making the text up-to-date, with practical, field-proven component configuration and the execution of mechanical drive steam turbines discussed at length. The emphasis was to be on the technology of the principal machine, but we did not want to overlook auxiliaries such as fixed-ratio gear transmissions, variable-speed transmissions, overrunning clutches, and couplings. With experience showing that machinery downtime events are often linked to malfunction of the support equipment, we decided to include governors, lubrication and sealing systems, overspeed trip devices, and other relevant auxiliaries. All of these are thoroughly cross-referenced in the index and should be helpful to a wide spectrum of readers.

While compiling this information from commercially available industry source materials, we were again impressed by the profusion of diligent effort that some well-focused companies expended to design and manufacture more efficient and more reliable turbomachinery. With much of this source material dispersed among the various sales, marketing, design, and manufacturing groups, we set out to collect the data and organize it into a text that acquaints the reader with the topic by using overview and summary materials. The information progresses through more detailed and more design-oriented write-ups and on to scoping studies and application and selection examples. Some of these are shown in both English and metric units; others were left in the method chosen by the original contributor.

The reader will note that we stayed away from an excessively mathematical treatment of the subject at hand. Instead, the focus was clearly on giving a single-source reference on a wide range of material that will be needed by the widest possible spectrum of machinery users. These users range from plant operators to mechanical technical support technicians, reliability engineers, mechanical and chemical engineers, operations superintendents, project managers, and even senior plant administrators.

Finally, the publishers and coauthors wish to point out that this book would never have been written without the full cooperation of a large number of highly competent steam turbine manufacturers in the United States and overseas. It was compiled by obtaining permission to use the direct contributions of companies and individuals listed in the figure sources and bibliography. These contributions were then structured into a cohesive write-up of what the reader should know about mechanical drive steam turbine technology as of 2008 and beyond. The real credit should therefore go to the various contributors and not the coauthors, who, in some instances, acted only as compiling editors. In line with this thought, we would be most pleased if the entire effort would serve to acquaint the reader with not only the topic, but also the names of the outstanding individuals and companies whose contributions made it all possible.

We wish to give special thanks to Seema, Reshma, and Masuma Singh for much-valued assistance. Their proofreading efforts and resulting suggestions greatly improved this text.

Heinz P. Bloch
Murari P. Singh

Acknowledgments

Special thanks are extended to

ABB Power Generation, Inc., and Asea Brown-Boveri, North Brunswick, N.J.—Mr. Sep van der Linden

Advanced Turbomachine, LLC, Wellsville, N.Y.—Steve Rashid

Dresser-Rand Steam Turbine, Motor and Generator Division, Wellsville, N.Y.—Messrs. R. J. Palmer, B. M. Oakleaf, and M. Singh

Elliott Company, Jeannette, Pa.—Messrs. Ross A. Hackel and D. Mansfield

General Electric Company, Industrial and Power Systems Division, Fitchburg, Mass.—Messrs. Donald R. Leger, Richard K. Smith, and Raymond B. Williams

Gulf Publishing Company, Houston, Tex.—Mr. Robert W. Scott

IMO Industries, De Laval Steam Turbine Division, Trenton, N.J.—Mr. Roy J. Salisbury

Lufkin Gear Company, Lufkin, Tex.—Mr. James R. Partridge

Mechanical Technology Inc., Latham, N.Y.—Dr. James F. Dill

Murray Turbomachinery Corporation, Burlington, Iowa—Mr. Douglas G. Martin

Philadelphia Gear Corporation, Philadelphia, Pa.—Mr. Robert J. Cox

RMT, Wellsville, N.Y.—John C. Nicholas

Salamone Turbo Engineering, Inc., Houston, Tex.—Mr. Dana J. Salamone

Saudi Aramco Oil Company, Dhahran, KSA—Mr. Abdulrahman Al-Khowaiter and Mr. Nayef Al-Otaibi

Siemens Power Corporation, Milwaukee, Wis.—Mr. Gary M. Cook

Sulzer Brothers, Inc., New York, N.Y.—Mr. Bernhard Haberthuer

Voith Transmissions, York, Pa.—Mr. David Pell

Woodward Governor Company, Loveland and Fort Collins, Col.—Dr. Ron Platz

Chapter

1

Introduction

1.1 Why Mechanical Drive Steam Turbines Are Applied

Dependability and versatility of equipment are vital to today's process plants, to pharmaceutical producers, mining interests, and a host of other users including, of course, petroleum, petrochemical, and chemical-process industries. Operating pressures and temperatures are constantly rising; single-train capacities grow by leaps and bounds; continuity of service becomes the vital force; and economics demands longer and longer periods between overhauls.

Steam turbines are faithful partners to the process industries. They have proved their basic reliability and today are showing a *new* versatility by keeping pace with every demand for higher capacity, speed, and reliability.

Wherever you look in the process industries, there are more mechanical drive turbines; wherever you look, both horsepower and speed go up, year after year. And wherever you look, technology advances are being incorporated into modern steam turbines. Many manufacturers deserve to be recognized for their ability to solve the tougher steam turbine application problems. Through advanced planning, imaginative research, persistent development, and painstaking evaluation, engineers have in the last quarter of this century created a whole new turbine generation: machines of sizes and speeds that were only dreamed of a few decades ago. Multiflow exhausts, solid rotors, high-speed bearings, taller last-stage blades ("buckets"), cam-operated valve gear and controls, and other highly sophisticated control systems and computerized designs are a few of the innovations that helped make this progress.

Knowledgeable manufacturers have available a wide selection of steam end designs, either single or multivalve, to meet any specific pressure and temperature conditions. In the overwhelming majority of cases both industrial and cogeneration systems designed for electric power generation use a simple, single-casing steam turbine. These turbines (Fig. 1.1) can be designed to provide the operating flexibility to economically utilize steam from a variety of sources to supply:

- Direct or geared power input for compressors, pumps, or other driven equipment
- Steam at the pressures and quantities required for integrated processes or lower pressure turbines
- The electric power desired
- Cogenerated power for sale to the local utility

1.2 Overview of Steam Turbine Fundamentals

Before discussing turbine selection, let's review how a steam turbine converts the heat energy of steam into useful work. The nozzles and

Figure 1.1 Straight noncondensing steam turbine (14,700 hp), pedestal-mounted, with electronic valve position feedback. *(General Electric Company, Fitchburg, Mass.)*

diaphragms in a turbine are designed to direct the steam flow into well-formed, high-speed jets as the steam expands from inlet to exhaust pressure. These jets strike moving rows of blades mounted on the rotor. The blades convert the kinetic energy of the steam into rotation energy of the shaft.

There are two principal turbine types: reaction and impulse (Fig. 1.2). In a reaction turbine, the steam expands in both the stationary *and* moving blades. The moving blades are designed to utilize the steam jet energy of the stationary blades and to act as nozzles themselves. Because they are moving nozzles, a reaction force—produced by the pressure drop across them—supplements the steam jet force of the stationary blades. These combined forces cause rotation.

To operate efficiently the reaction turbine must be designed to minimize leakage around the moving blades. This is done by making most internal clearances relatively small. The reaction turbine also usually requires a balance piston (similar to those used in large centrifugal compressors) because of the large thrust loads generated.

Because of these considerations, the reaction turbine is seldom used for mechanical drive in the United States, despite its occasionally higher initial efficiency. Reaction turbines are, nevertheless, in widespread use in Europe and the rest of the world. They deserve to be discussed and will be dealt with later.

The impulse turbine has little or no pressure drop across its moving blades. Steam energy is transferred to the rotor entirely by the steam jets striking the moving blades (see Fig. 1.3).

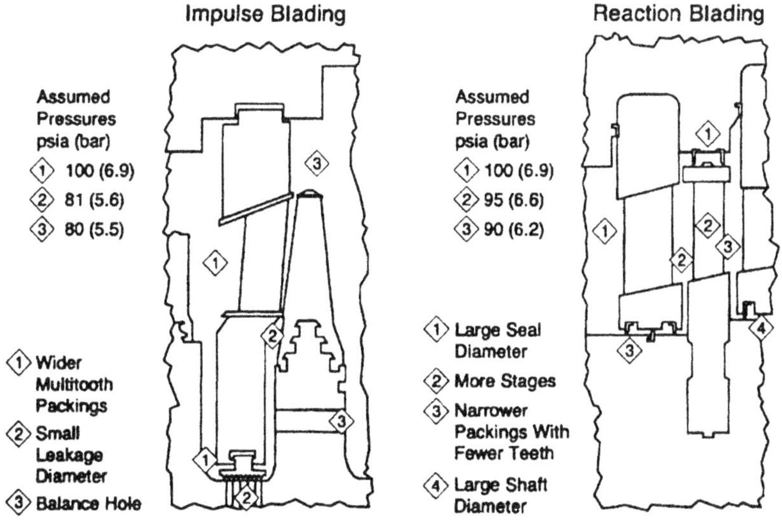

Figure 1.2 Impulse and reaction blade features. *(General Electric Company, Fitchburg, Mass.)*

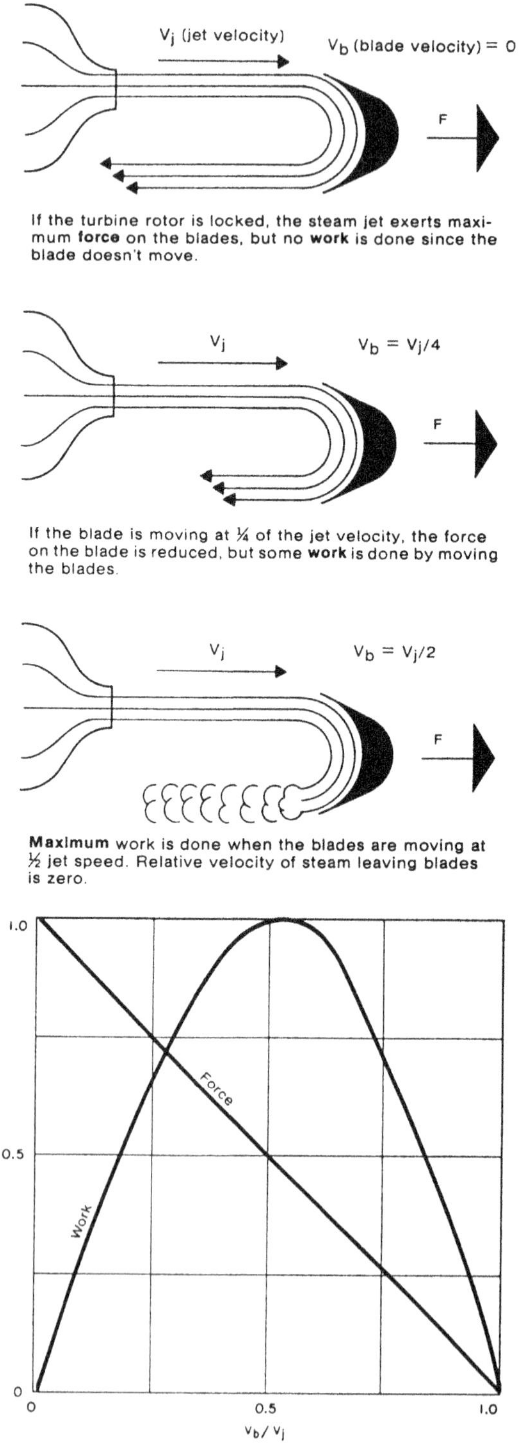

Vj (jet velocity)
Vb (blade velocity) = 0
F
If the turbine rotor is locked, the steam jet exerts maximum force on the blades, but no work is done since the blade doesn't move.
Vj
Vb = Vj/4
F
If the blade is moving at ¼ of the jet velocity, the force on the blade is reduced, but some work is done by moving the blades.
Vj
Vb = Vj/2
F
Maximum work is done when the blades are moving at ½ jet speed. Relative velocity of steam leaving blades is zero.
1.0
0.5
0
Force
Work
0
0.5
1.0
Vb/Vj

Figure 1.3 The impulse principle. *(The Elliott Company, Jeannette, Pa.)*

Since there is theoretically no pressure drop across the moving blades (and thus no reaction), internal clearances are large, and no balance piston is needed. These features make the impulse turbine a rugged and durable machine that can withstand the heavy-duty service of today's mechanical drive applications.

1.2.1 Steam turbine staging can vary

First, let's consider velocity-compounded (Curtis) staging. A Curtis stage consists of two rows of moving blades. Stationary nozzles direct the steam against the first row; reversing blades (not nozzles) then redirect it to the second row.

The large pressure drop through the nozzle produces a high-speed steam jet. This high velocity is absorbed in a series of constant pressure steps (see Fig. 1.4). The two rotating rows of blades make effective use of the high-speed jet, resulting in small wheel diameters and tip speeds, fewer stages, and a shorter, more rugged turbine for a given rating.

In pressure-compounded (Rateau) staging, the heat energy of the steam is converted into work by stationary nozzles (diaphragms) directing the steam against a *single* row of moving blades. As in a Curtis stage, pressure drops occur almost entirely across the stationary nozzles.

1.2.2 Modern impulse design

The importance of steam turbine efficiency has continued to increase over the last decade. Today, there is no pure impulse turbine. Manufacturers are using a combination of reaction and impulse design features to further improve turbine efficiency. The traditional impulse turbine manufacturers, who utilize the basic wheel-and-diaphragm construction, have been able to meet, and many times exceed, the performance of a pure reaction turbine. This is done on high-pressure stages by adding a small amount of reaction to improve the performance, without the need for tight leakage controls or increasing thrust forces. Tall, low-pressure buckets are designed with more reaction than ever before using advanced aerodynamic codes for these complex blade forms. The generous clearances of the wheel-and-diaphragm construction decrease the dependence on tight leakage control. Field data have shown that these modern impulse turbines will sustain their high level of performance over time and are much more tolerant to fouling, which can have a significant impact on thrust loads.

1.2.3 Single-valve vs. multivalve construction

Single-valve units (Fig. 1.5) are available when justified by plant economics. When used, individual nozzle ring segments are controlled by hand-operated shutoff valves. Hand valves (arrow, Fig. 1.6) may be specified for reduced steam consumption at part load or overload, or for design load with reduced steam pressures. Hand valves are not automatic and are only of value when manually operated as needed.

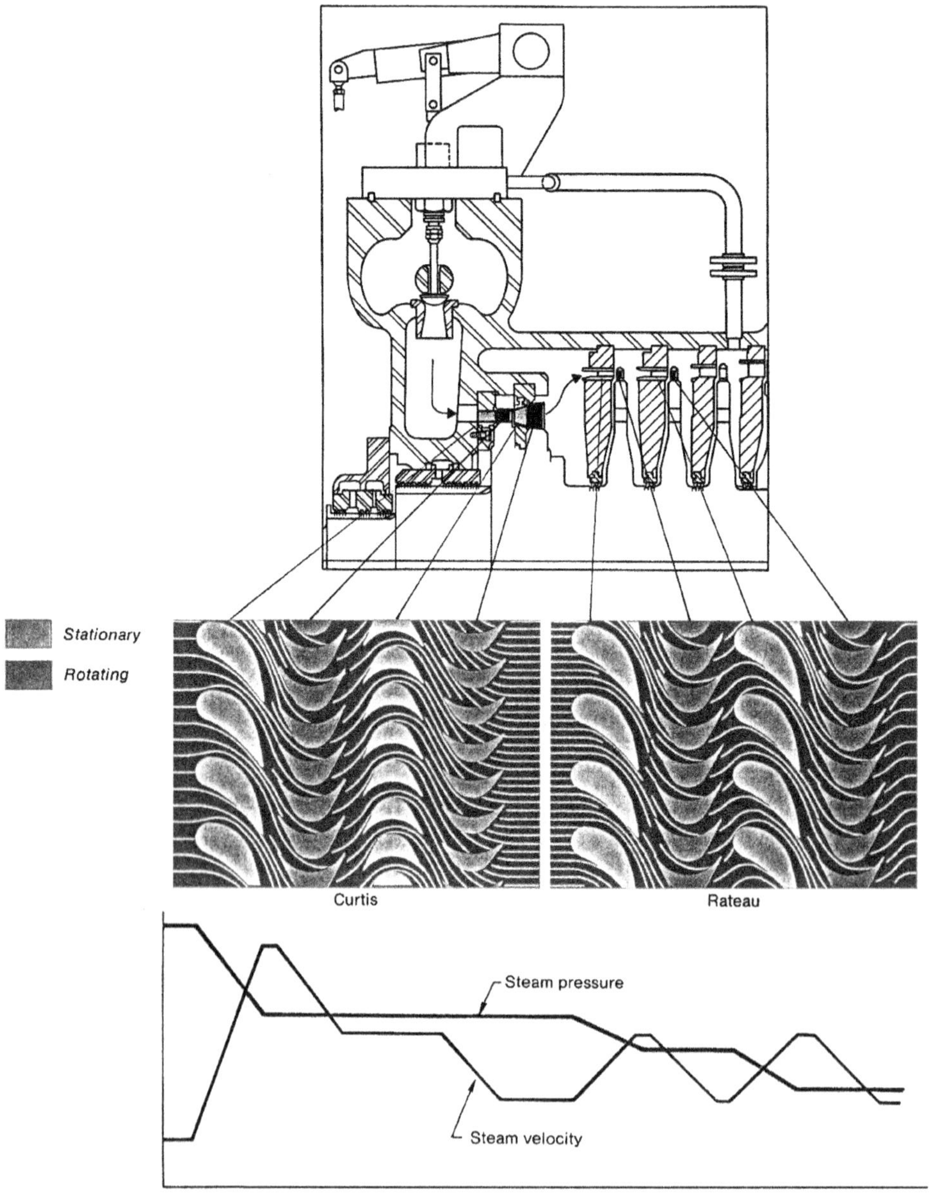

Figure 1.4 Steam flow through turbine stages. *(The Elliott Company, Jeannette, Pa.)*

Figure 1.5 Single-valve steam turbine. *(The Elliott Company, Jeannette, Pa.)*

Multivalve turbines (Fig. 1.7) automatically limit pressure drop across the governing valves, thereby minimizing throttling loss.

The prime benefit of a multivalve turbine is the fact that the nozzles forming a short arc are fed by a single valve, which will allow a better velocity ratio than would result if all available nozzles were fed with the same amount of steam. Valve gear designs will sequence valve opening so that subsequent valves will only open when the previous valve is wide open. Multivalve turbines are the wise choice if frequent load changes or varying outputs are anticipated or when inlet volume

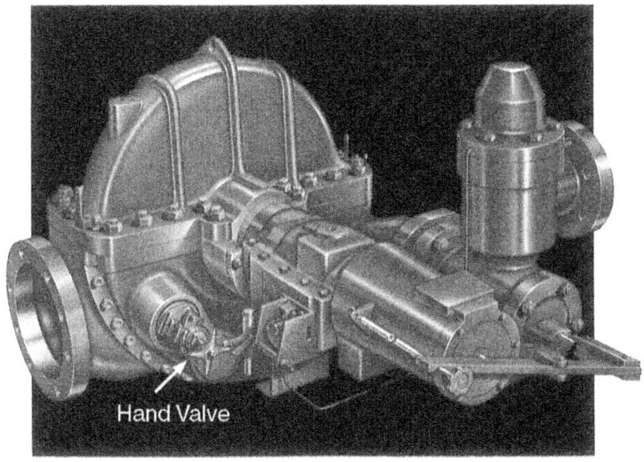

Figure 1.6 Single-valve steam turbine with hand valve indicated by arrow. *(Dresser-Rand Company, Wellsville, N.Y.)*

flows will be high. The multivalve arrangement usually improves efficiency over the full operating range of a steam turbine.

Single-stage turbines are available in six classes of construction. Class 1 (cast iron) is suitable for pressures not exceeding 250 psig (17.2 bar) and for temperatures not exceeding 500°F (260°C). If either one of these limits is exceeded, steel construction is required.

Classes 2 and 3 (carbon steel) incorporate construction features suitable for a maximum pressure of 700 psig (48.3 bar). Temperature limit for Class 2 is 650°F, 750°F for Class 3 (343 and 399°C, respectively).

For pressures exceeding 700 psig (48.3 bar), the casting is formed from a different pattern and otherwise utilizes construction features suitable up to a maximum pressure of 900 psig (62 bar). Class 4, 5, or 6 is required, depending on temperature. Class 4 (carbon steel) is suitable to a maximum temperature of 750°F (399°C). Alloy steels are required for temperatures exceeding 750°F, or 399°C. Class 5 (carbon-moly steel) can be used to 825°F (440°C), Class 6 (chrome-moly steel) to 900°F (482°C).

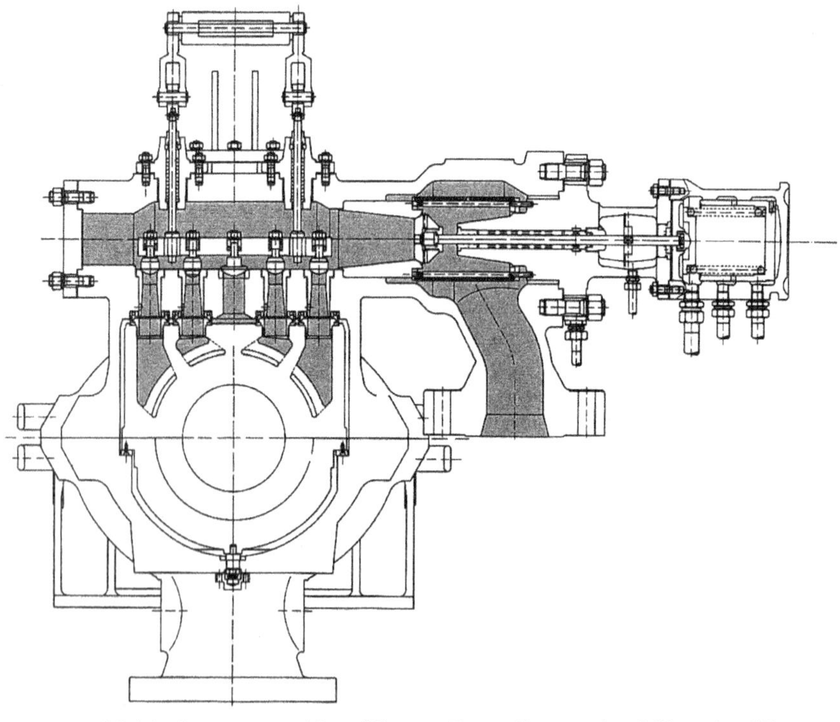

Figure 1.7 Multivalve steam turbine. *(Siemens Power Corporation, Milwaukee, Wis. and Erlangen, Germany)*

Note that these material classes do not define the situation in which the operating pressure is 700 psig (48.3 bar) or less, with an operating temperature exceeding 750°F (399°C). For this combination of operating limits, Class 3 construction, *with the appropriate material,* is utilized. In other words, 700 psig (48.3 bar) construction is utilized with the parts cast in the appropriate steel alloy (carbon-moly steel to 825°F, chrome-moly steel to 900°F (440 and 482°C, respectively).

1.2.4 Steam balance considerations

The steam balance of a process plant can be quite complicated because of the multiple steam pressure levels often required.

Selecting a turbine to complement a particular steam balance is made easier, however, by the wide variety of turbines available. Condensing, back-pressure or extraction/induction turbines can be used, as required, in designing both new plants and additions to existing plants.

Steam for process use, for example, can be supplied from the exhaust of a back-pressure turbine or from an extraction turbine. The choice would depend on the number of pressure levels involved, the design of the remainder of the plant, number of turbines required, etc. This versatility simplifies the job of optimizing your steam balance.

The steam balance diagrams in Figs. 1.8 through 1.11 illustrate how various types of turbines have been used to supply both shaft power and steam for other uses.

1.3 Overview of Steam Turbine Types and Controls

Figures 1.12*a* through 1.12*h* illustrate the types of turbines most frequently used in industrial and cogeneration applications. Figures 1.12*a* through 1.12*d* show noncondensing designs that exhaust to a header from which the steam is used for process or supply to a lower pressure turbine. Figures 1.12*e* through 1.12*h* represent condensing units that exhaust at the lowest pressure obtainable using water or air as a heat sink.

Figures 1.12*a* and 1.12*e* illustrate straight noncondensing and straight condensing turbines, simple types in which no flow is removed from the turbine between its inlet and exhaust.

Figures 1.12*b* and 1.12*f* show the next simplest variations, in which steam is made available for process from an uncontrolled, or nonautomatic, extraction. The extraction pressure is proportional to the flow passing beyond the extraction through the unit to its exhaust and is thus related to the inlet steam flow and the extraction itself. Variations may include two or more such uncontrolled extractions.

TWO 1500 PSIG/900°F (103.4 bar/482°C) turbines drive the large compressors in this application. Two different extraction pressures were used (400 and 255 PSI/27.6 and 17.6 bar, gage) with lower-pressure steam being supplied to a process and to the smaller turbine. Exhaust steam from all three turbines is then condensed at 4" HGA (135 mbar).

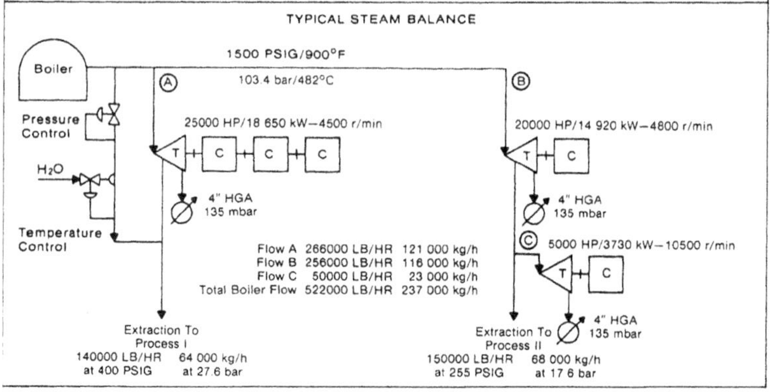

THREE TURBINES all use steam from the 900 PSIG/850° F (62 bar/455°C) boiler. Steam is extracted from both larger units at 410 PSIG (28.3 bar) for process I, and the remainder is condensed at 4" HGA (135 mbar). Smaller back-pressure turbine exhausts steam at 190 PSIG (13.1 bar) for use in process II.

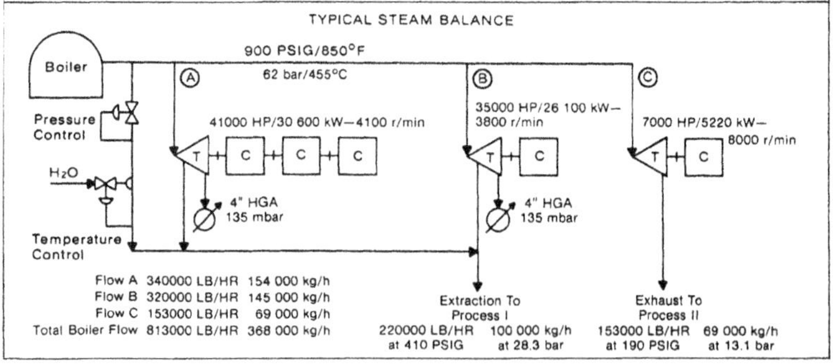

TOPPING TURBINE concept was used in this plant, where exhaust from high-pressure turbine is used to supply process demands as well as the low-pressure turbines. Exhaust steam from these low-pressure units is then condensed at 3.5" HGA (120 mbar).

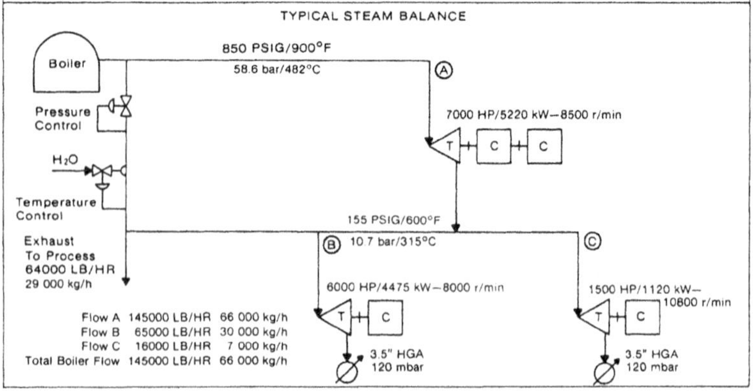

Figure 1.8 Steam balance representations incorporating mechanical drive steam turbines. *(The Elliott Company, Jeannette, Pa.)*

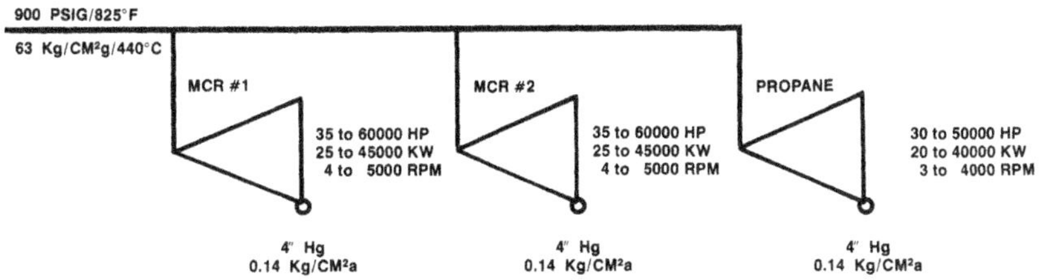

Figure 1.9 Typical heat balance for liquefied natural gas plants. *(General Electric Company, Fitchburg, Mass.)*

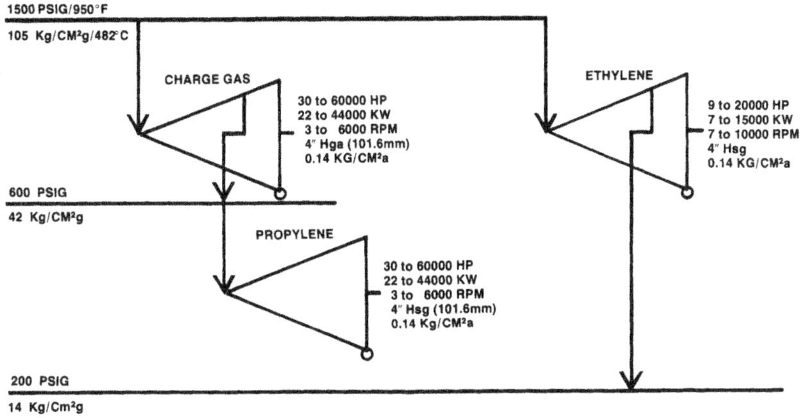

Figure 1.10 Typical heat balance for ethylene plants. *(General Electric Company, Fitchburg, Mass.)*

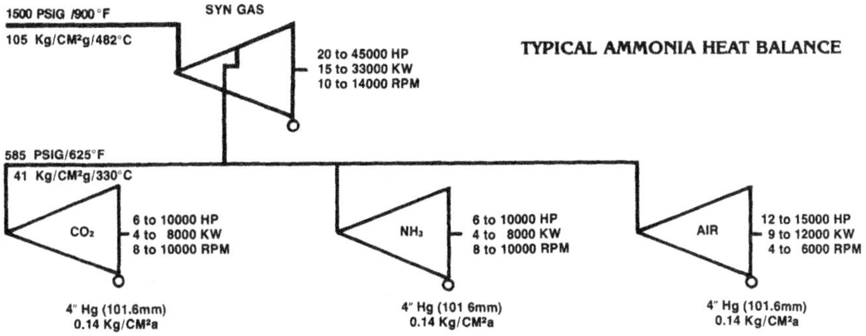

Figure 1.11 Typical heat balance for ammonia plants. *(General Electric Company, Fitchburg, Mass.)*

Figures 1.12*c* and 1.12*g* illustrate automatic extraction units providing process steam at one controlled pressure. The extraction control valves regulate the flow to the exhaust section of the turbine. Should an increase in process demand cause the extraction pressure to fall below the set value, the valve closes, reducing exhaust section flow, raising the extraction pressure, and diverting additional flow to the extraction.

Figures 1.12*d* and 1.12*h* show double automatic extraction units in which a second set of internal extraction control valves allows controlled extraction at two pressures. Although not shown here, triple automatic extraction is a further variation.

Figure 1.13 is a schematic map indicating where these various turbine types are applied. The pressure ranges selected for the boilers and the process lines are chosen for illustration. Topping turbines, high-pressure (HP) boiler to medium-pressure (MP) boiler, are used where

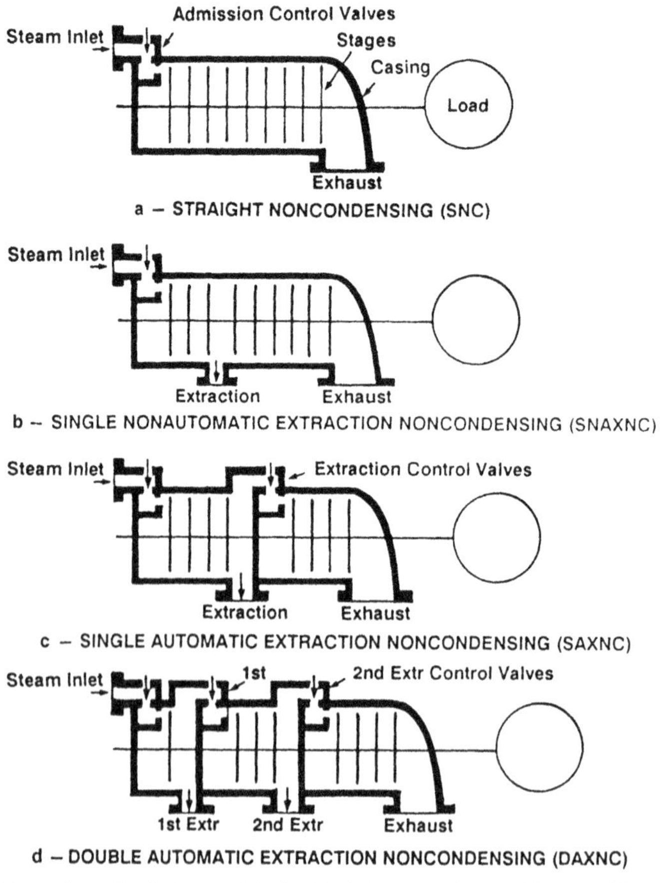

Figure 1.12 Basic types of industrial steam turbines. *(General Electric Company, Fitchburg, Mass.)*

an existing boiler is replaced with one at higher pressure to gain additional generation and improved efficiency. A bottoming turbine, from a process pressure to the condenser, is used to recover energy when process heat needs are reduced either permanently or seasonally.

The representative units of Figs. 1.12*a* through 1.12*h* are shown taking steam from the MP boiler, exhausting to process in the noncondensing cases, and to the condenser in the condensing designs. Any of these types can be designed for use with HP boilers where needed. Automatic extraction units can be designed to accept steam from a process line (admission or mixed pressure) when the steam available from other sources exceeds the process needs.

Four of the turbine types described earlier are illustrated as cross sections in Fig. 1.14.

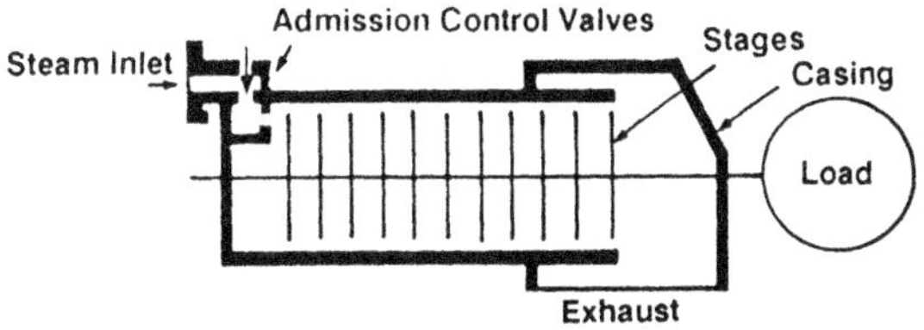

e – STRAIGHT CONDENSING (SC)

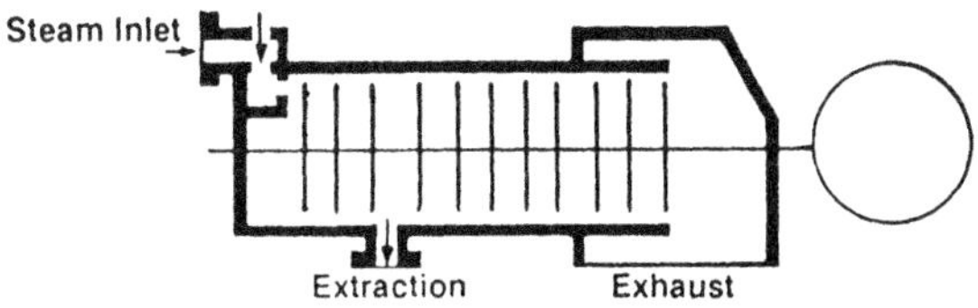

f – SINGLE NONAUTOMATIC EXTRACTION CONDENSING (SNAXC)

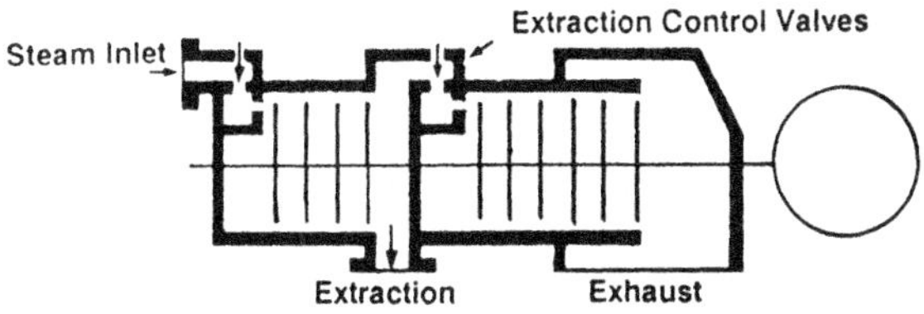

g – SINGLE AUTOMATIC EXTRACTION CONDENSING (SAXC)

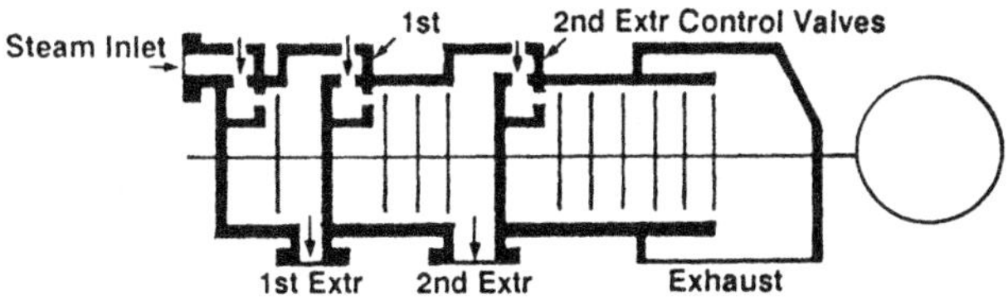

h – DOUBLE AUTOMATIC EXTRACTION CONDENSING (DAXC)

Figure 1.12 *(Continued)*

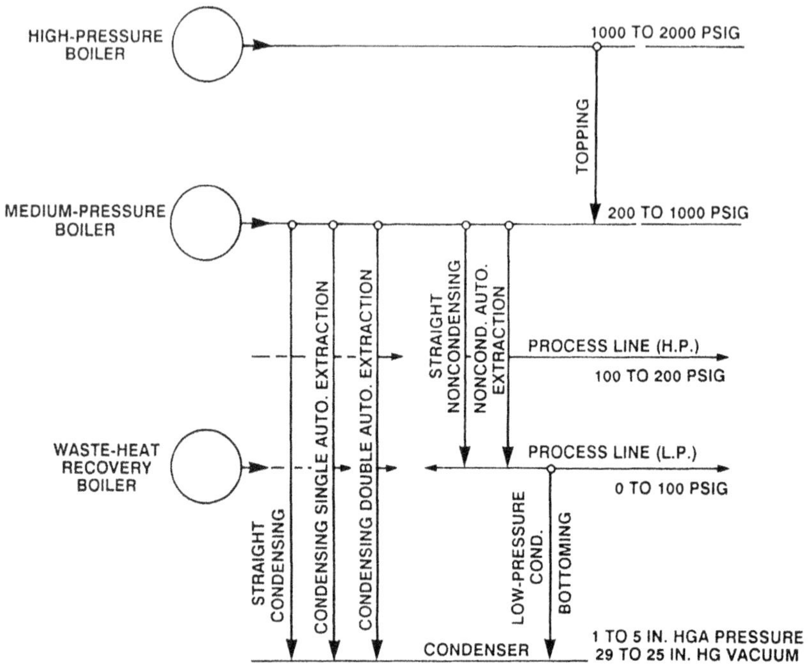

Figure 1.13 Schematic diagram illustrating regimens of application of various types of steam turbines. *(General Electric Company, Fitchburg, Mass.)*

1.3.1 Straight noncondensing

The most simple steam turbine configuration is the straight noncondensing design. The output of the turbine is a function of the initial steam conditions, the turbine exhaust pressure, and the process steam demand. The power production of this unit type is limited by the process demand, unless an artificial demand is created by the use of a steam vent at the exhaust.

A cross section of a typical noncondensing steam turbine is shown in Fig. 14*a*. The cam-positioned poppet inlet valves are shown on the upper casing. The casing is in two halves, each made from a single steel casting. The front standard, shown to the left, contains the thrust bearing, first journal bearing, and control devices. The turbine is anchored at its exhaust end. Thermal expansion of the casing is accommodated by the flexible support under the front standard. The steam path is of the impulse, wheel-and-diaphragm type, in which the moving buckets are carried on the periphery of wheels machined from a solid forging. The packing diameters between the wheels are made small to minimize the interstage packing leakage. A small-diameter shaft acts to minimize transient thermal stresses, optimizing starting and loading characteristics.

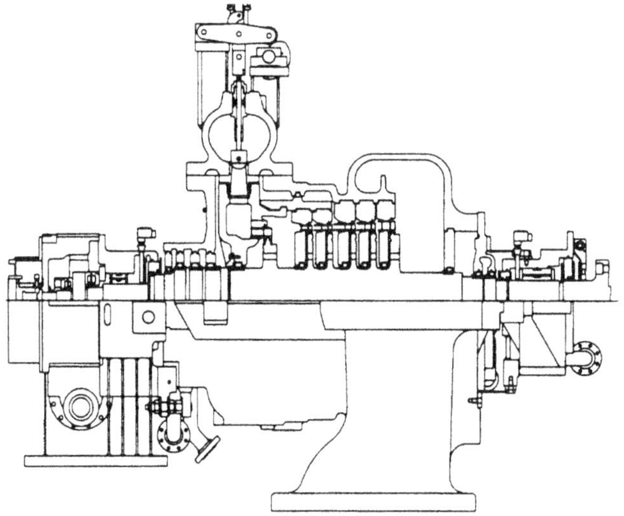

(a)

Figure 1.14 Cross-sectional views of typical mechanical drive steam turbines. *(General Electric Company, Fitchburg, Mass.)*

Most of the stage pressure drop is taken across the nozzles in the stationary diaphragms. Spring-backed packing rings seal against the rotating shaft.

The solid coupling to the driven load is shown to the right.

1.3.2 Automatic extraction noncondensing

Industrial plants having steam demands at two or more pressure levels can benefit from the use of these turbines. They provide the flexibility to automatically respond to variations in steam demands at the extraction and the exhaust.

In recent years, typical mechanical drive noncondensing turbines have been rated 10 to 40 MW with inlet steam conditions of 600 psig/750°F (41 bar/400°C) to 1450 psig/950°F (100 bar/510°C). Extraction pressures of 150 to 650 psig (10.3 to 13.8 bar) for the high pressure and 25 to 200 psig (1.7 to 13.8 bar) for the low pressure are typical of the broad spectrum of steam conditions used in industrial plants where these units are applied.

1.3.3 Automatic extraction condensing

These units provide additional operating flexibility and the ability to control power generation, as well as process header pressures. They are well suited to third-party cogeneration systems because of their ability to handle variations in the steam host's steam requirements while maintaining electric power delivery to the utility. They can be sized for electrical generation considerably in excess of that associated with the extraction steam flows.

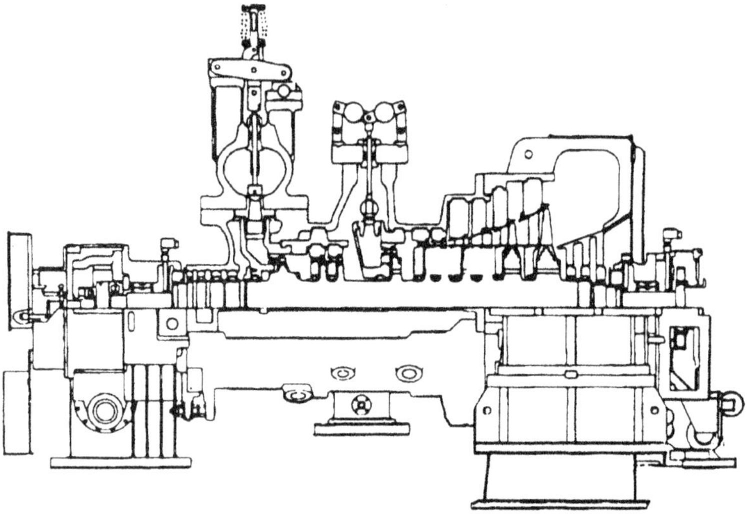

Single Automatic Extraction Condensing

(b)

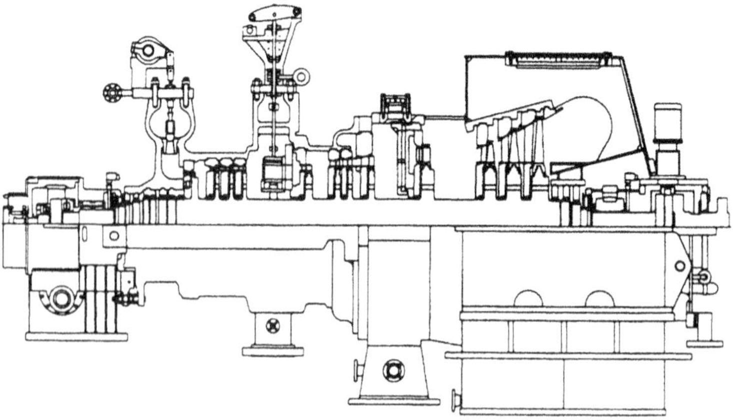

Double Automatic Extraction Condensing

(c)

Figure 1.14 *(Continued)*

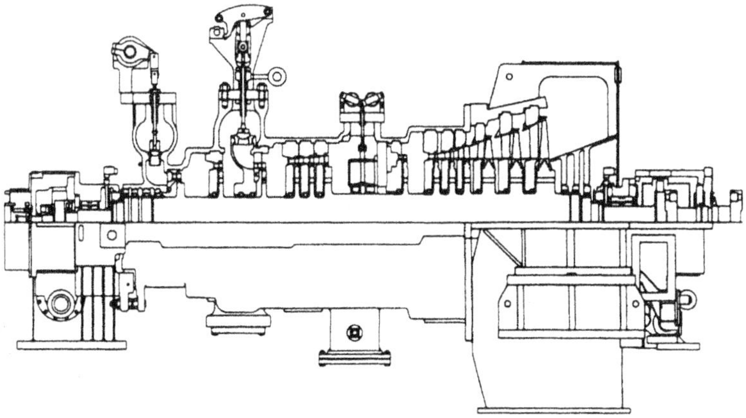

Double Automatic Extraction Condensing

(d)

Figure 1.14 *(Continued)*

Automatic extraction/admission valve gear is normally used when the extraction or admission flow is greater than 25 percent of the flow through the turbine.

A single automatic extraction condensing steam turbine is shown in Fig. 14*b*. It is a single-casing, single-flow machine with two bearings. This machine utilizes cam-lift valves and a 360° nozzle box. This is typically applied on turbines with inlet pressures over 900 psig (62.0 bar). This turbine also utilizes double-shell construction between the inlet nozzle box and the extraction valves. This double-shell design limits the steam pressure and temperature to which the outer casing is exposed. This design provides the greater flexibility in the turbine casing required by machines that must be tolerant to process variations. The extraction valve gear configuration in Fig. 14*b* is also a poppet valve design with a bar lift mechanism. The chest is a fabricated design that is partitioned to pass flow from each valve to a specified number of nozzles in the extraction diaphragm. The control range for these valves is ±10 percent of the normal pressure at that stage. The major benefit of the internal extraction valve gear is its ability to control the flow from wide open to only a small amount to the condensing section as required to keep that section cool.

A double automatic extraction condensing steam turbine is shown in Fig. 14*c*. It is a single-casing, single-flow machine with two bearings. The casing comprises a cast-steel shell down to the second extraction using a vertical joint. The upper and lower casing components are bolted together with a horizontal joint. The inlet valves are poppet valves with a bar lift mechanism. The extraction valves illustrated for the first extraction are internal spool valves. With spool valves, a horizontal, external cam shaft lifts four vertical stems. Each stem positions two internal spool valves, one each in the upper and lower halves of the turbine. The valves are designed to open sequentially, providing the efficiency advantage of multiple partial arc admissions to the immediate downstream stage group. An axial-flow extraction valve is shown for the second extraction. Axial-flow or *grid valves* are appropriate for low-pressure applications and high-volume flows. The additional benefit is the relatively short span that is required. Shown on the low-pressure end, which is the drive end, is a turning (barring) gear used to rotate the shaft when the turbine is cooling down. These turning devices are often utilized on high inlet temperature turbines over 850°F (455°C) and when the bearing span is over 150 inches (3810 mm).

The double automatic extraction turbine shown in Fig. 14*d* has the same inlet valve gear as the previous machine but has a cam-lifted valve gear with a cast nozzle box for the first extraction. This is the design typically used for moderate extraction pressures ranging from 650 to 250 psig (45 to 17 bar). The second extraction is a bar lift, spool valve design.

Steam volume increases rapidly as the steam expands to condenser pressure. Thus the length of the buckets (blades that make up the turbine rotor) increases rapidly between the inlet of the LP section and the last-stage buckets.

Today there is a wide range of bucket designs applied to steam turbines. Traditional tangential dovetails may not provide adequate centrifugal capability on high-speed machines such as those applied on syn-gas turbines. Here, axial entry dovetails and features such as integral and double-covered designs provide the long-term reliability and wide speed range necessary for mechanical drive applications. Tall condensing section buckets have seen the evolution from aircraft engine and gas turbine technology. Features such as axial entry dovetails and *Z* lock covers are used to dampen vibration and lower response factors to allow greater bucket loads with higher reliability. Examples of impulse buckets are shown later in this text.

Operation over a wide speed range adds a significant degree of complexity for the turbine designer. Accurate prediction of blade frequency and stress is necessary to ensure the level of reliability expected today.

1.3.4 Basic steam control considerations

The selection of a particular turbine type is influenced by the nature of the driven load as well as the need for power and process heat. One set of valves can control only one parameter at a time: speed/load, inlet pressure, extraction pressure, or exhaust pressure. The control of a second parameter requires the use of a second set of valves, and so on. This is illustrated in Fig. 1.15.

Modern control systems use electronic speed and pressure sensors, digital processing and logic, and hydraulically operated valves. Figure 1.15 shows direct mechanical devices to help visualization of principles, not to represent the actual implementation. Thus, speed control is modeled by a flyball governor. A speed increase in response to load decrease causes the flyballs to move outward, acting to close the associated steam valves. Pressure control is modeled by a mechanical bellows. An increase in the controlled pressure causes the bellows to expand, closing inlet valves or opening extraction valves, as appropriate to the application.

Figure 1.15*a* represents a unit with a simple speed governor. The exhaust pressure is established independently of the turbine, either by a condenser or other steam sources to an exhaust header. This arrangement typifies a straight condensing unit. The valves control speed and load only, and the turbine can operate in an isolated application or synchronized in parallel with other generating units or electrical systems.

Figure 1.15*b* shows a unit maintaining exhaust pressure through the action of inlet valves. (An associated speed governor takes over control when speed limits are exceeded. Pressure control is lost in the process.) This would be an impractical arrangement for an isolated load requiring accurate speed regulation. It works well when the turbine drives an electrical generator that operates in parallel with one or more other generators, which provide system speed control. Under normal operation one valve regulates one parameter: exhaust pressure. This arrangement typifies the straight noncondensing turbine. A drop in pressure at the turbine exhaust indicates an increased demand for exhaust steam. The pressure governor acts to increase inlet steam flow to match the changing exhaust requirements. The generator output varies with use of process steam.

Figure 1.15*c* represents a single automatic extraction unit. The speed governor responds to speed/load changes. The pressure governor regulates extraction pressure as process needs change. Simply changing the position of the extraction valve at a constant inlet valve position would also change the speed/load as well. Therefore, the two systems interact to maintain constant load. For example, a reduction in process steam requirements leads to an extraction reduction and an extraction pressure rise. The pressure governor opens the extraction valve, increasing low-pressure section flow and power. At the same time, it acts to proportionally close the inlet valve to reduce inlet flow, holding constant speed/load.

The two valves of Fig. 1.15*c* control two parameters: speed/load and extraction pressure. This unit is incapable of independently controlling its exhaust pressure as configured. If it were to drive an electrical gen-

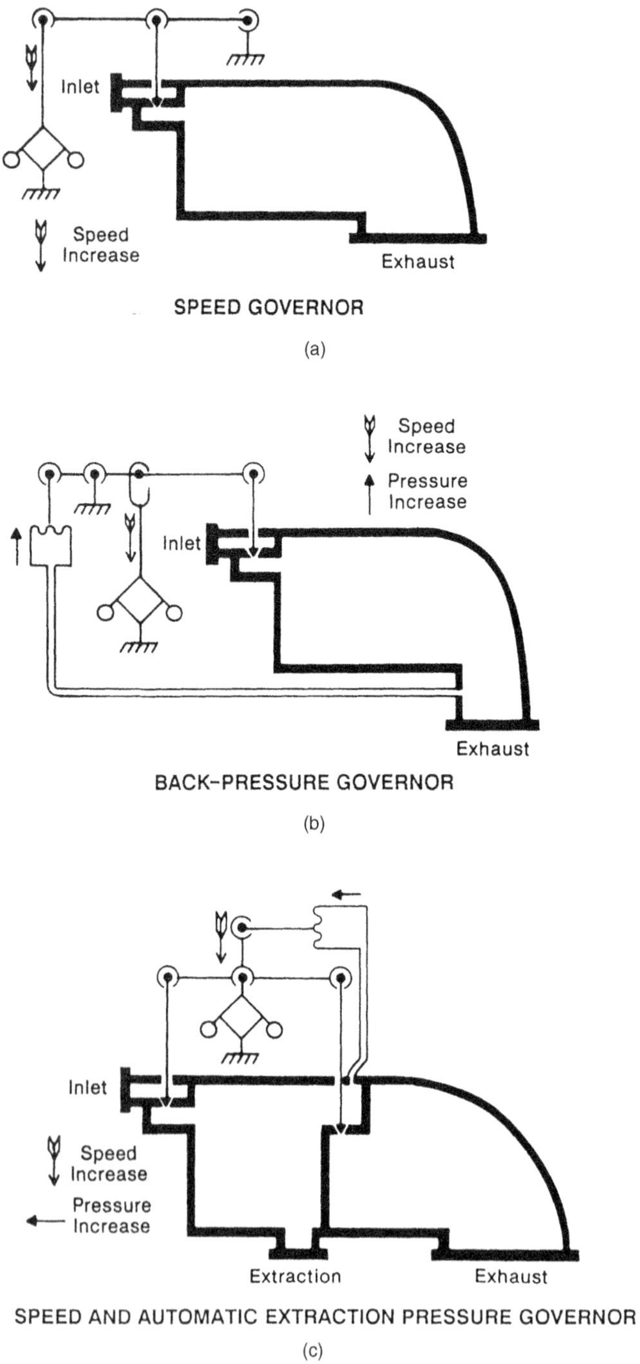

Figure 1.15 Schematic diagrams of steam turbine governors. *(General Electric Company, Fitchburg, Mass.)*

erator in parallel with others, then the two valves could be used to regulate extraction and exhaust pressures with a speed override.

1.3.5 Automatic extraction condensing controls

The turbines shown in Figs. 1.14*b* through *d* are automatic extraction condensing turbines commonly applied in industrial plants and third-party cogeneration systems. A typical governing system would permit simultaneous control of extraction pressure (with varying extraction flow) in one or more steam systems at the same time that the speed governor is maintaining frequency, even though the requirements for extraction steam and generated power may be varying. Thus, the turbine-governing system for either a single automatic, double automatic, or triple automatic extraction condensing turbine can respond to varying demands for steam in one-, two-, or three-process steam systems in addition to frequency or load.

Turbine inlet and extraction valve gears have the ability to control turbine section flows as required to accomplish these many tasks. This turbine can operate as an isolated unit or in parallel (synchronized) with other generating units or the utility tie. If the required kilowatt output for an isolated unit should ever be lower than the power produced by extraction steam (plus turbine exhaust cooling steam), then the speed governor would automatically decrease the controlled extraction flow to maintain system frequency. The deficit in extraction steam could then be made up by pressure-reducing valves throttling steam to the process system in parallel with the turbine extraction steam.

Thus, available steam turbine control systems provide a wide variety of system control functions to simplify plant operation and increase plant energy system reliability.

1.3.6 Geared and direct-drive types

Geared and direct-drive turbine-generator sets are a cost-effective and flexible means of generating power. With today's competitive economic climate, geared sets, with their lower first cost and competitive efficiency levels, are finding extensive application in the 15,000- to 20,000-kW range. A packaged gear-drive turbine is shown in Fig. 1.16. Direct-drive turbine-generator sets can be designed for very large ratings. Packaged direct-drive units, as depicted in Fig. 1.17, are available up to approximately 60,000 kW depending on the shipping limitations associated with last-stage bucket length and exhaust hood size.

Packaged equipment sets offer the benefits of single-point responsibility for the assembly of the turbine-generator or other driven equipment package, which includes all major equipment as well as the

Figure 1.16 Geared turbine-generator set (21.5 MW); straight condensing unit, fully base-mounted. *(General Electric Company, Fitchburg, Mass.)*

Figure 1.17 Single automatic extraction, condensing turbine, with base-mounted oil system (24 MW). *(General Electric Company, Fitchburg, Mass.)*

control system, the lube and hydraulic system, baseplate, and total system design, manufacture, and integration.

This design offers maximum factory assembly and testing to ensure quick and reliable field installation and start-up. Factory packaging includes the installation and alignment of the fully assembled turbine, oil system, and gear (if applicable) on a baseplate, and the generator, compressor, or pump on a separate baseplate, depending on the size of the unit. All major components are piped for the lube and control oil as required, steam seal piping, gland leakoff piping, instrumentation, and control wiring, all within the confines of the baseplate(s). Terminations consist of conveniently located pipe flanges at the edge of the base and electrical terminal boxes.

The rugged baseplates used to integrate the set should be designed to control deflection and stress during a simplified four-point lift. This type of a lift at the site is most reliable and permits rapid installation of the package. This base design avoids the use of costly and risky six-point lifts and would permit a range of customer foundation designs due to its rigidity.

For units whose overall lengths are compatible with shipping and manufacturing limitations, a single package including the turbine, gear (if applicable), and lube oil system module is provided. Otherwise, two modules are provided, one for the base-mounted turbine and the other for the lube oil system module. An example of this configuration is shown in Fig. 1.18. In this case, the lube oil system is supplied on a separate skid that can be located conveniently near the unit for interconnection to the turbine.

A photograph of the installation of an actual turbine-generator is shown in Fig. 1.19. The turbine is condensing and rated 28 MW. Its size permits mounting both the turbine and its combined lube/hydraulic system on a common base. The turbine base has been set in the photograph, while the base-mounted generator is being lowered into place.

Installation of the modules in the field consists of mounting each module on its foundation, shimming and grouting the turbine-generator bases, connecting the couplings, checking the final alignment, and flushing the oil system. Once the electrical connections are complete and the steam, service water, and other miscellaneous connections are made, the package is ready to produce power or to impart energy to fluid streams. The packaged unit concept results in significant savings when compared to the longer installation cycles of a nonpackaged design.

1.3.7 Modular design concepts

The design philosophy of the major international manufacturers of steam turbines recognizes the unique requirements of industrial appli-

Figure 1.18 A separate base is provided for this turbine module. *(General Electric Company, Fitchburg, Mass.)*

cations. These applications span an extremely wide range of turbine design parameters, such as inlet steam conditions, extraction and exhaust conditions, and turbine speeds. The wider energy range and widely differing applications have always complicated the normal design process. This design process involves many interactions among mechanical, thermodynamic, and application factors to achieve a truly optimum design.

To manage this process, many manufacturers depend on the building block principle. The major components of the turbine are designed with a well-planned structure. Figure 1.20 shows the higher-level division of the turbine. The front standard, for example, consists of a family of standard components of increasing size as bearing loads increase. Other major families of components include inlet and extraction valve gears, high-pressure casings, exhaust casings, and bearing assemblies. Each of these components has well-established application ranges, and operating experience can be readily identified. From these families of components, the engineer can select the optimum components to integrate with a customized steam path to meet the specific requirements of the application.

Capitalizing on this broad component experience coupled with recent developments in computer-aided drafting, the major manufacturers have been extending the frontiers in computer-aided design and computer-aided manufacturing for the heavy equipment industry. The goal of the program was to provide a firm turbine structure with enough flexibility to address industry needs while maintaining a strong base of successful operating experience.

The key to the structure concept was to develop a family of components with common interface points so that a turbine of any combination of steam conditions, ratings, and configuration could be customized to meet the unique requirements of each user. The various components contained within this structure are:

- A family of front standards that support the high-pressure casing and house the thrust and journal bearings
- An array of valve gear assemblies, each designed to cover a range of pressures, temperatures, and throttle flows

Figure 1.19 Turbine-generator set being installed. *(General Electric Company, Fitchburg, Mass.)*

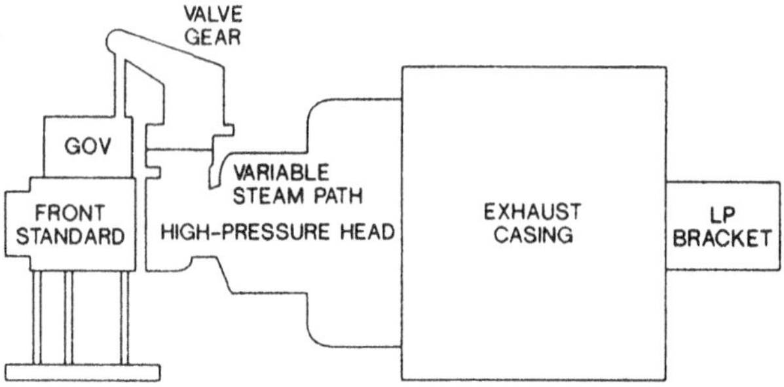

Figure 1.20 Building block concept of standardization by components. *(General Electric Company, Fitchburg, Mass.)*

- A family of high-pressure casings
- A family of extraction/admission valve gears and associated pattern sections
- A complete line of single- and double-flow exhaust casings for both condensing and noncondensing applications
- Microprocessor-based control systems from simple, single-variable controllers to triple-redundant, fully integrated monitoring and multivariable control function systems
- Structured steam and oil piping brought out to standard customer interface points
- A customized steam path consistent with overall thermodynamic requirements

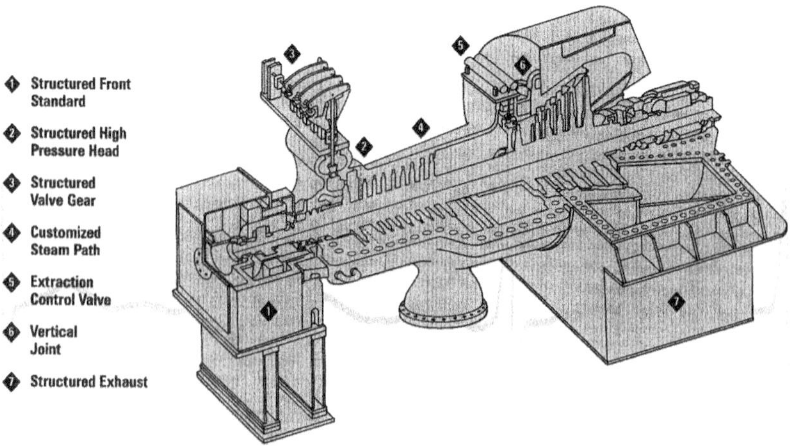

Figure 1.21 Structured or building block approach to modern steam turbine design. *(General Electric Company, Fitchburg, Mass.)*

An example of a structured turbine is presented in Fig. 1.21. Since variations within the industrial market make it impossible to structure groups of stages to cover every conceivable application, the primary challenge for this steam path structure is generally to create a matrix of nozzles, buckets, wheels, and casing pattern sections on a component basis, each with predefined interface points.

In this manner, it is possible for major manufacturers to offer the demonstrated experience of standard components together with the flexibility required for a wide range of industrial applications.

Chapter

2

Turbine Casing and Major Stationary Components

Mechanical drive steam turbines for compressors and pumps, as well as cogeneration turbine-generator sets, can be designed for inlet steam conditions ranging from 2000 psig/1005°F (138 bar/541°C) down to nearly atmospheric pressures with saturated inlet temperatures. These turbines can also have multiple provisions for the admission or extraction of steam for various process uses. Uncontrolled steam extractions are often used for feedwater heating where the pressure and energy of the extracted steam is approximately proportional to the flow of steam through the turbine. When controlled steam pressure or flow is required, automatically regulated extraction valve gear assemblies are provided within the turbine casings to maintain a constant pressure or flow at the extraction point.

The various steam inlet and exhaust schemes obviously call for different casing configurations. Associated with casings are a number of different valve geometries, stationary dividing walls, or guide blades directing steam flow to rotating blades, etc.

For purposes of clarity of presentation, our text deals with this collectively under the umbrella heading *stationary components* as opposed to the turbine rotor that rotates.

2.1 Casing Design

Based on capable manufacturers' experience in developing casing materials and construction for high inlet steam conditions, designs employ either a single- or double-shell construction. Both of these shell configurations have been used on many applications and have accumulated years of operation. These construction methods facilitate the accommodation of transient thermal stresses and provide a design that resists shell cracking and alignment changes during operation.

The double-shell construction prevents initial steam being in direct contact with the outer casing joint. A double-shell back-pressure turbine for high inlet steam conditions is shown in Fig. 2.1.

Adjusting devices allow all parts of the turbine to be aligned to each other, and the turbine to be aligned to the foundation, without the need to machine shims, etc.

The parts of the turbine that control the position of the rotating components in relation to the fixed components are supported and located precisely at shaft height: they move independently of each other. On large turbine casings distortion cannot be transmitted to the bearings. On these turbines, the bearings are supported on adjusting devices in fixed bearing pedestals completely separate from the steam-carrying

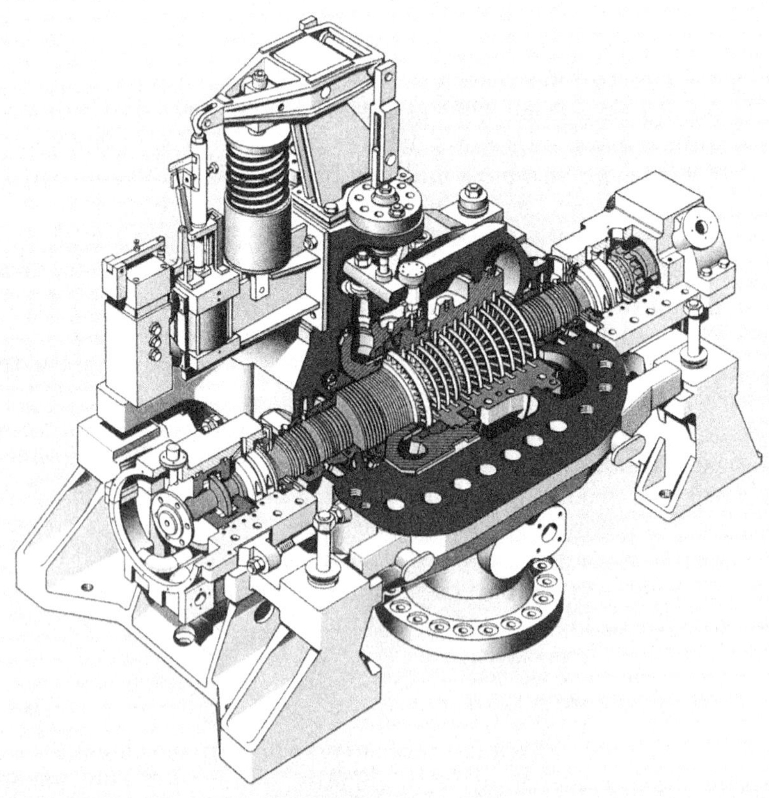

Figure 2.1 High inlet pressure reaction turbine, back-pressure type. *(Siemens Power Corporation, Milwaukee, Wis. and Erlangen, Germany)*

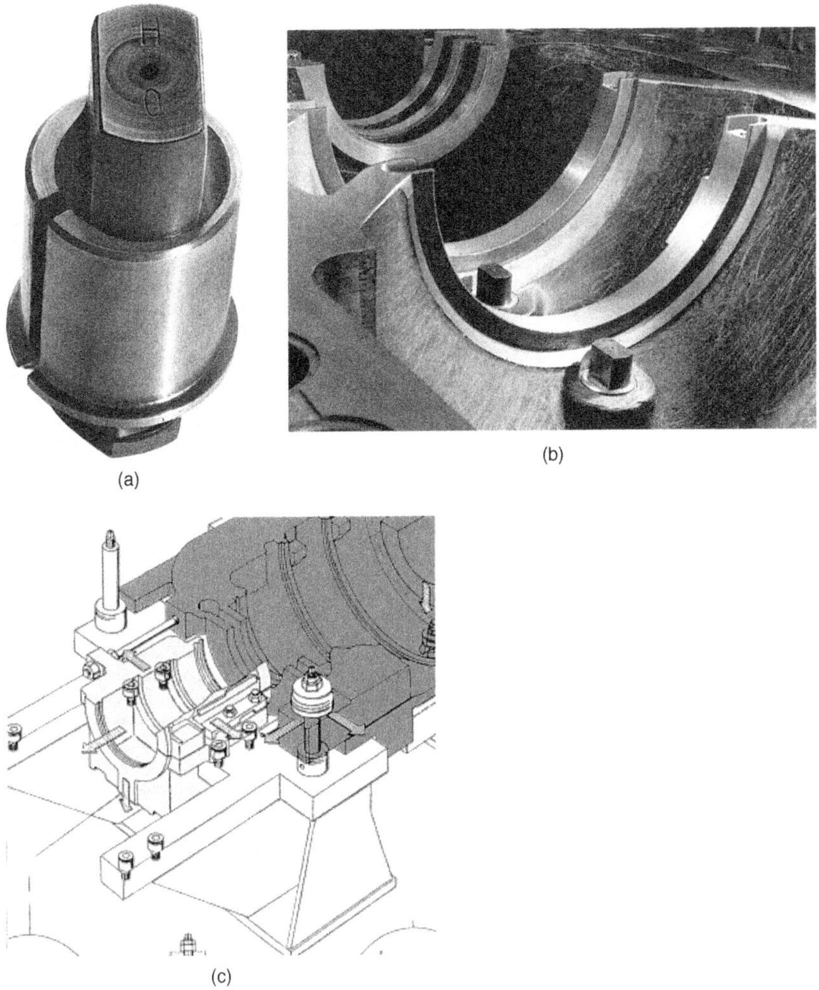

Figure 2.2 Adjusting devices for accurately locating turbine-internal components: (*a*) eccentric pin for aligning internal conponents; (*b*) fitted eccentric pin; (*c*) adjustment directions. *(Siemens Power Corporation, Milwaukee, Wis. and Erlangen, Germany)*

turbine casing. This is illustrated in Fig. 2.2. Except for a few occasional special designs, turbine casings are generally horizontally split and designed to provide reliable, leak-free operation with metal-to-metal joints, moisture drainage provisions, and multiple casing inspection openings.

For additional information on moisture drainage provisions, please refer to Sec. 12.5 on Erosion.

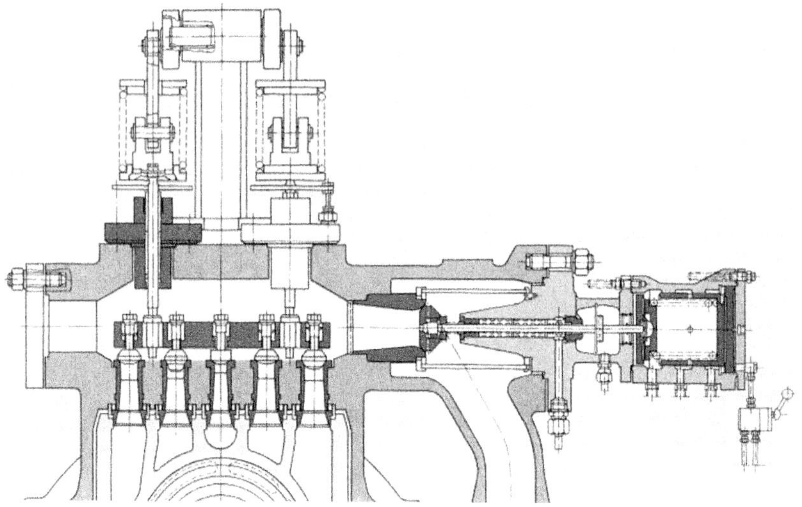

Figure 2.3 Bar lift mechanism arranged for sequential opening of five free-hanging inlet valves. *(Siemens Power Corporation, Milwaukee, Wis. and Erlangen, Germany)*

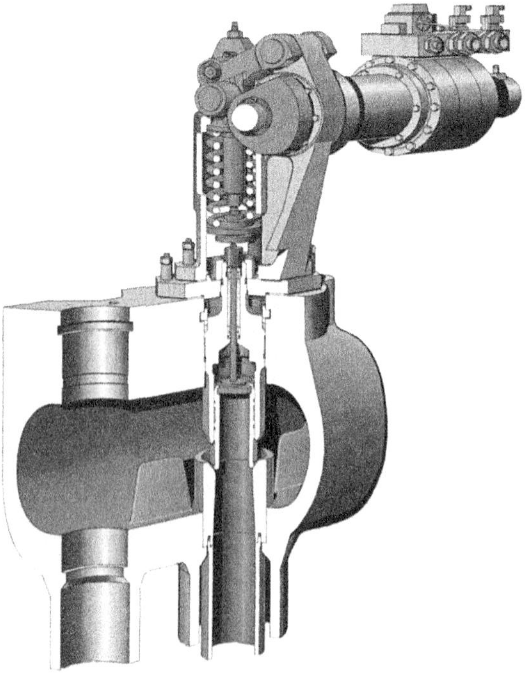

Figure 2.4 Control valve (admission valve) with cam drive. *(Asea Brown Boveri, Baden, Switzerland)*

2.2 Steam Admission Sections

As the term implies, steam enters the turbine at the admission section through one or more governing valves. Either bar lift or cam lift arrangements are common. These are shown in Figs. 2.3 and 2.4, respectively.

The bar lift valve gear shown in Fig. 2.3 is relatively simple. A servo motor operates the lever that raises or lowers the horizontal bar that is enclosed in the steam chest. The governing valves hang loosely on the bar and are adjusted for length of stem so that they will open sequentially. There are typically five valves in this design but as many as seven valves have been used.

The flow capacity of this type of design is limited to about 600,000 lb/h (273 T/H). Special precautions should be used on high-flow applications to prevent these free-hanging valves from spinning and wearing out in service. The single-valve sketch in Fig. 2.5 shows a design feature that minimizes valve action in a turbulent steam flow. The design with the sphere-shaped nut on the valve stem rests in a cone-shaped seat in the valve-lifting bar. This design not only reduces friction in any sideways motion of the valve but also reduces the bending stress in the valve stem.

The cam lift gear shown in the next two illustrations is used for very high flow or high pressures. Cams, Fig. 2.6, can be shaped for obtaining the best balance of lifting forces and flow-travel characteristics. The spring load on top of each valve stem is to oppose the blowout force of the valve stem when the valve is wide open. This spring force is also used as a safety measure to help make sure the valves close in case of emergency unloading of the turbine.

Shown in Fig. 2.7 is an arrangement using a bar lift valve gear with dual inlet to the steam chest. The dual inlet has the advantage of reducing steam velocity in the chest, thus reducing the probability of valve wear at a given set of inlet conditions. Note also that the dual trip valve arrangement would make it possible to test one of the two trip

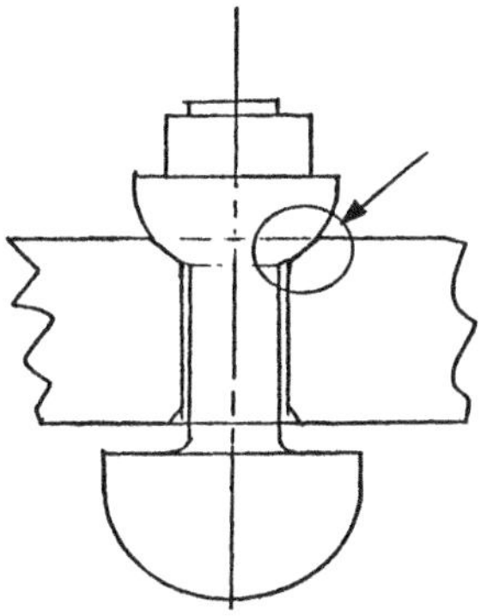

Figure 2.5 Contoured geometry reduces friction and bending stress in the valve stem. *(IMO Industries, Inc., DeLaval Steam Turbine Division, Trenton, N.J.*

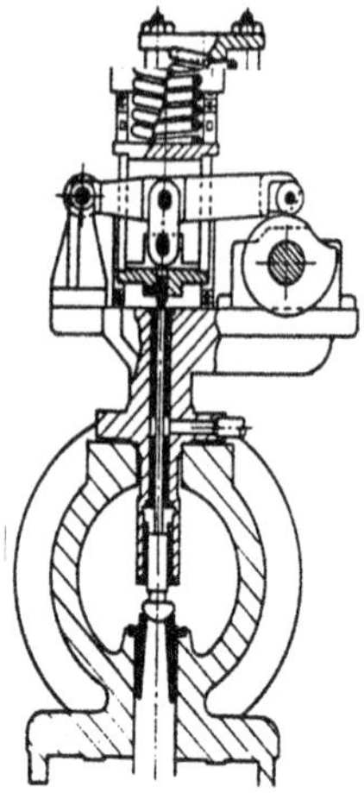

Figure 2.6 Cam lift detail. *(IMO Industries, Inc., DeLaval Steam Turbine Division, Trenton, N.J.)*

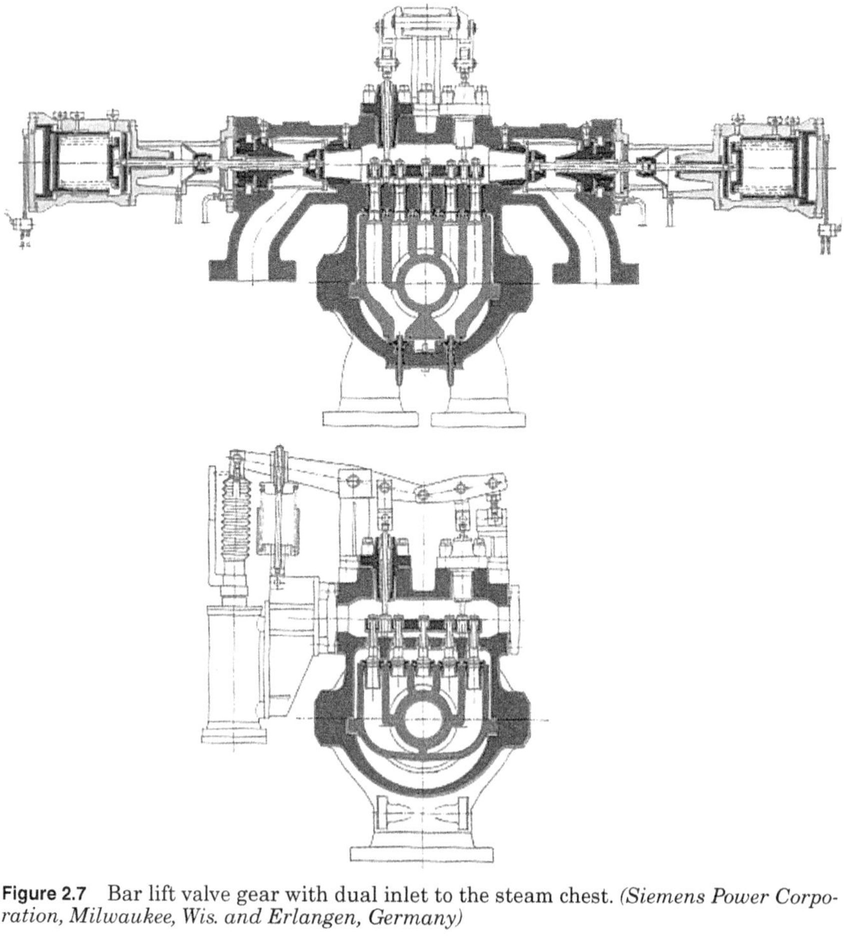

Figure 2.7 Bar lift valve gear with dual inlet to the steam chest. *(Siemens Power Corporation, Milwaukee, Wis. and Erlangen, Germany)*

valves while leaving the second one in operation. This could be an important reliability assurance task.

Figure 1.6, earlier, showed a hand-operated valve in a single-stage turbine. There may be as many as three hand-operated valves plus a governing valve in some single-stage turbines. Hand-operated valves are often used to obtain improved part load steam performance. These valves must be either closed or wide open. They are not used to throttle the steam. Their effect is shown in Fig. 2.8.

Referring back to the admission section of large multivalve steam turbines, we note that emergency stop valves, also called main shutoff or trip-throttle valves (Fig. 2.9, item 2) are often provided with steam strainers, item 5, that are built into the steam chest, item 1. They are typically removable without having to dismantle pipework. Figure 2.9 shows these control valves operated by servomotors, item 12.

All moving parts within the steam passage are allowed large clearances so that deposits cannot prevent their free movement. Angle-section rings or special linkage arrangements provide a kinematic connection between the outer casing and the steam chamber. The nozzles preceding the governing stage are usually replaceable although they are often welded in place and not bolted. The governing stage of either a reaction or impulse turbine is designed as an impulse wheel.

A stage in a multistage turbine consists of both rotating and stationary blades. The stationary blades can be part of a nozzle ring, Fig. 2.10, top, or a diaphragm, Fig. 2.10, bottom. In either case their function is to direct steam onto the rotating blades, turning the rotor, and producing mechanical work.

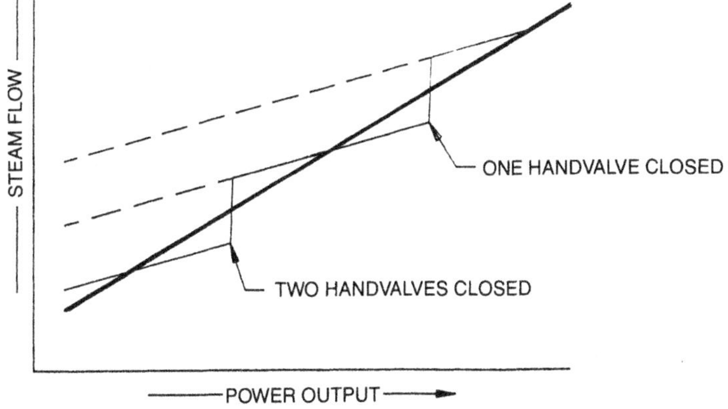

Figure 2.8 Improved part load performance (efficiency increase) achieved by closing hand valves. *(Coppus-Murray Turbomachinery Corporation, Burlington, Iowa)*

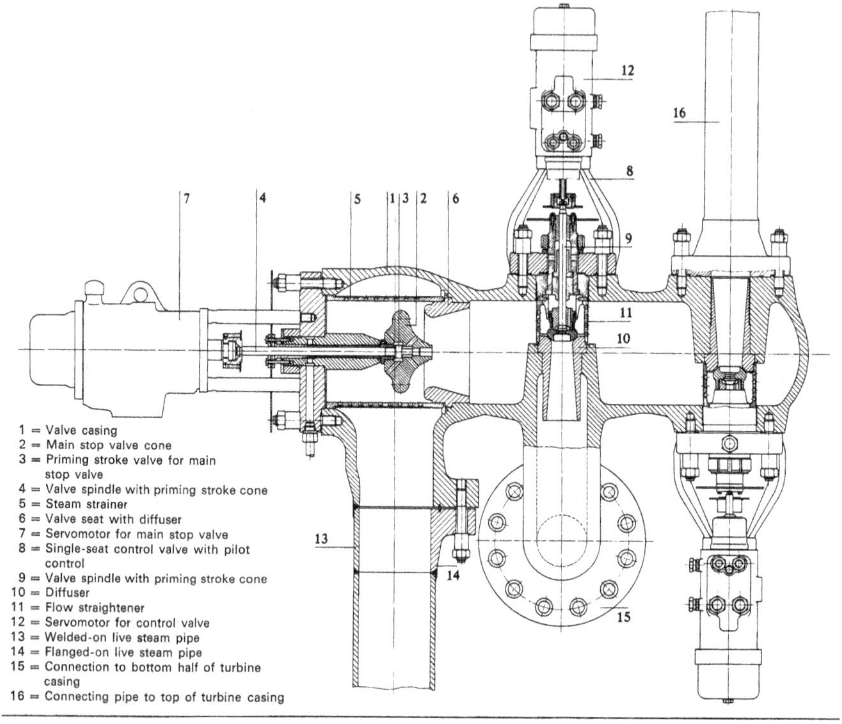

Figure 2.9 Valve casing (valve chest) assembly. *(Asea Brown Boveri, Baden, Switzerland)*

There are a number of fabrication methods available for nozzle rings. Elliott is one of several companies whose first-stage nozzle rings are often made by milling steam passages into stainless steel blocks, which are then welded together (Fig. 2.10).

The nozzles in the intermediate pressure stages are formed from stainless steel nozzle sections and inner and outer bands. These are then welded to a round center section and an outer ring.

Low-pressure diaphragms of condensing turbines are sometimes made by casting the stainless nozzle sections directly into high-strength cast iron. This design includes a moisture catcher to trap condensed droplets of water and keep them from reentering the steam path. These diaphragms can also be completely fabricated.

2.3 Steam Turbine Diaphragms and Labyrinth Packing

A diaphragm in an impulse turbine is a stationary partition located between each rotating wheel that

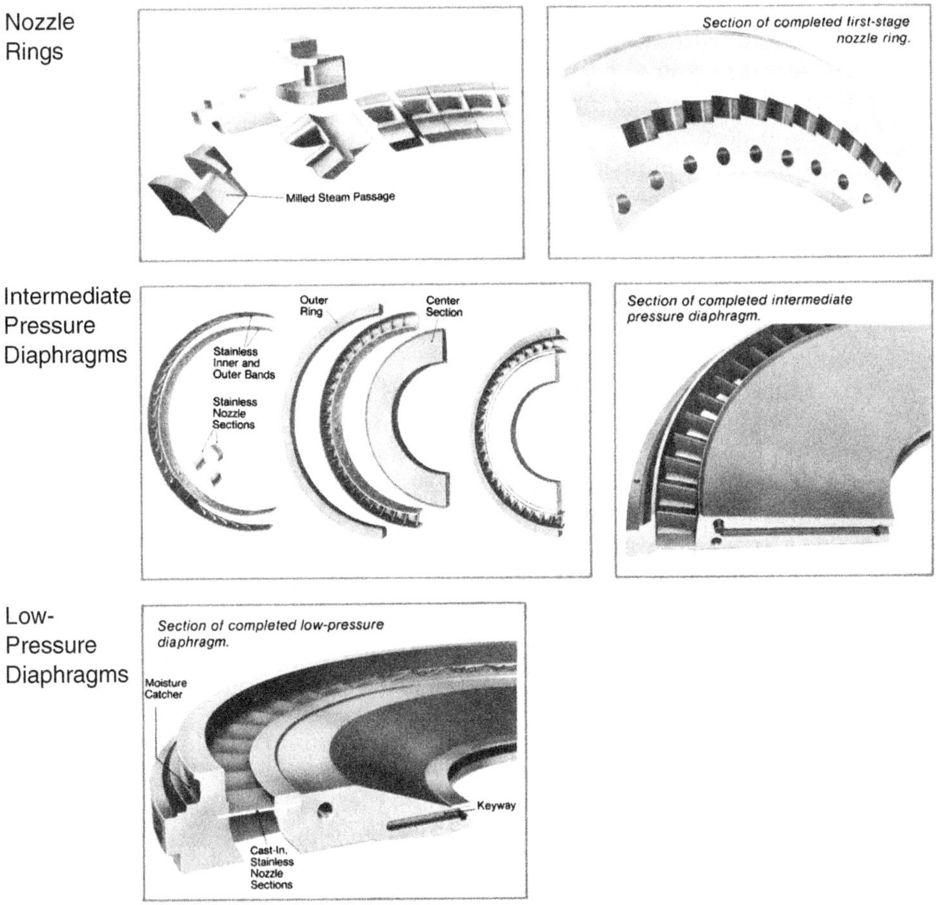

Figure 2.10 Nozzle rings and diaphragms direct steam onto the rotating blades (buckets) of a turbine. *(Elliott Company, Jeannette, Pa.)*

- Separates the turbine into a number of separate pressure stages of successively lower pressure
- Contains the nozzles that accelerate and direct the steam jet into and through the blading of the next succeeding rotating wheel
- Retains the packing through which passes the shaft containing the wheels of each stage

The nozzles contained in the diaphragms vary in size and passage area from stage to stage to efficiently handle the steam volume encountered at the various pressure levels throughout the turbine.

With few exceptions, modern, state-of-the-art diaphragms are typically manufactured in four styles or type configurations: conventionally cast, investment cast, milled and welded, and spoke type.

Conventionally cast diaphragms generally consist of preformed nozzle blade sections of cold rolled 405 stainless steel material cut to length corresponding to proper nozzle height. The individual nozzle sections are positioned in proper location and ductile cast iron material (ASTM A-536, Gr. 80-55-06) poured to form the diaphragm; the resultant steam path is not entirely of stainless steel construction.

The cast conventional diaphragm is limited to low-temperature applications (500°F and less) and stage pressure drops of approximately 25 psig. Its frequency of use is decreasing, being gradually phased out by the other basic diaphragm types especially in the small nozzle height range. While the basic diaphragm is less expensive than other types, some problem areas are associated with cast conventional diaphragms:

- The actual nozzle area is difficult to control. In many cases expensive hand filing is required to remove core sand from the nozzle passages and to obtain the proper areas.
- Scrap rates are high. The quality of the casting obtained from this method is difficult to control, in particular at the bore.

Investment cast diaphragms consist of individual nozzle blocks of 17 percent chrome material (17-4 PH) (Fig. 2.11), assembled to inner and outer steel rings (ASTM A-283 Gr. D) and welded together by the submerged arc welding process (Fig. 2.12). Note then this nozzle block has sidewalls included resulting in a total stainless steel steam path with the investment cast process. The nozzle blocks can be made to a much closer tolerance and with a smoother finish than with conventional casting methods.

With overall steel construction, the allowable stage temperature is 750°F (400°C), and the allowable pressure drop across the diaphragms is not a severe limitation because the actual thickness can be increased to that required.

The inside diameter of the nozzles of a family of diaphragms with approximately the same pitch diameter remains constant, so that the wheel outside diameter is also constant. This, many times, results in the possible use of a standard turbine wheel.

Recent uses of investment cast diaphragms include a 12.5-MW condensing turbine with 13.0-in (330-mm) nozzle height and with a 53.25-in (1353-mm) pitch diameter. The investment cast diaphragms have proven superior to the conventional cast type (closer area control, smoother finish, sidewalls) and, with mechanical design modifi-

Figure 2.11 Investment cast diaphragm. *(Dresser-Rand Company, Wellsville, N.Y.)*

cations, can yield performance comparable to the milled and welded design.

The milled and welded diaphragm consists of individual nozzle blocks assembled and welded together, similar to the investment cast diaphragms. The most significant difference is in the nozzle block

Figure 2.12 Nozzle block being submerged arc welded. *(Dresser-Rand Company, Wellsville, N.Y.)*

itself. The nozzle is milled to proper configuration from a solid block of stainless steel (hot rolled 405 ASTM A-276, Fig. 2.13). The result is a total stainless steel steam path. The individual nozzles are then positioned to steel inner and outer rings (ASTM A-283, Gr. D) and welded together by the submerged arc welding process. The finished assembly is shown in Fig. 2.14.

The stage temperature limit for this type of diaphragm is 750°F (400°C) but can be increased with the use of proper inner and outer ring materials. The allowable pressure drop across the milled and welded diaphragm is the same as across the investment cast diaphragms.

Advantages of the milled and welded type diaphragms include the achievement of close bucket-nozzle spacing (both entrance and exit) along with the use of constant pitch diameters from stage to stage. This results in "residual velocity pickup," which further improves the stage efficiency and is referred to as *flugelized* construction.

A wide range of nozzle heights can be machined from a standard block size.

The spoke type diaphragm (i.e., like spokes in a wheel, Fig. 2.15) is usually used when larger nozzle heights (greater than 3.25 in or 83 mm) are required. The nozzle blades (Fig. 2.16) are made of either 405 stainless steel material (sand cast, milled plate, or preformed) or 17-4 PH investment cast material.

The nozzle blades are fitted to steel inner and outer rings (ASTM A-283, Gr. D) located in proper position by ridges on the rings and pins. The assembly is then tack-welded and the ridges removed by machining. Each nozzle blade is then welded at all edges by the manual shielded electric arc method.

Figure 2.13 Impulse steam turbine nozzles milled from solid blocks of stainless steel. *(Mitsubishi Heavy Industries, Ltd., Hiroshima, Japan)*

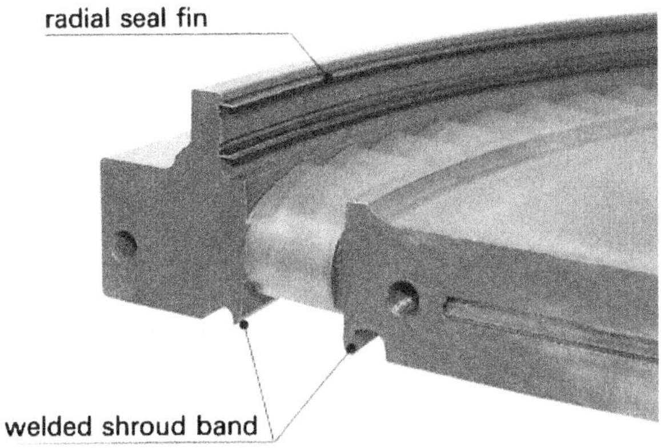

Figure 2.14 Welded steam turbine diaphragm assembly (shroud-band type diaphragm). *(Mitsubishi Heavy Industries, Ltd., Hiroshima, Japan)*

The construction is costly by virtue of the extensive manual welding operation.

Any unit in compliance with applicable standards of the American Petroleum Institute (API 612) will require use of labyrinth packing for interstage sealing. The features described below will center around the use of labyrinth packing:

Figure 2.15 Spoke type diaphragm. *(Dresser-Rand Company, Wellsville, N.Y.)*

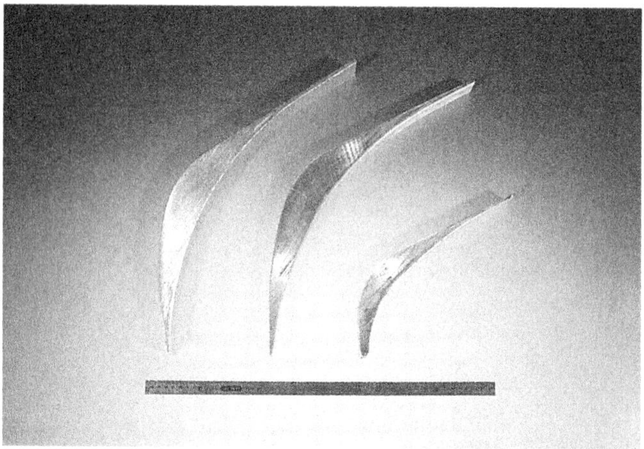

Figure 2.16 Nozzle blades for spoke type diaphragm. *(Dresser-Rand Company, Wellsville, N.Y.)*

- The diaphragms are horizontally split and keyed, the key serving principally as a seal but also as an axial locating device for upper and lower halves (Fig. 2.17). Note also the locating dowel. Locating dowels are mounted at the diaphragm split to ensure symmetry of the packing bore.
- The lower half of the diaphragm is secured to the casing after shimming by use of set screws (Fig. 2.18).

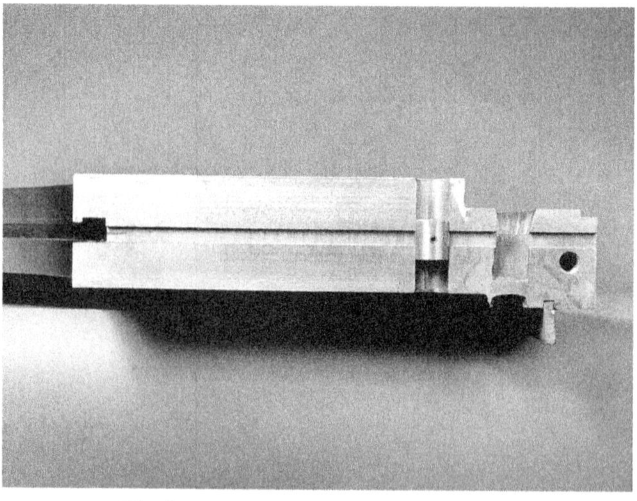

Figure 2.17 Diaphragm split line showing keyway and locating dowel. *(Dresser-Rand Company, Wellsville, N.Y.)*

Figure 2.18 Lower casing half showing diaphragms secured with set screws. *(Dresser-Rand Company, Wellsville, N.Y.)*

- The outer diameter is approximately 0.060 in (1.5 mm) smaller than the casing bore to allow shimming of the lower half for setting labyrinth-to-rotor clearances (Fig. 2.19).
- When lifting the casing upper half, the upper half diaphragm is prevented from dropping by use of stops located at the casing split. In normal fitup, there is clearance between the stop and the diaphragm so that the upper half may rest on its locating dowel.

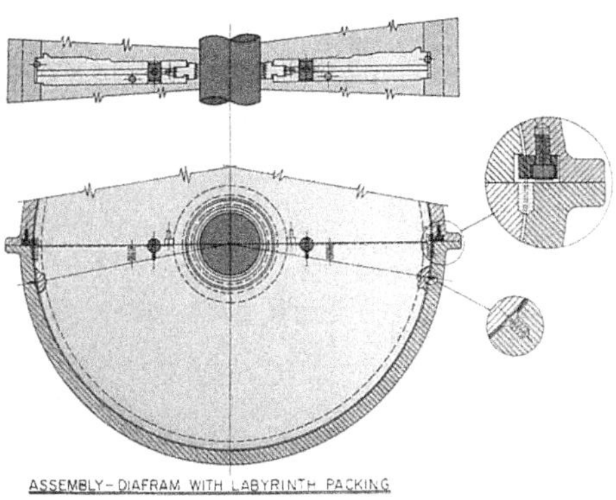

Figure 2.19 Diaphragm contour geometry. *(Dresser-Rand Company, Wellsville, N.Y.)*

Figure 2.20 Windage shields located a short distance from wheel periphery. *(Dresser-Rand Company, Wellsville, N.Y.)*

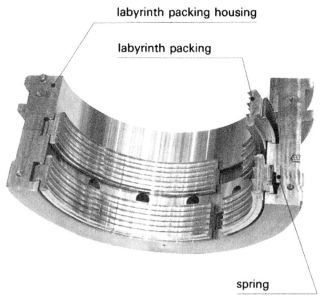

Figure 2.21 Labyrinth packing assembly. *(Mitsubishi Heavy Industries, Ltd., Hiroshima, Japan)*

- All diaphragms are fitted with windage shields located a short distance from the wheel outside diameter. The windage shields also serve as additional casing protection (Fig. 2.20), plus stage moisture removal.
- A special arrangement of diaphragm, similar to that used on high-temperature service, has the diaphragm halves bolted together. This eliminates the need for upper half stops and allows for a better horizontal flange casing bolt pattern, i.e., the bolt line closer to the internal pressure source, a concern in metal-to-metal applications.

The next six illustrations, Figs. 2.21 through 2.26, show various kinds of packing (seals) used in steam turbines. Solid packing rings incorporate a spring arrangement that pushes the packing toward the

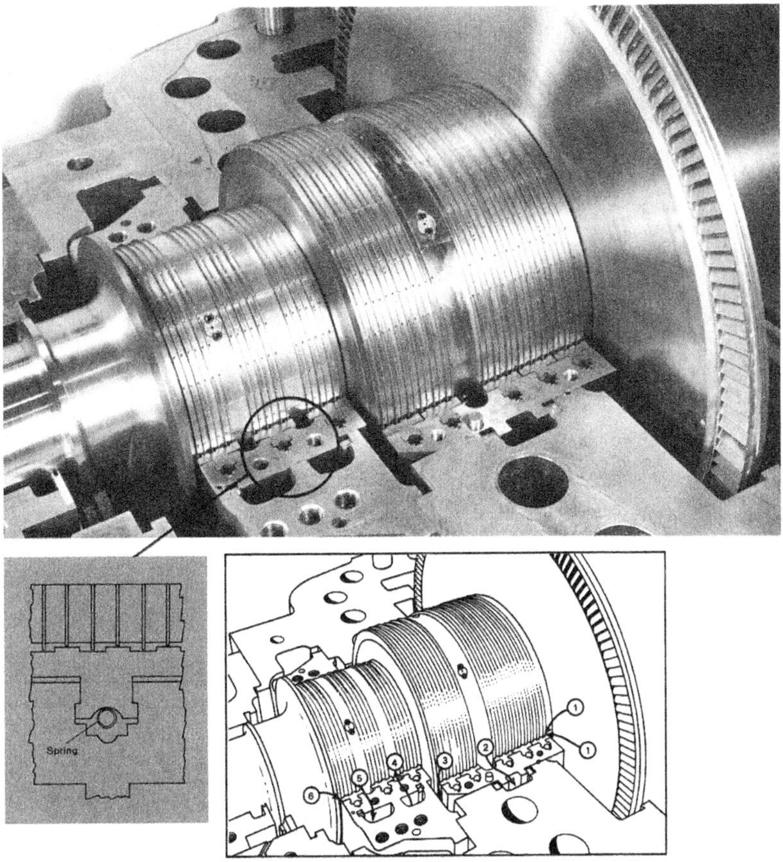

Figure 2.22 Labyrinth seal. *(Elliott Company, Jeannette, Pa.)*

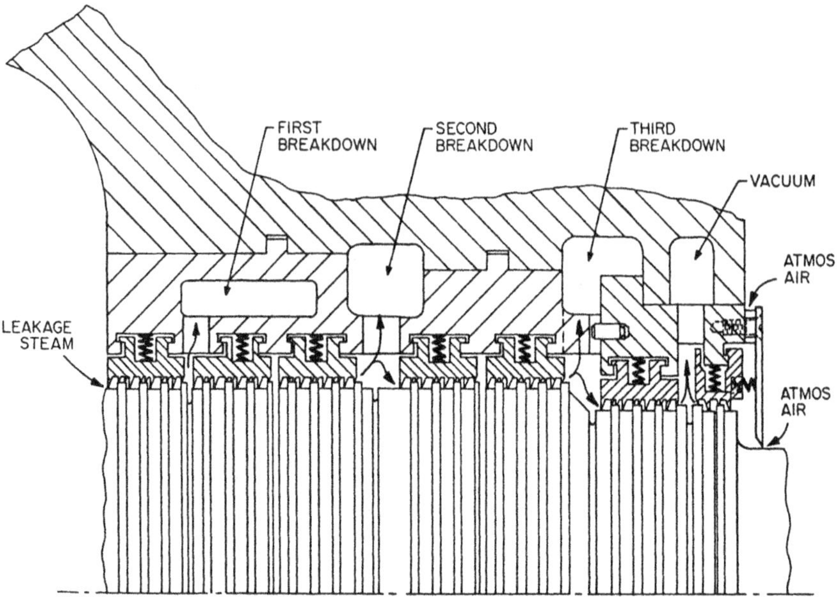

Figure 2.23 Spring-loaded labyrinth segments reduce steam leakage to atmosphere. *(Transamerica DeLaval, Engineering Handbook, McGraw-Hill, 1983)*

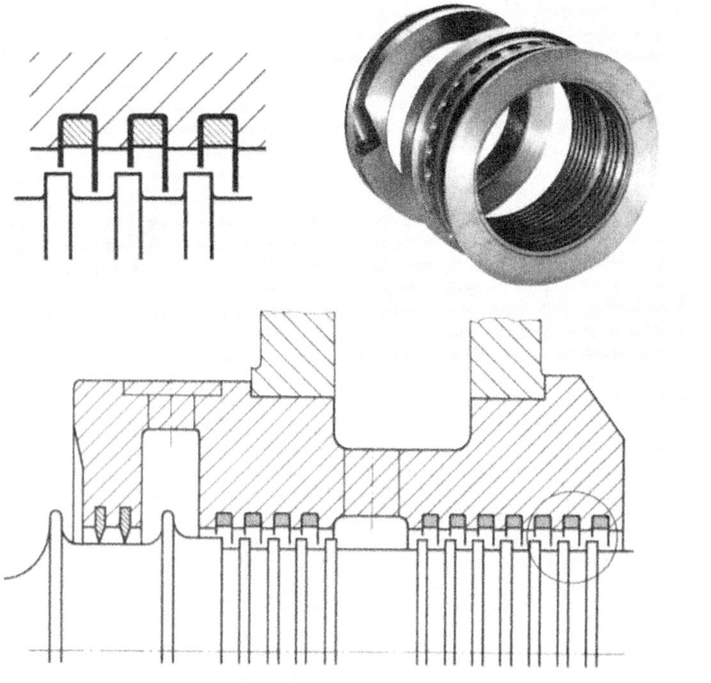

Figure 2.24 Sealing strips (*J*-strips) in the stationary part of packing glands are used in reaction turbines. They project into grooves in the rotor shaft. *(Siemens Power Corporation, Milwaukee, Wis. and Erlangen, Germany)*

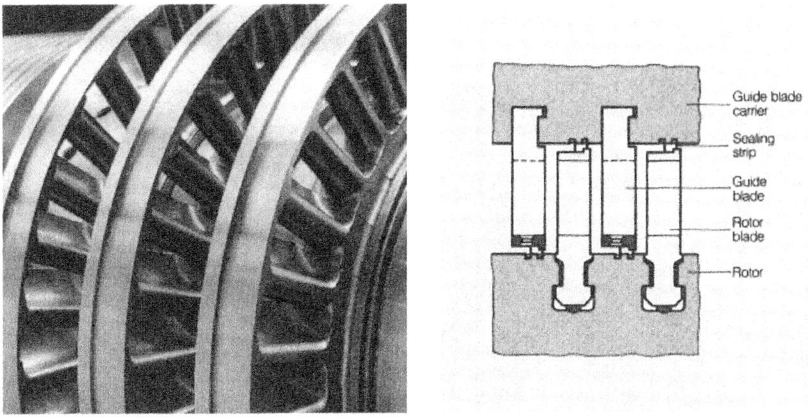

Figure 2.25 Sealing strips rolled into guide blade carrier and rotor of reaction type turbines. *(Siemens Power Corporation, Milwaukee, Wis. and Erlangen, Germany)*

shaft. The packing rests on a small lip in the packing holder that prevents it from normally touching the shaft. If the shaft does touch this packing because of thermal or other changes in the parts, the packing is expected to move radially, thus compressing the loading springs. Two views, Figs. 2.24 and 2.25, are shown of packing that is made by rolling sealing strips into a casing or rotor. This type of seal design is commonly used at the blading tips on reaction type turbines.

Figure 2.26 Carbon ring packing segments. *(Dresser-Rand Company, Wellsville, N.Y.)*

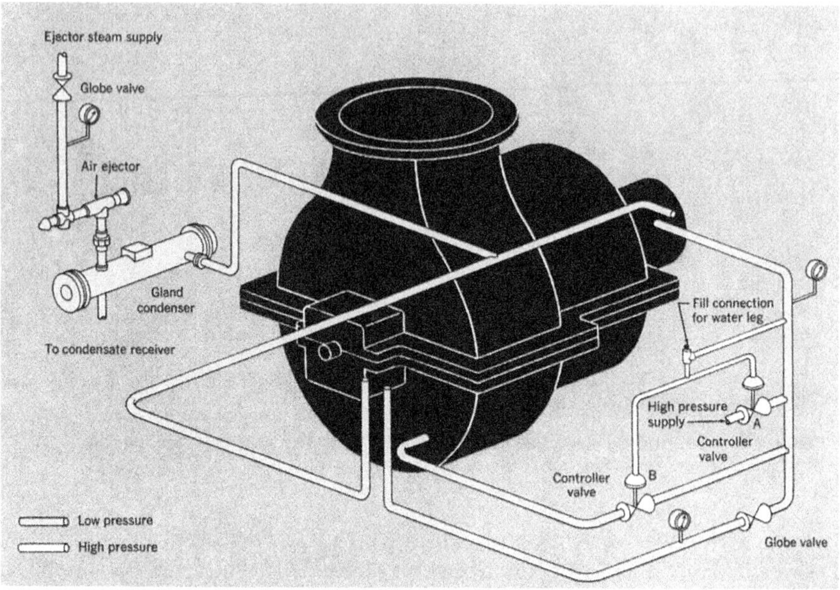

Figure 2.27 Steam leakoff and sealing principle. *(Dresser-Rand Company, Wellsville, N.Y.)*

The last illustration, Fig. 2.26, depicts carbon packing (see also Fig. 1.5, earlier). In this design, floating carbon rings are assembled with very close clearance to the shaft. The packing assembly is held in place and against a sealing surface by a garter spring that goes 360° around the assembly.

Although carbon rings can be designed and manufactured for minimum leakage, they are more maintenance-intensive than either the labyrinth or dry gas seal design. For a detailed description of dry gas seals, see an up-to-date text on compressor technology.

Finally, Fig. 2.27 depicts a schematic of the steam sealing method typically employed with the larger (1000 hp and up) steam turbine sizes. Steam leakage is led to successively lower pressure regions in the turbine or, ultimately, to a gland steam condenser.

Turbine manufacturers are using commercially available brush seals to enhance turbine sealing. These seals have been used in gas turbine technology for many years. Brush seals are circumferential seals comprised of densely packed, fine nickel chromium alloy wires (approximately 0.006-in diameter), arrayed on a 45° angle against the tangent of a shaft surface over which they are configured to run. These are very-high-tip-speed, high-temperature-seals and are capable of surface speeds up to 1100 ft/s (feet per second) and 1200°F.

Brush seals not only provide reduced original leakage, but tolerate rubs and transients much better than labyrinth seals. This eliminates the needs for spring-back labyrinths and other complicated systems.

Figure 2.28 Improved brush seals.

Brush seals can be used instead of, or in addition to, the existing labyrinth seals. One or more labyrinth teeth are removed, and the brush seals are installed in their place. The typical clearance between the brush seal ID and the rotating shaft is 0.000 in to 0.005 in. This reduces the steam leakage by about 30 percent with a reduction in the size of the gland condenser and associated systems. See Fig. 2.28 for a typical brush seal installed with labyrinth seals.

Similar designs are being tried by utilizing brush seals in conjunction with carbon rings to make use of the reduced leakage of carbon rings and the wearability of brush seals.

But post-1980 and even newer developments supersede the old seal design and are well worth considering in both new as well as retrofit situations. One such design is Wavy Face Dry Running Seals. A detailed description of this type of seal and its use is given in Section 16.1.2.

Chapter

3

Bearings for Mechanical Drive Turbines

3.1 Journal Bearings for Industrial Turbomachinery

The journal bearing designs described in this chapter are common to many types of industrial turbomachinery, including steam turbines. The types of bearings most commonly found in turbomachinery include:

- Plain cylindrical
- Axial groove
- Pressure dam
- Lemon bore or elliptical
- Offset half
- Multilobe
- Tapered land
- Tilting-pad

The reason for such a large selection of bearings is that each of these types has unique operational characteristics that render it more suitable for one particular application compared to another.

The fundamental geometric parameters for all journal bearings are diameter, pad arc length, length-to-diameter ratio, and running clearance. For the bearing types consisting of multiple pads, there are also variations in the number of pads, pad preload, pad pivot offset, and bearing orientation (on or between pads). In addition to the geometric parameters, there are several important operating parameters. The key operating parameters are oil type, oil inlet temperature, oil inlet pressure, rotating speed, gravity load at the bearing, and applied external loads. Volute loads in pumps, mesh loads in gearboxes, and asymmetric loads in steam turbines are examples of external loads.

The vibration attenuation for any steam turbine rotor/bearing system is significantly influenced by the journal bearing design. A good bearing design will provide sufficient bearing damping for vibration suppression plus operate at reasonable metal temperatures. A poor design will lead to vibration problems or high-temperature operation, both resulting in reduced bearing life and increased downtime.

Steam turbine journal bearing design has evolved over the years from primarily fixed-geometry bearings to tilting-pad bearings. Fixed-geometry or sleeve bearings have the annoying property of creating an excitation force that can drive the turbine unstable by creating a subsynchronous vibration. This phenomenon usually occurs at relatively high rotor speeds and/or light bearing loads. This inherent bearing-induced instability can be completely eliminated with the tilting-pad bearing.

Further enhancements in tilting-pad steam turbine bearing design over the years include switching almost exclusively to four pad bearings loaded between pivots. The tilting-pad axial length has also increased from around 50 percent of the journal diameter to typically 75 percent of the journal diameter. Both of these enhancements provide increased effective damping for vibration suppression for most steam turbine designs.

Most recently, reduced temperature tilting-pad bearing designs have evolved to allow larger-diameter bearings to run at faster surface velocities. These designs utilize the cool inlet oil more efficiently, thereby reducing the operating metal temperatures.

The following sections address all of these advancements, with the advantages of each design feature clearly outlined. The disadvantages of these advancements, when applicable, are also addressed.

3.1.1 Fixed-geometry journal bearing stability

Fixed-geometry or sleeve bearings have the annoying property of creating an excitation force that can drive the rotor unstable by creating a subsynchronous vibration. This phenomenon usually occurs at relatively high rotor speeds and/or light bearing loads, or, more generally, at a high Sommerfeld number. The Sommerfeld number S is defined as follows:

$$S = (\mu NLD)/(60W)(D/C_d)^2$$

where μ = average fluid viscosity, lbf·s/in^2
N = rotor speed, r/min
L = bearing length, in

D = bearing diameter, in
W = bearing load, lbf
C_d = bearing diametral clearance, in

The problem with sleeve bearings (i.e., all journal bearings excluding tilting-pad bearings) is that they support a resultant load with a displacement that is not directly in line with the resultant load vector but is at some angle with rotation from the load vector. This angle can approach 90° for light loads and high speed.

The specific case of a vertically downward gravity load is illustrated in Fig. 3.1 for a two-axial groove bearing. This sleeve bearing supports the vertically downward load with a displacement that is not directly downward but at some angle with rotation from bottom dead center. This angle is defined as the attitude angle ψ, as shown in Fig. 3.1.

Figure 3.2 illustrates the hydrodynamic pressure distribution for a high speed, lightly loaded, unstable journal bearing. Note that the bearing eccentricity ratio $\varepsilon = e/c$ (Fig. 3.1), is very small and the attitude angle ψ approaches 90°. In this manner, a light Y direction load is supported by a $+X$ displacement. This occurs because the load is so light and the resulting pressure profile becomes very small, with little change from the maximum film to the minimum film locations. For equilibrium, the summation of all vertical components of the hydrody-

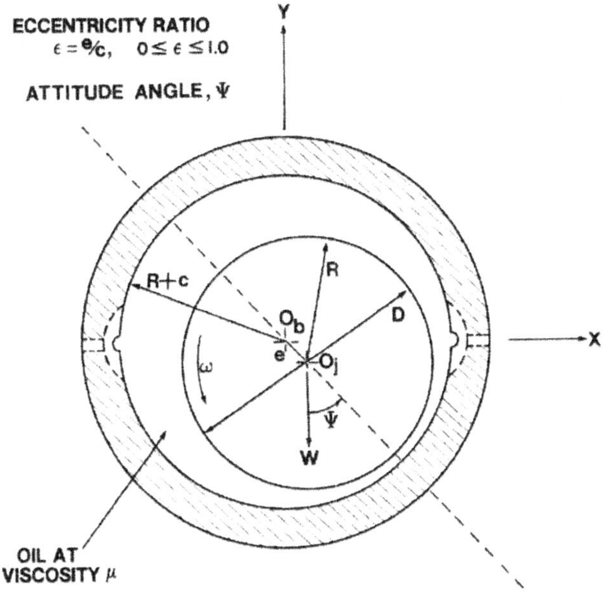

Figure 3.1 Fixed-geometry bearing schematic. *(RMT, Wellsville, N.Y.)*

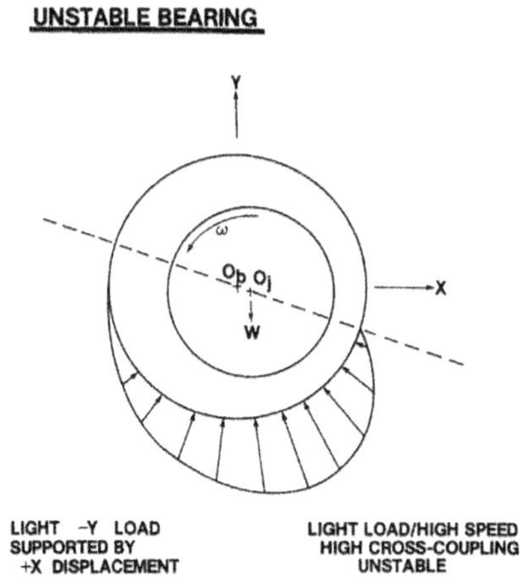

Figure 3.2 High-speed, lightly loaded, unstable bearing. *(RMT, Wellsville, N.Y.)*

namic forces multiplied by the area must be equal and opposite to the external load *W*. Likewise, the sum of all horizontal forces must be zero, which can occur only for attitude angles that approach 90°.

Since a downward load is supported by a horizontal displacement, any downward force perturbation will result in a horizontal displacement, which will result in a horizontal force, which in turn produces a vertical displacement, and so on. Thus, the bearing generates unstable cross-coupling forces that actually drive the rotor and cause it to vibrate at a frequency that is normally in the range of 40 to 50 percent of running speed.

Conversely, a relatively low speed, heavily loaded (i.e., low Sommerfeld number), stable journal bearing is illustrated in Fig. 3.3. Note that

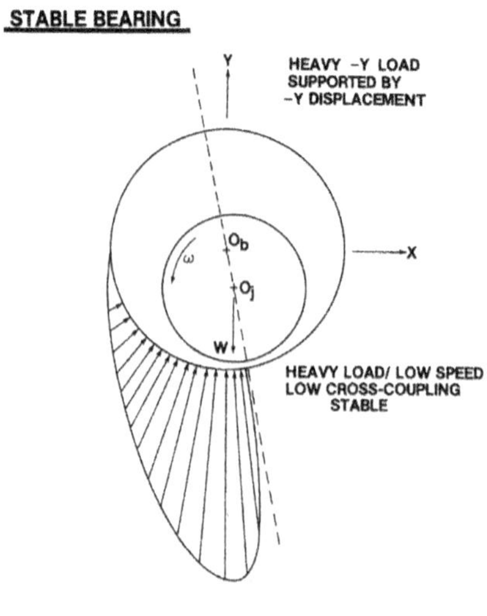

Figure 3.3 Low-speed, heavily loaded, stable bearing. *(RMT, Wellsville, N.Y.)*

the bearing eccentricity ratio ε is very large, and the attitude angle ψ approaches 0°. In this manner, a heavy $-Y$ direction load is supported by a Y displacement. This occurs because the load is so heavy and the resulting pressure profile becomes very large, with large gradients from the maximum film to the minimum film locations. From a force summation $\Sigma F_x = 0$ and $\Sigma F_y = W$. This can occur only for attitude angles that approach 0°. Since a downward load is now supported by a vertical displacement, cross-coupling forces are at a minimum and the bearing is stable.

Sleeve-journal-bearing-induced oil whirl and shaft whip are illustrated in Fig. 3.4. The 1× or synchronous vibration line is clearly indicated on the plot. Oil whirl, caused by the destabilizing cross-coupling forces produced by high-speed, lightly loaded sleeve bearings (i.e., a high Sommerfeld number), manifests itself as approximately 50 percent speed as of running speed frequency (shaft vibrates approximately once per every two shaft revolutions). This can be seen in Fig. 3.4 at speeds below approximately 7500 r/min (below twice the rotor's first critical).

Above 7500 r/min, the instability frequency locks onto the rotor's first fundamental natural frequency, which is at approximately 3800 c/min. This reexcitation of the rotor's first natural frequency is sleeve-bearing-induced shaft whip which shows up as a vibration component that is below 50 percent of running speed and occurs at speeds that are above twice the rotor's first critical speed.

Usually, sleeve bearings are designed not to be unstable until the rotor speed exceeds twice the rotor's first critical speed. Thus, an approximate 0.5× is a rare occurrence and bearing-induced instabili-

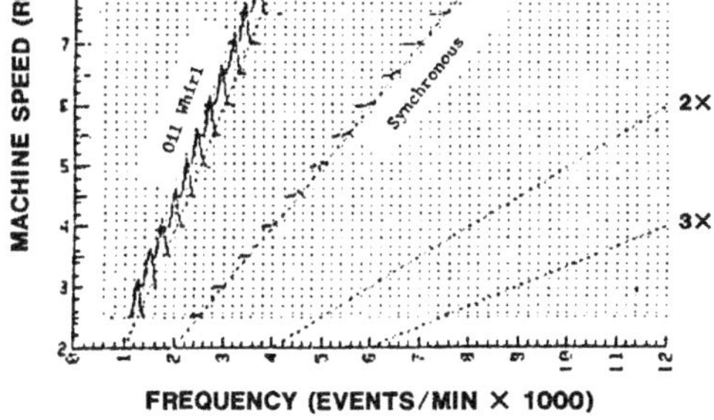

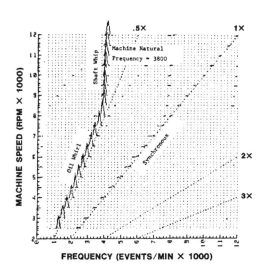

Figure 3.4 Bearing-induced shaft whip and oil whirl. *(RMT, Wellsville, N.Y.)*

ties usually show up as shaft whip at frequencies less than 50 percent of synchronous speed.

Axial groove bearings have a cylindrical bore with typically two to four axial oil feed grooves. These bearings have been very popular in relatively low-speed steam turbines. For a given bearing load magnitude and orientation, the stability characteristics of axial groove bearings are controlled primarily by the bearing clearance. Tight clearances produce higher instability thresholds but tight clearance bearings present other problems that make them undesirable. For example, as clearance decreases, the bearing's operating oil temperature increases. Furthermore, babbitt wear during repeated start-ups will increase the bearing's clearance, thereby degrading stability. In fact, many bearing-induced instabilities in the field are caused by bearing clearances that have increased due to wear from oil contamination, repeated starts, or slow-rolling with boundary lubrication.

Because of these limitations, other fixed-bore bearing designs have evolved to counteract some of the poor stability characteristics of axial groove bearings. Some antiwhirl sleeve bearing examples include pressure dam bearings, offset half bearings, and multilobe bearings. These bearing designs have been successful in increasing the instability threshold speed compared to axial groove bearings. The pressure dam bearing is probably the most popular with steam turbine designers and is still being used today in some lower-speed applications.

3.1.2 Tilting-pad journal bearings

Even though they are costlier than fixed-geometry bearings, tilting-pad journal bearings have gained popularity with steam turbine designers because of their superior stability performance. Unlike fixed-geometry bearings, tilt-pad bearings generate very little destabilizing cross-coupled stiffness regardless of geometry, speed, load, or operating eccentricity. However, turbines supported on tilting-pad bearings are still susceptible to instabilities due to other components within the machine such as labyrinth seals and/or the turbine blades (steam whirl).

Figure 3.5 is a schematic of a five-pad tilting-pad bearing loaded between pivots. Note that the journal center O_j is directly below the bearing center O_b, and the downward load is supported by a vertically downward displacement. Because of the pad's ability to tilt, the attitude angle is zero and thus the cross-coupling forces are zero. Figure 3.6 illustrates a typical shaft centerline plot for a tilting-pad bearing during a run-up to high speed. The attitude angle is very small and thus the cross coupling produced by the bearing is essentially zero.

Another advantage of tilting-pad bearings is the possibility of many variations in design parameters. Some examples include the number of

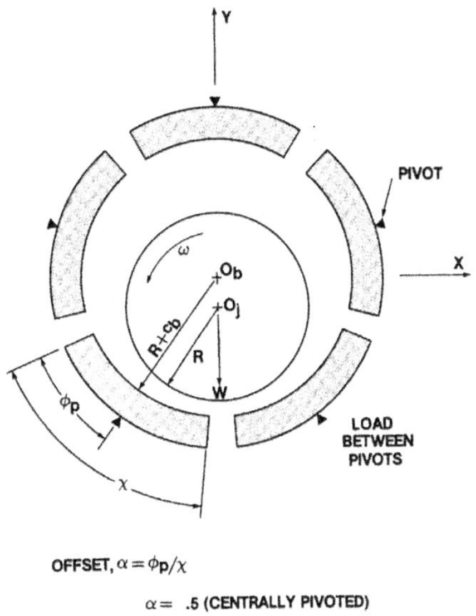

Figure 3.5 Tilting-pad bearing schematic. *(RMT, Wellsville, N.Y.)*

pads—four-pad designs are very popular with steam turbine designers, followed by those using five pads. The pivot loading is another design parameter to consider; load between pivots is used almost exclusively with steam turbines, as opposed to load on pivot. A typical steam turbine tilting-pad bearing is shown in Fig. 3.7 with four tilting pads loaded between pivots.

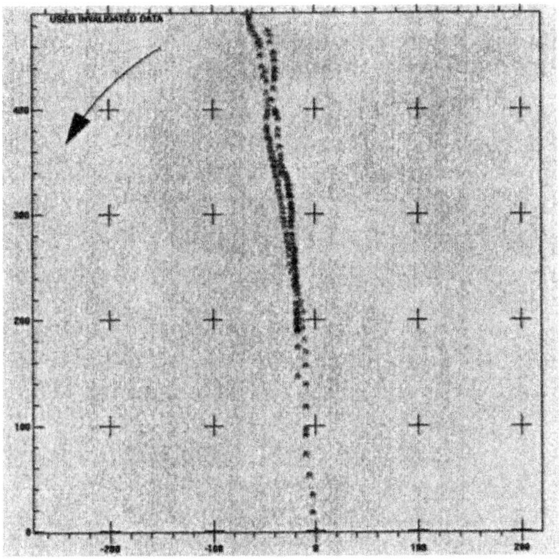

Figure 3.6 Tilting-pad bearing shaft centerline position. *(RMT, Wellsville, N.Y.)*

Figure 3.7 Four-pad tilting-pad bearing in steam turbine application. *(RMT, Wellsville, N.Y.)*

Between-pivot loading provides more symmetric stiffness and damping coefficients compared to on-pivot loading. Furthermore, a four-pad tilting-pad bearing loaded between pivots produces stiffness and damping properties that are equal in the horizontal and the vertical directions. Symmetric support properties provide circular orbits, whereas asymmetric supports cause highly elliptical orbits. Circular orbits are preferable for most steam turbine applications, since their vibration amplitudes are smaller going through the rotor, which is critical compared to the major axis of an elliptical orbit.

Pad pivot offset is another parameter available to a tilting-pad bearing designer. Pad pivot offset is defined as:

$$\alpha = \phi_p/\chi$$

where α = pad pivot offset

ϕ_p = angle from pad leading edge to pivot, degrees

χ = pad arc length, degrees

The pad pivot is offset if it is moved in the direction of rotation by some angle from the centered position. Centrally pivoted pads are more popular with steam turbine designers, but recently more turbine bearings are being designed with offset pivots, typically around 60 percent. Offset pivots increase load capacity and thereby reduce bearing operating temperatures.

Possibly the most important tilting-pad bearing parameter that the designer must consider is pad preload. Tilting-pad bearing preload is defined as:

$$m = 1 - (C_b/C_p)$$

where m = pad preload
C_b = bearing assembled clearance, mils
C_p = pad clearance, mils

The zero preload case is illustrated in Fig. 3.8, where the pad's radius of curvature equals the pivot radius ($R_p = R_b$) and thus the pad clearance equals the bearing clearance ($C_p = C_b$).

A preloaded pad is illustrated in Fig. 3.9. The pad clearance is greater than the bearing clearance. Typical preload values for steam turbine bearings range from 0.2 to 0.5. When a pad is preloaded, a converging film section always exists and the pad will produce a hydrodynamic force even if the bearing load approaches zero.

In many cases, the biggest advantage of light preloaded pads for steam turbine applications is that bearing damping tends to increase while bearing stiffness remains approximately constant. Both of these trends help to increase *bearing effective damping* (the amount of bearing damping that is effective in rotor vibration suppression) and decrease rotor vibration levels.

One disadvantage to decreased pad preload is that the top pads become unloaded and may be susceptible to pad flutter. Fluttering

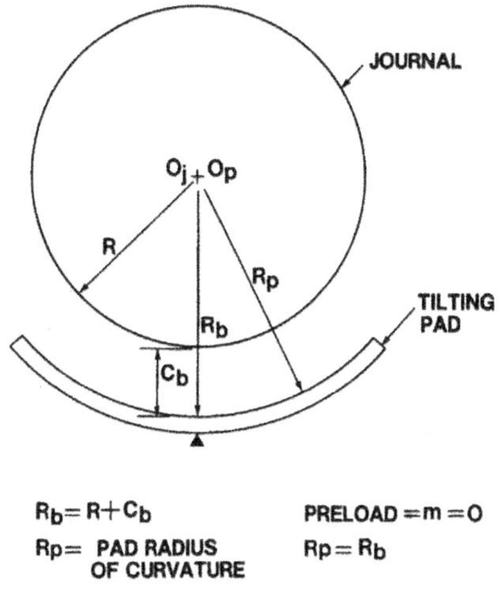

Figure 3.8 Zero preloaded pad. *(RMT, Wellsville, N.Y.)*

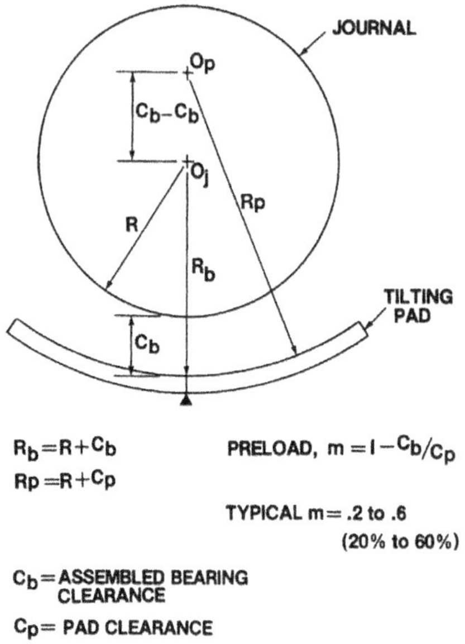

Figure 3.9 Preloaded pad. *(RMT, Wellsville, N.Y.)*

pads may cause babbitt damage near the leading-edge corners, as illustrated in Fig. 3.10.

Another powerful design parameter available to the tilting-pad bearing designer is pad axial length. In many steam turbine applications, increasing pad axial length increases effective damping by increasing

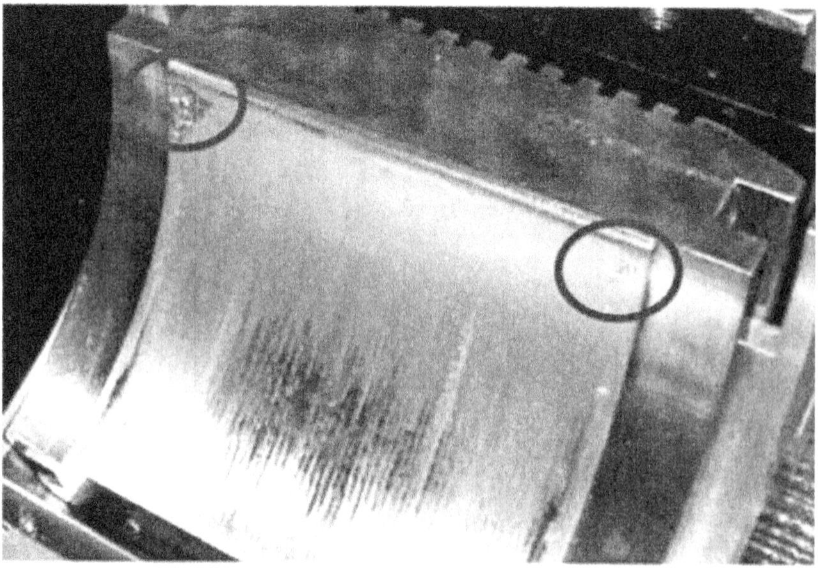

Figure 3.10 Pad flutter damage. *(RMT, Wellsville, N.Y.)*

bearing damping and decreasing bearing stiffness. For this reason, tilting-pad bearing L/D ratios have increased in recent years from a standard of 0.5 to 0.75.

3.1.3 Advanced tilting-pad journal bearings

In recent years, the focus of tilting-pad bearing designers has been on reduced-temperature designs. This has been driven partly by the stringent rotor dynamic specifications that steam turbine manufacturers must adhere to in order to sell their rotating equipment. Large journal diameters result in less severe critical speeds. However, larger-diameter journals result in high surface velocities, often exceeding 90 m/s (300 ft/s). Furthermore, for increased efficiencies, steam turbines are required to run at even higher speeds, again increasing surface velocity. At these high surface speeds, special care must be taken to cool the bearing.

Figure 3.11 illustrates a vintage pressurized housing design with oil inlet nozzles and drain holes between each set of pads from the 1980s. At the time, this was the best tilting-pad bearing available for cool operation. As speeds increased, increasing the bearing oil flow often

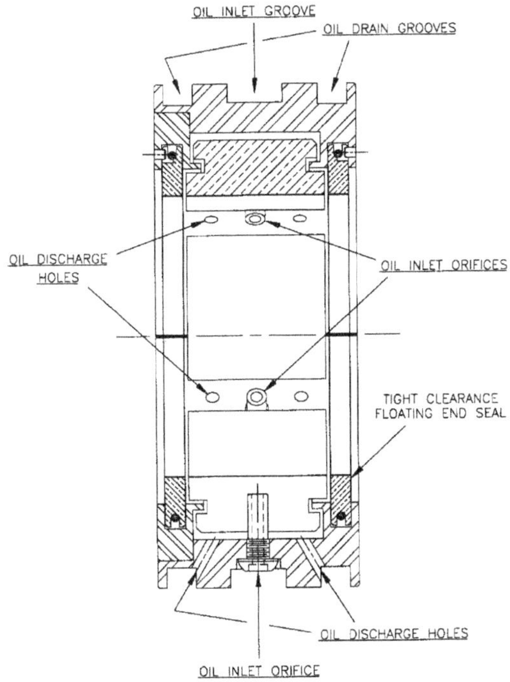

Figure 3.11 Pressurized housing tilting-pad bearing design. *(RMT, Wellsville, N.Y.)*

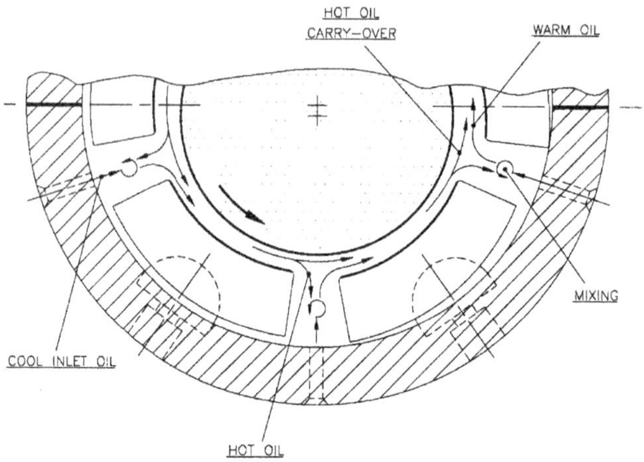

Figure 3.12 Pad-to-pad hot oil carryover and hot-cool oil mixing. *(RMT, Wellsville, N.Y.)*

produced only marginal decreases in bearing operating temperature. Two major problems associated with cooling a tilting-pad bearing are illustrated in Fig. 3.12. Unfortunately, a substantial percentage of hot oil is carried over by the journal from the trailing edge of the upstream pad into the leading edge of the next downstream pad. Furthermore, the cool inlet oil mixes with the discarded preceding pad's hot oil before it enters the downstream pad's leading edge.

It was soon discovered that better results were achieved by applying the cool inlet oil more effectively. One method of reducing the carryover and hot-cool oil mixing is to introduce cool inlet oil directly into the pad's leading edge, as illustrated in Fig. 3.13. This effectively prevents

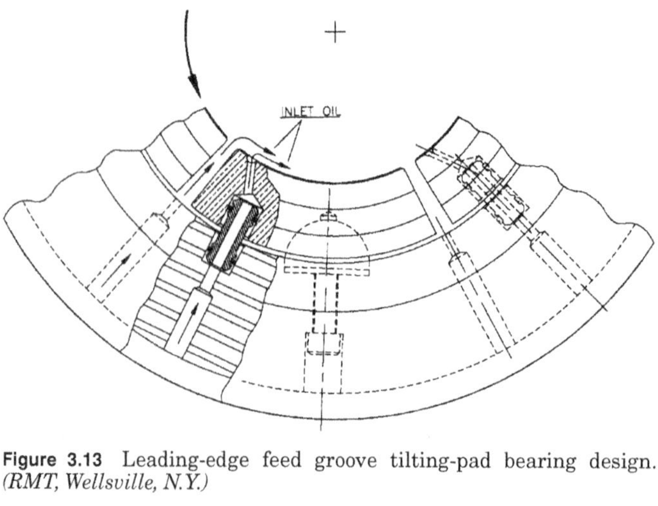

Figure 3.13 Leading-edge feed groove tilting-pad bearing design. *(RMT, Wellsville, N.Y.)*

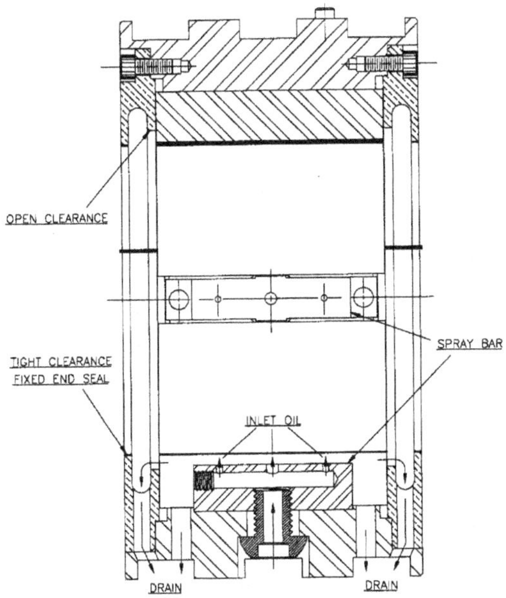

Figure 3.14 Evacuated housing spray bar tilting-pad bearing design. *(RMT, Wellsville, N.Y.)*

mixing and blocks some of the hot oil carryover. In 1991, Tanaka published results showing the cooling effect of removing the end seals of a bearing design similar to the one shown in Fig. 3.11. These results led to the design depicted in Fig. 3.14 featuring spray bars, open-end seals, and open housing drains.

Figure 3.15 shows the effectiveness of this evacuated housing design compared to the pressurized housing design illustrated in Fig. 3.11. This test was performed on a steam turbine with a 4.0-in four-pad tilting-pad bearing loaded between pivots. The evacuated housing design, which employs offset pivoted pads to further reduce pad oper-

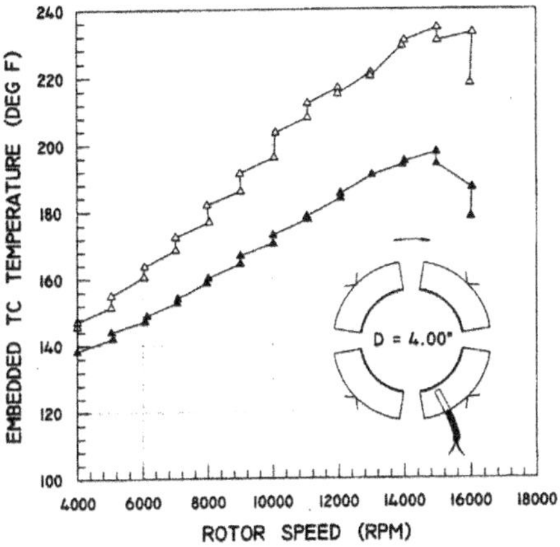

Figure 3.15 Evacuated vs. pressurized housing metal temperature comparison. *(RMT, Wellsville, N.Y.)*

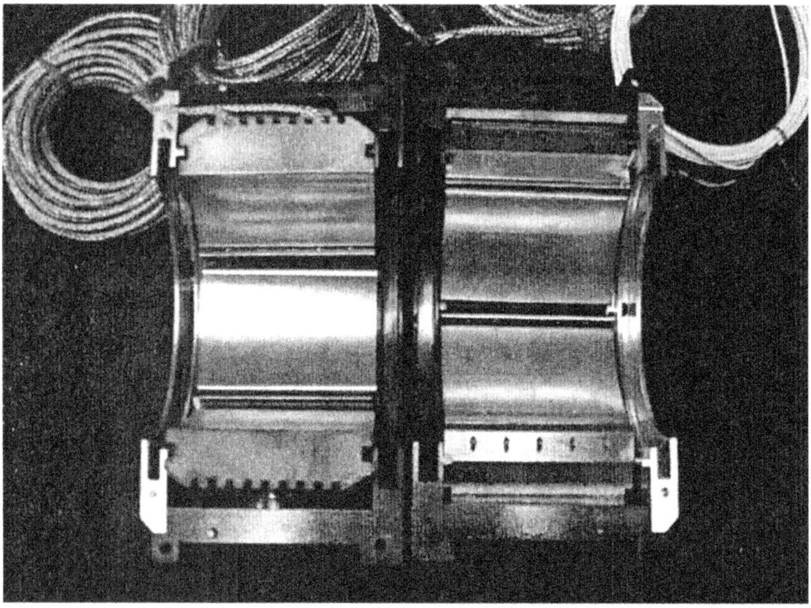

Figure 3.16 Advanced tilting-pad bearing design with Spray Bar Blockers® and Bypass Cooling®. *(RMT, Wellsville, N.Y.)*

ating temperatures, was effective in reducing bearing metal temperatures from 235 to 200°F. This design also employs offset pivoted pads to further reduce pad operating temperatures.

Figure 3.16 is a photo of an advanced evacuated housing design with babbitted chrome-copper pads. Since the chrome-copper material's heat transfer properties are a factor of six greater than those of steel, they are very effective in taking the heat away from the babbitt. This effect is amplified by the heat transfer grooves located on the back of the pads. Cool inlet oil is injected into these grooves, carrying the heat away from the pad body. This ByPass Cooling® oil does not participate in lubricating the bearing. Instead, it is routed out of the heat transfer grooves and directly into the oil drain.

Another feature of the bearing shown in Fig. 3.16 is the Spray Bar Blocker®. This device sprays cool inlet oil directly into the pad's leading edge, shielding it from the hot oil that exits the preceding pad's trailing edge. It also effectively blocks much of the hot oil carryover from the upstream pad, preventing it from entering the leading edge of the downstream pad. The hot oil is then allowed to exit the bearing. This advanced evacuated housing bearing is extremely effective in reducing bearing operating temperatures, enabling operation at extremely high surface velocities of up to 400 ft/s, and thus allowing steam turbines to operate at much higher speeds.

3.1.4 Lubrication-starved tilting-pad bearings

While the advanced tilting-pad bearing designs with evacuated housings have been successful in reducing the operating temperature of the pad by 10 to 20 percent, care must be taken when implementing the design. The wide-open end seals and housing drains result in low housing cavity pressures, often below 1.0 psig. Conversely, conventional flooded housing designs have typical housing pressures that range from 5 to 15 psig for a 20-psig oil inlet. These low housing pressures can lead to oil starvation and subsynchronous vibration if improperly applied.

Figure 3.17 illustrates one misapplication example in which open-end seals and open housing drains will not work in a high-speed balance vacuum. Even though the inlet oil is introduced to the housing through a spray bar at approximately 20 psig, the housing cannot maintain a positive pressure, and the oil immediately atomizes, resulting in oil starvation. This manifests itself as a subsynchronous rotor vibration.

The solution is to use dummy end seals with a reasonable clearance and to temporarily block the open housing drains. Dummy end seal clearances of approximately twice the bearing clearance will produce a reasonable housing pressure of approximately 10 psig. In this case, the inlet oil exiting the spray bars will not atomize even in a high-speed balance vacuum bunker.

Another example of the misapplication of the evacuated housing design is depicted in Fig. 3.18. In this case, the oil is introduced with a single housing hole between each set of tilting pads at the pad's axial centerline. The pad length-to-diameter ratio is 1.0, resulting in a rela-

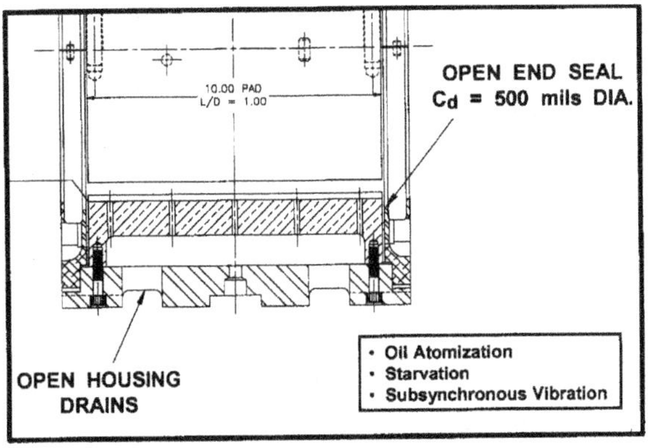

Figure 3.17 High-speed balance vacuum pit oil atomization resulting in subsynchronous vibration. *(RMT, Wellsville, N.Y.)*

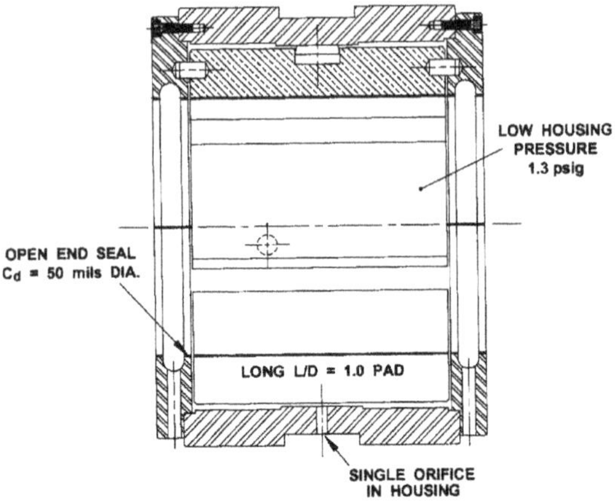

Figure 3.18 Single housing orifice design resulting in subsynchronous vibration. *(RMT, Wellsville, N.Y.)*

tively long pad. The designer, desiring lower pad operating temperatures, used open-end seals that resulted in a relatively low housing pressure of 1.3 psig. This low housing pressure coupled with the long pad caused oil starvation and a subsynchronous rotor vibration. The spray bar design shown in Fig. 3.19 distributed the oil along the full axial length of the pad and was successful in eliminating the subsynchronous rotor vibration.

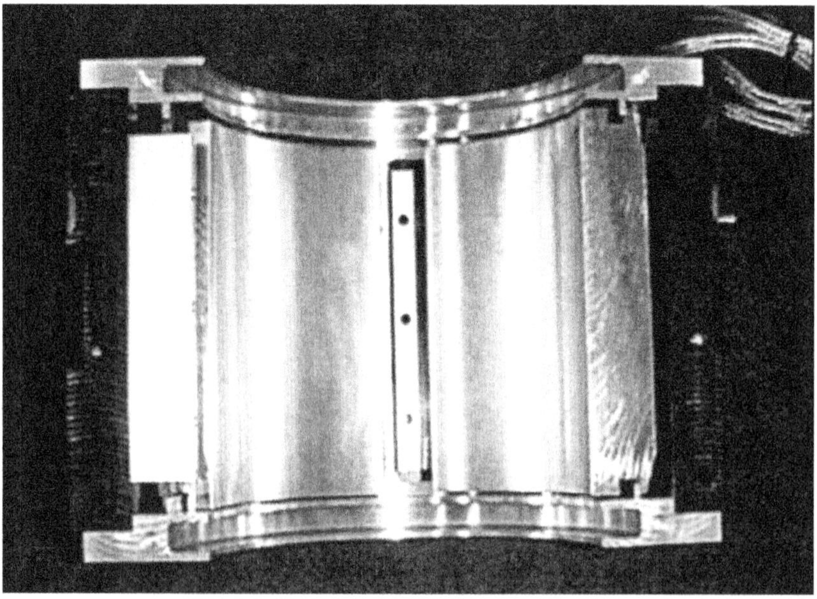

Figure 3.19 Multiorifice spray bar design. *(RMT, Wellsville, N.Y.)*

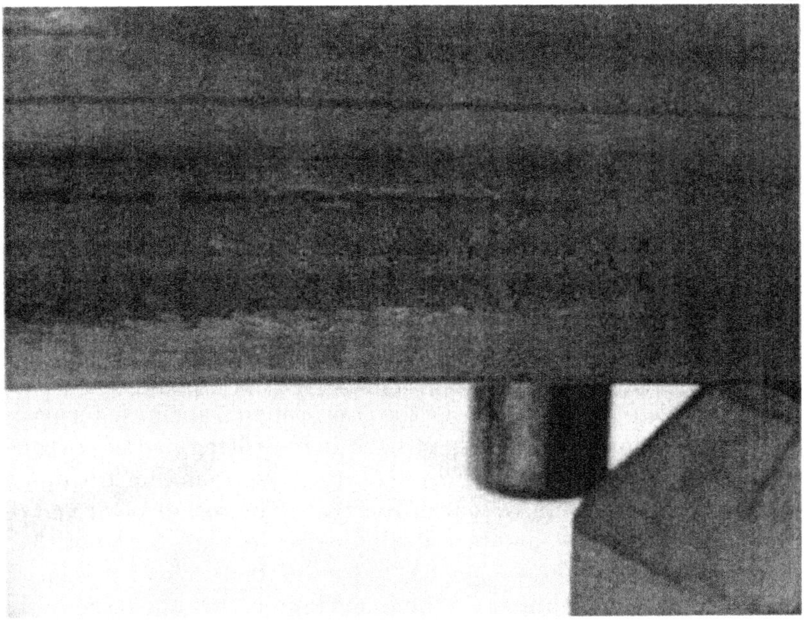

Figure 3.20 Dry friction rub caused by oil starvation. *(RMT, Wellsville, N.Y.)*

A typical dry friction rub on a tilting pad from a steam turbine is shown in Fig. 3.20. This rub was caused by insufficient lubrication oil starvation in an advanced evacuated housing design. The rub caused a subsynchronous vibration that was eliminated by increased oil flow.

Up to this point, discussion has addressed several styles of fixed geometry, or fixed pad, bearings. Each of these bearings has specific advantages in different applications, but they all have a characteristic called cross-coupled stiffness, which creates an out-of-phase force to the displacement and couples the equations of motion for the lateral degrees of freedom. Under certain conditions, this cross-coupling can cause the bearing to be unstable and an oil whirl will result. Oil whirl is essentially the same as rotation of the oil wedge at a frequency roughly one-half that of shaft r/min.

The tilting-pad journal bearing consists of several individual journal pads that can pivot in the bore of a retainer. The tilting pad is like a multilobe bearing with pivoting lobes, or pads. The same concept of preload applies to the tilting-pad bearing. The pads have a machined bore, and these pads can be set into a retainer to achieve a particular bearing set bore. The primary advantage of this design is that each pad can pivot independently to develop its own pressure profile. This independent pivoting feature significantly reduces the cross-coupled stiffness. In fact, if pad pitch inertia is neglected and the bearing is

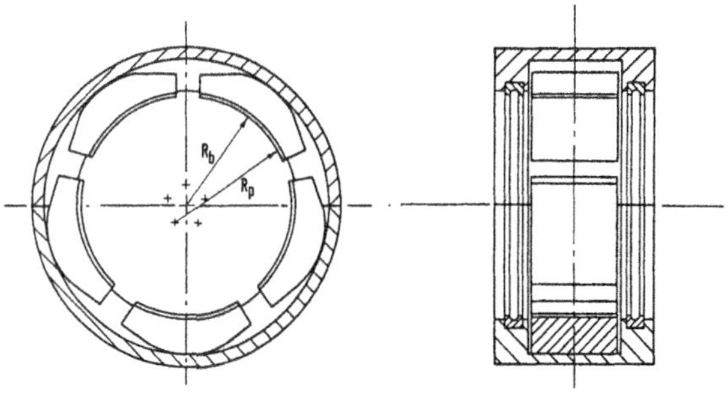

Figure 3.21 Rocker pivot tilting-pad journal bearing. *(Salamone Turbo Engineering, Houston, Tex.)*

symmetric about the vertical axis, the cross-coupled stiffness terms are eliminated. The number of pads utilized in the tilting-pad bearing can be three, four, five, or seven. However, the most common tilting-pad bearing arrangements have four or five pads. The rocker pivot and the spherical pivot arrangements are illustrated in Figs. 3.21 and 3.22, respectively. The rocker design has a line-contact pivot between the pad and the bearing retainer. As the name implies, the spherical design has a semispherical surface-contact pivot. Both of these designs allow the pads to pitch in the conventional manner. However, the spherical design has the additional ability to accommodate shaft misalignment.

3.2 Key Design Parameters

The discussion of preload brought out the distinction between two different bearing bores. These are pad machined bore and bearing set bore. Note that the plain journal, axial groove, and pressure dam bearings have only one bore (the set bore is the same as the pad bore). The difference between the pad machined bore radius and the journal radius is the pad machined clearance. The difference between the bear-

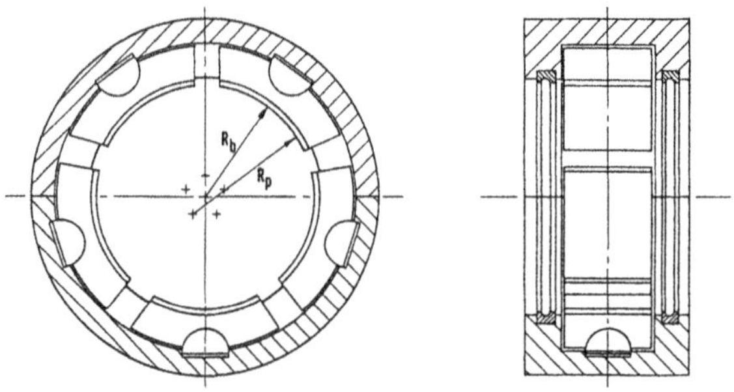

Figure 3.22 Spherical pivot tilting-pad journal bearing. *(Salamone Turbo Engineering, Houston, Tex.)*

ing set bore radius and journal radius is the bearing set clearance. The set clearance is the same as the running clearance, which is often specified as a clearance ratio of mils (milli-inches) per inch of journal diameter. Some typical values of clearance ratio are between 1.5 and 2.0 mils/in. Obviously, there are some applications where these values do not apply. The manufacturer will specify the recommended clearances for the particular bearing application.

Slenderness ratio is also referred to as L/D ratio. This is the ratio of the bearing length to the shaft diameter. This ratio typically varies between 0.2 and 1.0. However, some plain journal bearings have slenderness ratios above 1.0. The bearing length affects the stiffness and damping characteristics of the bearing. In the selection of a bearing length, one must consider the bearing unit loading. The unit load is the bearing load divided by the product of the bearing length and the shaft diameter; therefore, the units are psi. Typical values of unit loading are between 150 and 250 psi (10 to 17 kg/cm^2).

3.3 Thrust Bearings for Turbomachinery

The thrust bearing has two functions in a turbomachine: It constitutes the axial reference point for locating the rotor in the casing and it carries the axial thrust (Fig. 3.23).

The axial thrust can originate from steam thrust generated by the rotor parts subject to steam pressure or from the thrust forces developed in flexible couplings (gear tooth or diaphragm couplings).

A coupling thrust can always be expected when two shafts, each of which is located by an axial bearing, are connected via a flexible coupling. If one or both the rotors change their length because of temperature changes, forces are developed in the coupling that counteract the thermal movement. With gear tooth couplings the friction between the teeth has to be absorbed; with diaphragm couplings the spring force of the deflecting diaphragm comes into play.

The thrust bearing of many modern industrial turbines consists of the bearing collar (integral part of the shaft) and two rings of thrust bearing pads, each of which is provided with a tilting edge (Fig. 3.24). By tilting the pads a wedge-shaped gap can form between the bearing collar and the bearing pads. The space between bearing collar and pads is filled with oil.

Because of its viscosity the oil is forced from the surface of the rotating collar into the wedge gap.

Because oil is practically incompressible, the decreasing flow section in the wedge gap must effect an increase of the oil pressure in the wedge gap. This oil pressure is balanced by the axial force of the rotor via the bearing collar.

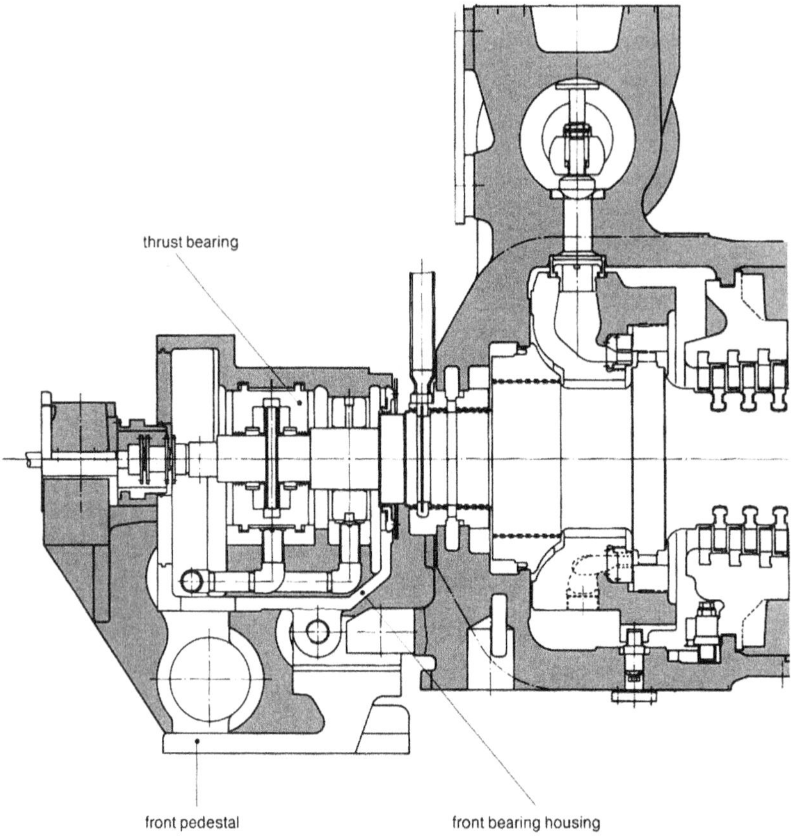

Figure 3.23 Thrust bearing location in front bearing housing of reaction turbine. *(Siemens Power Corporation, Milwaukee, Wis. and Erlangen, Germany)*

The bearing is therefore a thrust bearing with a hydrodynamically generated load bearing oil film. At constant axial load the thickness of the oil film increases with rising speed by the hydrodynamic effect on the loaded side.

Consequently, because of the constant axial clearance of the bearing collar the thickness of the oil film must decrease on the nonloaded side.

Within the normal speed range of the turbine there is no metal-to-metal contact between bearing collar and pads. Provided clean lubricating oil is used, the bearing is free from wear. Metal-to-metal contact is only possible on starting up. However, at this stage the axial rotor forces are very low. To avoid mechanical damage to the thrust bearing surfaces also under the condition of metal-to-metal contact, suitable materials must be selected for the bearing collar and the tilting-pad lining. In both turbine and compressor construction the use of low-alloy

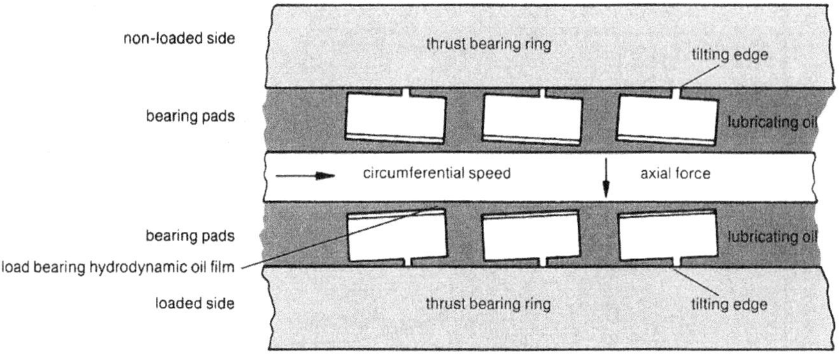

Figure 3.24 Functional representation of tilt-pad thrust bearing used in major turbomachinery. *(Siemens Power Corporation, Milwaukee, Wis. and Erlangen, Germany)*

carbon steel for the shaft (bearing collar) and babbitt metal for the surfaces of the tilting pads has proved very successful.

The load limit of a thrust bearing is defined by three criteria:

1. The minimum permissible thickness of the oil film between bearing collar and pads
2. The maximum temperature of the babbitted surface of the pads
3. The maximum permissible surface pressure with respect to the fatigue limit of the babbitt metal

At normal operating speed the thickness of the oil film must not fall below a certain minimum value to avoid metal-to-metal contact due to marginal deformation or overloading of the bearing pads.

The thickness of the oil film is a function of the circumferential speed and the axial force of the bearing collar. If a particular thickness of the oil film is stipulated, a permissible axial force and mean surface load is related to each circumferential speed. The higher the circumferential speed, the higher the permissible load for a given thickness of oil film.

These conditions are illustrated by limiting curve 1 in Fig. 3.25. Because a thrust bearing of twice the dimensions does not require double the minimum permissible oil film thickness, the load capacity of large bearings is higher than that of small ones at equal circumferential speeds.

However, with constant thickness of the oil film the oil temperature in the wedge gap increases with rising circumferential speed so that the temperature limit of the babbitt metal is reached (point A in Fig. 3.25). With a further increased circumferential speed therefore limit curve 2 applies for the permissible load, for which the maximum oil film temperature equals the temperature limit.

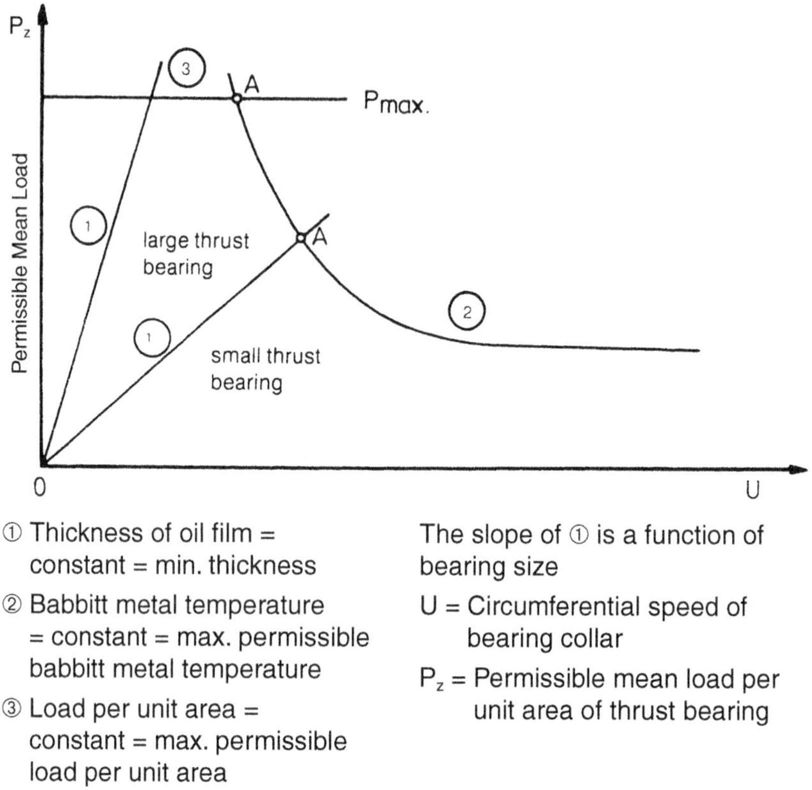

Figure 3.25 Typical operating parameters for turbomachinery thrust bearings. *(Siemens Power Corporation, Milwaukee, Wis. and Erlangen, Germany)*

The thickness of the oil film increases with rising circumferential speed, since the permissible load is now below that for constant oil film thickness.

For thrust bearings with mean diameter of 250 mm and up, the load permitted for constant oil film thickness reaches the load limit set by the fatigue limit of the babbitt metal prior to approaching the temperature limit. Therefore, the load must be maintained constant with rising circumferential speed (limit curve 3 in Fig. 3.25). The thickness of the oil film will now increase and the temperature rises. At point A of the diagram the temperature limit is again reached. From here on limit curve 2 applies again and the load decreases to a value below the fatigue limit.

In theory, the babbitt surface can stand temperatures of up to approximately 145°C (293°F). If a maximum operating temperature of 110°C (230°F) is selected, the safety margin up to 145°C (293°F) would correspond to a further 100 percent load increase.

Experience shows, however, that manufacturing tolerances of bearing pads as well as minor deformations of the thrust bearing housing effect different loading of the individual pads. This can result in possible temperature differences of the hottest spots of the individual pads of approximately ±20°C (36°F). A temperature limit of 90°C (194°F) is thus assigned to the hottest spot of the babbitt lining of thrust bearings used by such experienced manufacturers as Siemens. The permissible load is often defined for this conservative temperature.

The design specification for thrust bearings is therefore aimed at keeping the maximum temperature of the babbitt lining as low as possible. Also, the bearing design must be so rigid to keep any deformations leading to uneven load distribution as small as possible.

The temperature rise of the oil in the thrust bearing, i.e., the difference in temperature of the oil at inlet and outlet should be approximately 15°C (27°F). Considerable friction is generated at high speeds and high-axial forces and a large oil flow quantity is thus required. When operating at lower speeds a reduced flow of oil is adequate for the temperature difference of 15°C (27°F). The oil flow required to produce this temperature rise is obtained by plugging an appropriate number of oil drain holes. Too low a temperature rise results in unnecessarily high friction due to steeply increasing oil viscosity at low temperatures. Oil that is too hot causes excessive babbitt metal temperatures. Exact dimensioning of the axial clearance between the bearing collar and the two tilting-pad rings is also extremely important: Too much clearance will incorrectly locate the shaft and can produce axial vibration of the rotor upon thrust reversal.

Too little clearance, on the other hand, produces both unnecessarily high friction and thrust bearing loads. If a double-acting thrust bearing is operated without an axial force present, the oil films between the bearing collar and the two pad rings have the same thickness. The bearing collar locates itself exactly in the center of the bearing clearance. The oil films, however, already generate hydrodynamic forces against the two faces of the bearing collar. These forces are equal and balance each other. They already constitute a load for the axial bearing and frictional heat is generated. The smaller the axial clearance selected, the smaller the thickness of the oil film becomes at no-load operation and high loads act on the bearing pads already at no-load operation. In that case the no-load friction heat is also very high. If one of the two rings of pads is now subjected to an external axial load, the thickness of the oil film on the loaded side is decreased. This causes the hydrodynamic axial forces to increase until equilibrium with the external axial force is achieved.

Simultaneously the thickness of the oil film at the unloaded side increases (thickness of oil film = axial clearance – oil film thickness of loaded side), so that the hydrodynamic forces and thereby the frictional forces decrease on this side.

A thrust bearing can be monitored satisfactorily only by temperature sensors (thermocouple, resistance thermometer) near the babbitt lining in the area of highest oil film temperatures. A thermometer in the oil discharge of the thrust bearing cannot indicate overheating of a bearing pad, because the additional friction heat generated by the endangered pad is so small in relation to the total friction heat of the thrust bearing that temperature rise in the early stages of incipient bearing failure is not likely to be detected. A thermometer in the oil discharge can only monitor the bulk oil supply temperature of the thrust bearing.

Finally, some remarks on the application of thrust bearings in turbines: Well-engineered thrust bearings are often simple and robust and may incorporate a minimum of moving parts. The best proof of this statement is seen in Fig. 3.26*a* and *b*, which compare conventional rocking pivot tilt-pad bearings with the KMC flexure pivot configurations. Where the former types create a converging wedge through rocking motion of the pad pivot support, the latter form a wedge through flexure of the post support while eliminating the pivot wear and high-contact stresses. The same principles are embodied in the combination radial/thrust flexure pivot tilt-pad bearing depicted in Fig. 3.27.

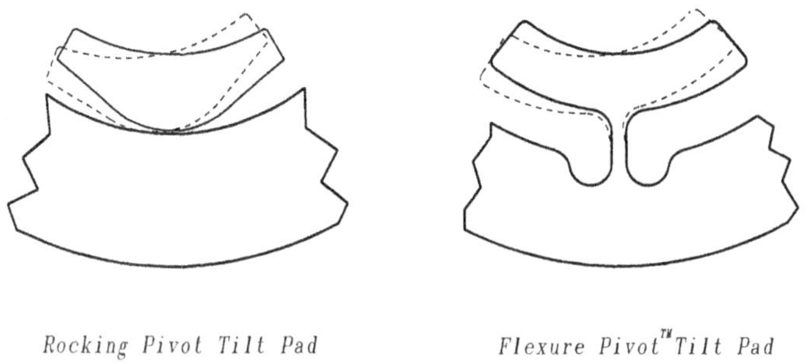

(a)

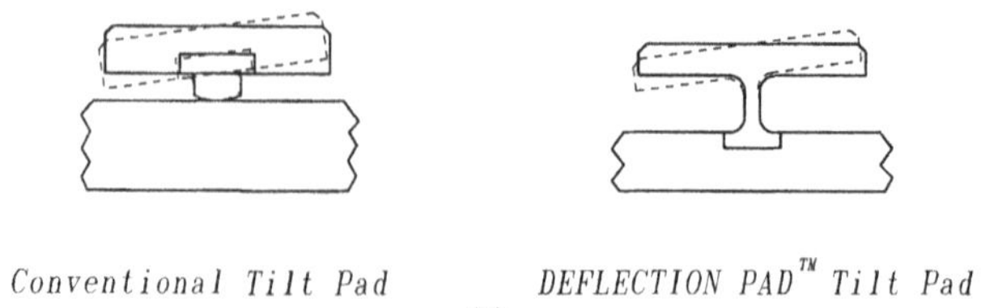

(b)

Figure 3.26 (*a*) Conventional rocking pivot tilt-pad radial bearing and modern flexure pivot™ tilt-pad principles; (*b*) conventional tilt-pad thrust bearing and modern flexure pivot™ tilt-pad principles. *(KMC, Inc., West Greenwich, R.I.)*

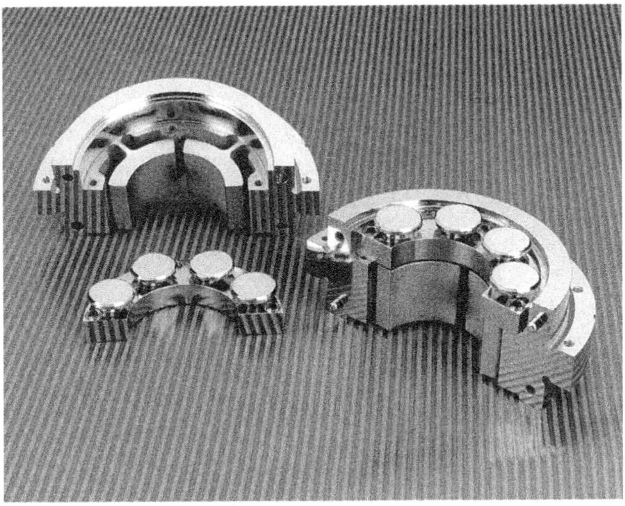

Figure 3.27 Combination radial thrust flexure pivot tilt-pad bearings. *(KMC, Inc., West Greenwich, R.I.)*

In many reaction turbines, the bearing housing is supported on adjusting elements in a pedestal that, in turn, is rigidly bolted to the concrete foundation. It is connected to the turbine merely by two thrust rods. With thermal expansion of the turbine outer casing, the bearing housing is shifted on the pedestal. It is guided so that it will slide without cocking or twisting. The outer casing itself is supported by its brackets on the pedestal. Any deflection of the turbine outer casing does not affect the position of the bearing housing.

For this type of mounting, thrust bearings with self-equalizing tilting pads (e.g., Kingsbury), are not required. However, for turbine designs where the bearing housing is affected by casing thermal expansion, or where compensation is achieved by bending of "wobble-plates," Kingsbury thrust bearings represent a better solution. Figure 3.28 depicts a Kingsbury bearing which accommodates considerable misalignment. This type of bearing is often found on impulse turbines and centrifugal compressors.

3.4 Active Magnetic Bearings

Active magnetic bearings (AMBs) are frictionless, unlubricated bearings that levitate moving machinery parts in an electromagnetic field. They are able to support large loads at speeds well beyond the range of conventional bearings with no wear and virtually no vibration. As of mid-1994, a few dozen major turbomachines (up to 7000 hp/approximately 5000 kW) and several hundred canned motor pumps and simi-

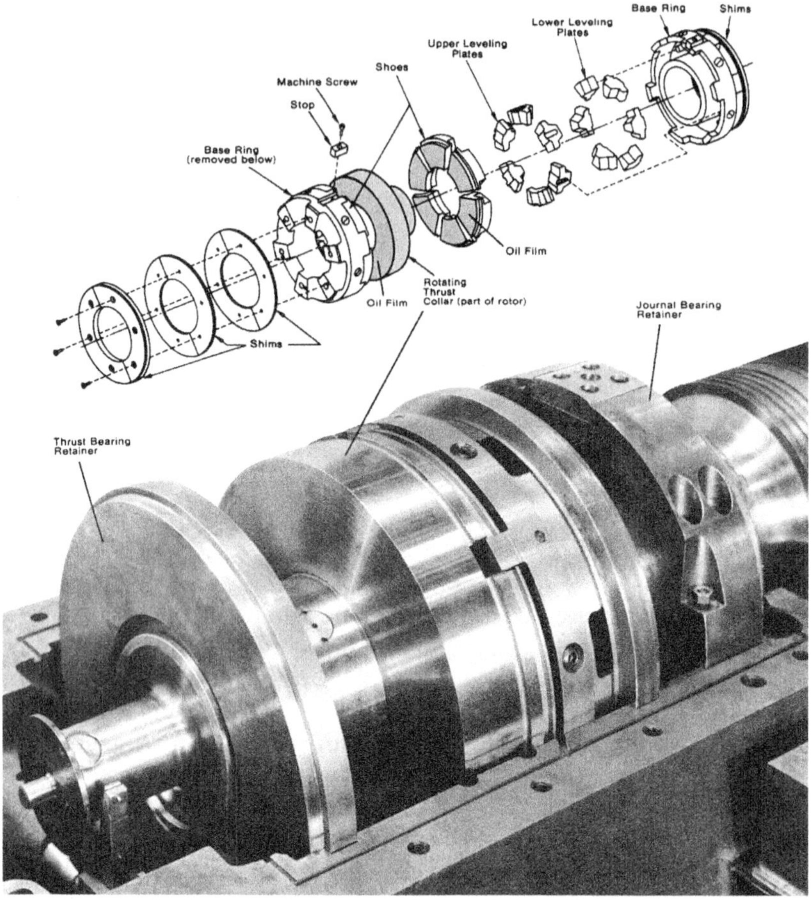

Figure 3.28 Kingsbury thrust bearing in a state-of-the-art impulse turbine. *(Elliott Company, Jeannette, Pa.)*

lar fluid machines have been equipped with these bearings. Bore ranges reached 1470 mm (58 in), speed bracketed 400 to 120,000 r/min, load capacities exceeded 80,000 N (17,600 lb), and temperature environments ranged from −185 to 480°C (−300 to 900°F).

As costs come down and more experience is gained, we will undoubtedly see more AMB applications in the future. The reason for this trend is best summarized in Table 3.1. For additional information, please refer to the section on Magnetic Bearings in the companion volume to this text, *A Guide to Compressor Technology.*

The operating principle of an AMB is simple: A rotating shaft is surrounded by electromagnetic coils that exert magnetic force to keep the shaft suspended. Unlike conventional bearings, magnetic bearings require no lubricants, seals, pumps, or valves.

The AMB typically includes a stator and a journal. The journal is assembled on a portion of the rotating shaft and is surrounded by the stator. In an eight-pole design, the stator is separated into four sectors. Current circulates to each magnetic coil, creating a flux within each bearing sector. This creates a magnetic force in each sector that can be varied without affecting the other sectors, allowing independent control and physical separation.

Each sector (two poles) exerts an attractive force on the journal. Two opposite sectors of poles at each axis are used to center the journal. These forces, which can be varied by current, provide the stability, stiffness, and damping control to the bearing.

TABLE 3.1 Comparison of Magnetic vs. Hydrodynamic Bearings

Requirement	Magnetic bearings	Hydrodynamic bearings
1. High loads	Load capacity low, but bearing area could be higher than with conventional bearings (see 3 below)	High load capacity (except at low speeds)
2. High speeds	Limited mainly by bursting speed of shaft; system response to disturbance must be considered carefully	Shear losses can become significant
3. Sealing	No lubricant to seal, and the bearing can usually operate in the process fluid	Seals must be provided
4. Unbalance response	Shaft can be made to rotate about its inertial center, so no dynamic load transmitted to the frame	Synchronous vibration inevitable
5. Dynamic loads	Damping can be tuned, but adequate response at high frequency may not be possible	Damping due to squeeze effects is high, and virtually instantaneous in its effect
6. Losses	Very low rotational losses at shaft, and low power consumption in magnets/electronics	Hydrodynamic and pumping losses can be significant, particularly at high speeds
7. Condition monitoring	Rotor position and bearing loads may be obtained from the control system	Vibration and temperature instrumentation can be added
8. Reliability and maintainability	Magnets and transducers do not contact the shaft so operating damage is unlikely; electronics may be sited in any convenient position	Very reliable with low maintenance requirements

SOURCE: MTI, Inc., Latham, N.Y.

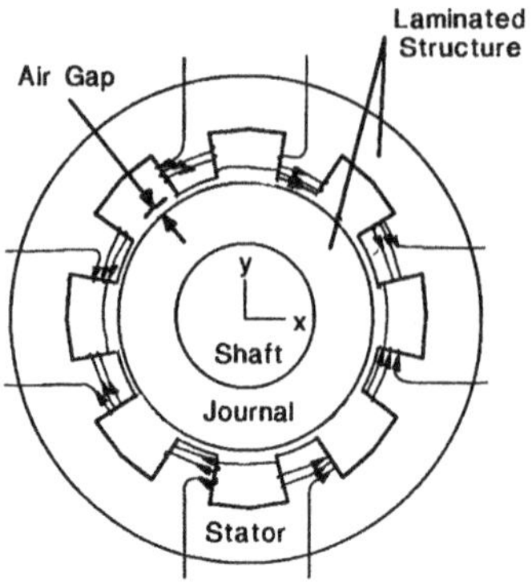

Eight-Pole Active Magnetic Bearing Configuration

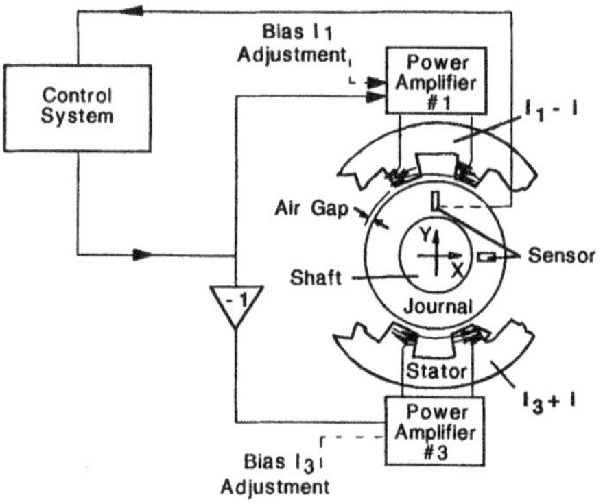

Basic Elements of Active Magnetic Bearing

Figure 3.29 Basic active magnetic bearing configuration. *(MTI, Inc., Latham, N.Y.)*

Both the stator and the journal are constructed of stacked, ferromagnetic laminations. Laminated materials reduce eddy currents that would cause power loss and degrade bearing performance.

The basic operation in an active magnetic bearing is:

$$\text{Magnetic force} \propto \left(\frac{\text{current}}{\text{air gap}}\right)^2$$

While this is a nonlinear relationship, linearization of bearing dynamics is achievable if the air gap is large, relative to the journal excursion, and the steady-state current supplied through each sector is large, compared to the dynamic control current. This steady-state or bias current produces the I^2R loss, the primary power loss in an active magnetic bearing. Because the total resistance of the coils and current path is quite small, the overall power loss is insignificant when compared to conventional bearings.

A journal that floats in a magnetic field produced solely by bias currents is not stable. To ensure stability, the journal motion must be monitored continuously and corrected instantaneously by superimposing a small control current on each bias current.

Figure 3.29 illustrates the basic active magnetic bearing configuration, showing the required sensors and control circuits in combination with the bearing elements. If the journal moves upward from its centered position, the current in the top sector is reduced by i. The current in the bottom sector is increased by the same amount. The control current produces a net downward force that pulls the journal back to the center. A modified version of this operating principle is incorporated in active magnetic thrust bearings.

The topic of magnetic bearings is given additional coverage in the companion volume to this text, *A Practical Guide to Compressor Technology.*

Chapter

4

Rotors for Impulse Turbines

Although rotors for impulse steam turbines exhibit great variety in physical size, wheel diameter, number of wheels and other construction features, they may all be conveniently classified in one of three basic categories:

1. *Built-up rotors:* Those rotors that are constructed by shrinking the wheels onto a shaft (Fig. 4.1)
2. *Solid rotors:* Those rotors in which the wheels and shaft are machined from a single, integral forging (Fig. 4.2)
3. *Combination solid and built-up rotors:* Those rotors in which some of the wheels are integral with the shaft and some are shrunk on (Fig. 4.3)

There are several factors that determine the type of construction that is utilized for any particular turbine rotor application. The most significant of these factors are:

1. Long-term operating experience
2. Pitch diameter
3. Maximum operating speed
4. Steam temperature

4.1 Long-Term Operating Experience

This selection factor is often influenced by user preferences based on maintainability criteria or design conservatism. An applicable American Petroleum Institute Specification (API 612) states:

Figure 4.1 Built-up rotor. At lower operating speeds, turbines are often supplied with built-up rotors in which the disks are shrunk and keyed to the forged alloy steel shaft. Disks are profiled to keep stresses at a minimum. *(Elliott Company, Jeannette, Pa.)*

> Purchaser approval is required for built-up rotors when blade tip velocities exceed 825 ft./sec. at maximum continuous speed or when stage inlet temperature exceeds 825°F.

Design conservatism relates to pitch diameter, operating speed, and steam temperature.

4.2 Pitch Diameter and Speed

A commonly used industry rule-of-thumb to determine if built-up construction is suitable for a particular application is when the product of r/min and pitch diameter does not exceed 160,000. This is easily kept in

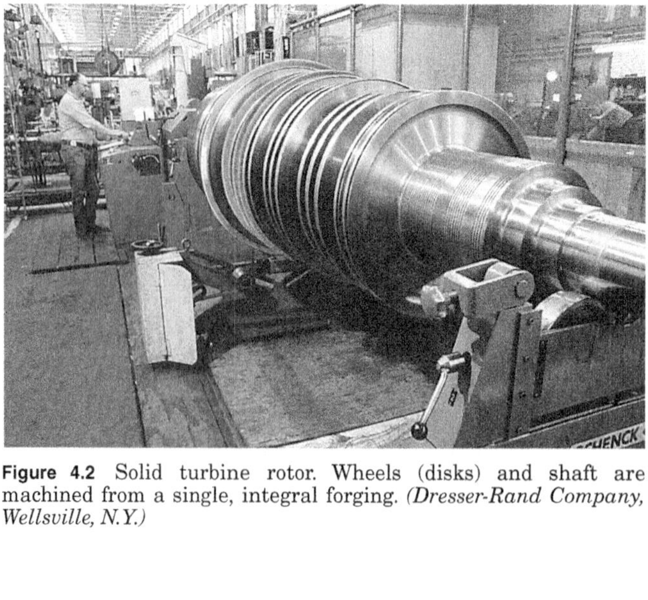

Figure 4.2 Solid turbine rotor. Wheels (disks) and shaft are machined from a single, integral forging. *(Dresser-Rand Company, Wellsville, N.Y.)*

Figure 4.3 Combination solid and built-up rotor. *(Dresser-Rand Company, Wellsville, N.Y.)*

mind by remembering that shrunk-on wheels (20 in or 508 mm pitch diameter) are suitable for a maximum speed of about 8000 r/min (20 × 8000 = 160,000). This is equivalent to a tip velocity limitation in the range of 700 to 800 ft/s (213 to 244 m/s), which is lower than the API 612 requirement.

4.3 Steam Temperature

Because the need for solid rotor construction is so readily related to tip speed (r/min and pitch diameter), special caution is required to ensure that the last of the four factors listed above, temperature, is not ignored. The relationship between steam temperature and the effectiveness of shrunk-on wheels is fairly straightforward. The wheels are held in place by an interference fit, which is achieved when the heated wheel is placed in position and allowed to cool to the same temperature as the shaft. As long as the wheel and shaft are at the same temperature, the desired interference fit or shrink is sustained. However, if during any transient operating condition a positive temperature differential develops between wheel and shaft, the desired shrink is reduced and may even be entirely lost. If this happens, the wheel is constrained from turning on the shaft by the key, but it may move axially with rather disastrous results. It should be evident that the danger of incurring a large temperature differential between wheel and shaft becomes more real as the temperature of the steam to which the wheel is subjected increases. For this reason, an integral wheel is utilized whenever the maximum temperature for the stage exceeds 750°F (400°C) or stage inlet temperature exceeds 825°F (440°C). Since stage temperatures diminish progressively from the first stage to the exhaust of a multistage turbine, a situation is frequently encountered in which stage temperatures dictate the need for integral wheels in some of the head end stages but become sufficiently low to permit shrunk-on wheels in the later stages. It is this situation that most commonly results in the combination solid and built-up type of rotor construction. This type of construction seems to be adaptable to conditions that are most commonly encountered on large condensing generator drive applications.

4.4 Built-Up Construction

Now that the basic categories of steam turbine rotor construction and the uses of each have been discussed, perhaps some of the differences in the manufacturing of the various types can be best pointed out by attempting to trace very briefly a rough step-by-step manufacturing sequence for each of the rotor styles.

After being received, the machining of the rough forging for a built-up rotor shaft begins on an engine lathe where all facing and turning operations are accomplished. In the turning of any critical shaft diameters, such as journals, shaft ends, and underwheel diameters, approximately 0.015 to 0.020 in (0.35 to 0.5 mm) is left for grinding to final dimensions. Provision is made for the location of the wheels by machining circumferential grooves in the shaft. These narrow shrink ring grooves are located axially on both sides of each intended wheel hub location. With turning and finish grinding operations completed, the next step in the machining sequence is normally the completion of all necessary milling operations. These include the milling of a keyway for each of the wheels and for any other keyed rotor components such as couplings, thrust collars, and governor drive worm gears.

Concurrently with the machining of the rotor shaft, the turbine wheels and blading are also in the manufacturing process. The rough wheel forging is machined to the desired profile, and the machining of the bore is completed. After a final trimming cut on the wheel rim, the circumferential dovetail groove that is to receive the bucket roots is machined in the wheel rim. Special care is necessary to achieve the required fit of wheel dovetail to blade dovetail (Fig. 4.4). After completing the machining of the bucket airfoils and rivets, the roots on a set of buckets are custom machined expressly to fit the groove in the particular wheel that is to receive the blades. This is necessary to achieve the close tolerances that are required to ensure a satisfactory degree of load sharing among the four bucket locks. The blades are assembled in the wheel by inserting each blade individually through a radial slot that is milled at one point in the wheel rim to provide access of bucket

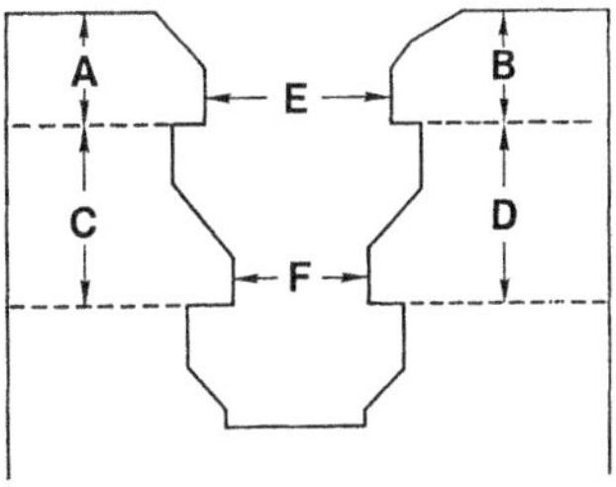

DIM	DWG	ACTUAL
A	.283	.283
B	.283	.283
C	.380	.380
D	.380	.380
E	.4605	.461
F	.322	.325

Figure 4.4 Wheel groove machining and inspection details are important to ensure turbine reliability. *(Dresser-Rand Company, Wellsville, N.Y.)*

root to dovetail groove. Once in the groove each bucket may be driven around to its ultimate position in the wheel. Buckets are, thereby, stacked in both directions from a point 180° opposite the access opening. The final bucket to be put in the wheel is a special locking bucket. This bucket has a specially formed root that is designed to fill the radial access slot in the wheel rim and lock all the other buckets in position. This locking bucket (Fig. 4.5) itself is held in the wheel by a locking pin (or pins) fitted into a drilled and reamed hole that passes axially through the blade root and both sides of the wheel rim. After the blades are completely assembled in the wheel, the bucket shroud is

Figure 4.5 Steam turbine locking bucket. *(Dresser-Rand Company, Wellsville, N.Y.)*

Figure 4.6 Bucket shroud fastened to bucket tips. *(Dresser-Rand Company, Wellsville, N.Y.)*

attached (Fig. 4.6). Each shroud segment is placed over a group of blades (normally five, six, or seven blades per group) with the rivet on the tip of each blade extending through a drilled hole in the shroud segment. The attachment is made by peening over the head of each blade rivet. After completion of the bucketing and shrouding procedures, each wheel is statically balanced with any necessary corrections being made by grinding material from the wheel rim.

There are special cases where integrally shrouded blades are used in built-up rotor construction. These are essentially lightly loaded blades (as used in geothermal applications), and the feature is to eliminate stress risers (riveted junction) in the anticipated corrosive environment (Fig. 4.7). In addition, the integral shroud will aid in thermodynamic performance. Integrally shrouded blades are not standard practice.

In preparation for the rotor assembly the wheels are placed in a gas-fired furnace and heated as required to achieve the necessary bore expansion. The actual shrinking-on procedure is normally accomplished with the rotor supported in vertical position with the exhaust end down. Starting with the last stage each wheel is removed from the furnace in turn and lowered down over the governor end of the shaft to its proper position where it shrinks tightly on the shaft as the wheel cools. Each wheel must be turned to align the keyway in the bore with the key that is prepositioned in the shaft keyway. Keyways for adjacent wheels are oriented 180° apart on the shaft and this, in turn, establishes an oppositely oriented 180° alternate spacing of locking buckets.

Figure 4.7 Integrally shrouded blading used in special applications. *(Dresser-Rand Company, Wellsville, N.Y.)*

Each wheel is preceded and followed onto the shaft by a heated ring that shrinks into a previously machined shrink ring groove in the shaft to provide positive axial positioning of the wheel.

The shaft material for built-up rotor construction for engineered turbines is usually per ASTM A293, Cl. 3, a chrome-molybdenum-nickel alloy steel. The forging is purchased with a proper heat stability test per the requirements of ASTM A293. The commercial specification for these materials is given in Table 4.1.

TABLE 4.1 Typical Materials of Construction for Multistage Mechanical Drive Steam Turbines

	Material	Commercial specifications
Steam chest and casing		
600 psi—750°F/41 bar—399°C	Cast carbon steel	ASTM A-216 Grade WCB
600 psi—825°F/41 bar—440°C	Carbon-molybdenum steel	ASTM A-217 Grade WC1
900 psi—900°F/62 bar—482°C	Chromium-molybdenum steel	ASTM A-217 Grade WC6
2000 psi—950°F/138 bar—510°C	Chromium-molybdenum steel	ASTM A-217 Grade WC9
Exhaust casing		
Condensing and Non-Condensing (Cast)	High-strength cast iron	ASTM A-278 Class 40

TABLE 4.1 Typical Materials of Construction for Multistage Mechanical Drive Steam Turbines (*Continued*)

	Material	Commercial specifications
Non-Condensing (Cast)	Cast steel	ASTM A-216 Grade WCB
Fabricated	Steel	ASME SA-516 Grade 60 & 70
Nozzles	12% Chromium	AISI-405
Nozzle rings	Cast iron	ASTM A-278 Class 40
	Carbon steel	ASME SA-516 Grade 60
	12% Chromium-stainless steel	AISI 405
Diaphragm centers		
Fabricated	Steel	ASME SA-516 Grade 60 or A514-F
Cast	Ductile iron	ASTM A-536 Grade 60-45-12
Disks (built-up rotors)		
Forged	Chromium-nickel-molybdenum steel	AISI 4340 ASTM A471 Class 6
Cross-rolled plate	High-strength alloy steel	ASTM A517 Grade F
Blades	12% Chromium-stainless steel	AISI Type 403 or ASTM A565
Shroud Bands	12% Chromium-stainless steel	AISI Type 410
Damping Wire	15% Chromium steel	Inconel X750
Rotor	Nickel chromium-molybdenum steel	AISI 4340
Built-up	Chromium-molybdenum steel	AISI 4140
Integral	Chromium-nickel-molybdenum-vanadium steel	ASTM A-470 Class 4, 7 or 8
Bearing liners	Bonded tin-base babbitt	ASTM B-23 Alloy #2
Bearing housings	Cast steel	ASTM A-216 Grade WCB
Bearing retainers	Steel	ASME SA-516 Grade 60
Shaft end labyrinth seals		
Sealing strips	High lead bronze	ASTM B584
	12% Chromium-stainless steel	AISI 410
Stationary baffles	Nickel steel	ASME SA-516 Grade 60
	Chromium-molybdenum-steel forgings	AISI 4340
Governor valves	12% Chromium-stainless steel	AISI Type 410
Governor valve stems and seals	12% Chromium-stainless steel, nitrided	AISI Type 416
Governor valve seats	12% Chromium-stainless steel	AISI Type 416
Bar lift rods and bushings	12% Chromium-stainless steel, nitrided	AISI Type 416
Steam strainer screen	Stainless steel	AISI Type 321

4.5 Solid Construction

As might be expected, the manufacturing sequence for a solid rotor includes some operations or procedures that duplicate those employed in manufacturing a built-up rotor. However, there are some basic differences. Since both rotor shaft and wheels are machined from a single forging (Fig. 4.8), the turning of the shaft diameters and machining of wheels are combined into one integrated machining sequence. As is done with the shaft for a built-up rotor, the final 0.015 to 0.020 in is removed from all critical shaft diameters by grinding. Normally, the turning and grinding operations are followed by all necessary milling operations. When specified, axial balancing holes are drilled in some or all of the wheels. These holes serve to ensure equalization of pressure on both sides of a turbine wheel and thereby reduce steam thrust while making some contribution to stage efficiency. Unless stress considerations dictate otherwise, balancing holes are provided on all flugelized stages having little or no reaction and greater than 60 percent admission. The drilling of these holes in the wheels of a solid rotor is a more demanding machining operation than for separate turbine wheels because of access limitations imposed by the adjacent wheels.

The bucketing procedure for a solid rotor is essentially the same as for a separate turbine wheel. As in the separate wheel the buckets are inserted into the dovetail groove through a radial access slot and from

Figure 4.8 Solid rotor construction. *(Dresser-Rand Company, Wellsville, N.Y.)*

there the buckets are driven around in the groove as required to fill the wheel by stacking buckets in both directions from a point 180° opposite the access opening. The wheel is closed in the same manner as described previously by fitting and pinning a special locking bucket or locking piece to fill the bucket access opening. The choice of a locking bucket or locking piece (essentially a locking bucket root without any attached airfoil) is dictated by stress considerations. If the weight of an airfoil would tend to impose prohibitively high stresses on the locking pin (or pins), the substitution of a locking piece is indicated. The use of shroudless blades (Fig. 4.9) is among the design options open to the manufacturer.

The rotor material for solid rotor construction is either ASTM A-470 Class 4 (for temperatures to 900°F [482°C]), or ASTM A-470 Class 8 (for temperatures to 1050°F [565°C]). These are chrome-molybdenum-nickel-vanadium alloy steels. The forgings are typically purchased with a proper heat stability test and ultrasonic inspection per ASTM A-470.

4.6 Shaft Ends

In view of the rather stringent requirements imposed on coupling hub bores and shaft ends by the American Petroleum Institute Specification on special purpose couplings (API 671), there is now a progressive trend toward the use of integrally flanged coupling hubs on solid rotors. Figure 4.10 shows such a rotor.

Figure 4.9 Shroudless blading without lacing wires. *(Dresser-Rand Company, Wellsville, N.Y.)*

Figure 4.10 Integrally flanged coupling hub on a solid steam turbine rotor. *(Dresser-Rand Company, Wellsville, N.Y.)*

A number of manufacturers have applied this approach to large generator drive units and will recommend integral hubs on future critical service mechanical drive units (typically, ethylene or ammonia service). There may be other occasions where an integral flange will improve a rotor dynamics situation (typically lightweight, high-speed rotors) by reducing the overhang and overhung weight.

4.7 Turbine Rotor Balance Methods

A balance machine is used to detect the amount and location of the unbalanced masses on a rotor. In essence, it is simply a device that spins the rotor on a set of spring-mounted bearings. With the soft bearings, any imbalance will cause the rotor to move about as it spins. The machine measures the phase angle and amplitude of the movement, and computes the unbalance which must be present to cause the motion. Appropriate corrections can then be made by the operator.

The method used to assemble and balance turbine rotors may be divided into two general methods applied either to built-up or integral rotors.

With the built-up rotor, each wheel or disk is fully machined, with the blading installed and completed. The wheel is then fitted temporarily with a small shaft or balancing arbor, and statically balanced by grinding metal from a suitable location at the edge of the disk. If the turbine shaft has a relatively small diameter and is symmetrical, it is assumed to be in balance. The wheels are assembled to the shaft using keys and a shrink fit. When cool, the shaft is checked for straightness and placed in a balance machine for dynamic balancing. The correction required at this step is usually quite small since each wheel has already undergone static balancing. The remaining parts such as the thrust disk and overspeed cup assembly are then put on the shaft and a final correction made to the edge of the thrust disk if required.

The integral, or solid rotor, requires a somewhat different approach. Since the wheels are part of the shaft, they cannot be balanced individually. The shaft is therefore dynamically balanced after machining, but before any blades or buckets are installed. Since it is symmetrical, very little correction is usually required. Wheels are then bladed in small groups, or one at a time, with a balance performed between each group. In this way, if an unbalance is introduced by the blading procedure, it is corrected at the same general plane of imbalance rather than elsewhere on the rotor. As with the built-up rotor, the final balance is done after the smaller parts, thrust disk, etc., are installed.

4.7.1 At-speed rotor balancing

A preferred balance procedure is to perform balance corrections utilizing an at-speed facility. The rotor is to be assembled, given a preliminary low-speed balance, and then a final at-speed balance using either job bearings or those of the same type to be supplied in the unit.

Low-speed balancing of multimass rotors is an attempt to improve the mass distribution to more nearly coincide with the axis of rotation. This process only measures the sum of unbalance at two journal locations with the rotor in a rigid state. Unbalance forces can still exist at service speed, especially for flexible rotors where bending of the rotor can exaggerate a specific unbalance distribution. Incremental low-speed balancing can help alleviate the offset of unbalance distribution, as only two major mass components are summed in each balancing operation.

However, the most effective results can be obtained by measuring dynamic forces at the bearings at maximum service speed using bearings that closely simulate actual operating rotor/bearing dynamics. For flexible rotors, where bending deflections at speed modify the resulting forces from residual unbalance, balancing corrections can be made in any of several planes. Trial weights can determine the most effective permanent corrections, minimizing deflections and resultant bearing forces. With an at-speed balancing process, incremental rotor assembly and balance are thus not required. Only a preliminary three-plane, low-speed balance is needed for flexible rotors to limit initial rotor response, prior to making corrections using at-speed measurements.

Rotor response plots, along with those of bearing pedestals, can be generated for the rotor with initial low-speed balance corrections, and for final high-speed corrections. Typical final results give less than 0.5 mils rotor vibration and pedestal velocity equivalent to less than 10 percent of rotor static load at maximum continuous speed.

Some advantages of at-speed balancing

1. Compensation of unbalance distribution of flexible rotors and effect of oil film on spin axis results in lower vibration levels, permitting more margin for changes due to service conditions. This apples both to at-speed and while traversing through the first critical speed.
2. Residual stresses from assembly and temporary rotor bends due to internal moments can be alleviated by running rotor to overspeed limit, resulting in final conditions prior to final balancing.
3. Less metal removal is required since the process is more effective.
4. Sensitivity and balancing accuracy are actually increased due to pedestal flexibility and mass, as compared to more rigid and massive machinery supports.
5. Shipments of new equipment and start-up of repaired or rerated rotors are expedited.
6. Unbalance response calculations can be verified in the at-speed balance machine. Differences in the support stiffness of the balance machine and the actual turbine casing have to be taken into consideration for this comparison.

Some disadvantages of at-speed balancing

1. With pedestal stiffness and oscillating mass used to obtain a force-measuring system, rotor critical speeds are reduced to below field values. Sensitivity is increased even higher when second critical speed is reduced to below trip speed, and added care is required in rare cases whenever it is reduced to below maximum operating speed. This sometimes requires increasing allowable limits.
2. Due to drive coupling limitations, heavy rotors must be lifted for start-up using hydrostatic lift bearings. Lift grooves in bearings affect oil film stiffness and damping, which may further reduce critical speeds.
3. After at-speed balancing, the low-speed balance readings can be above standard tolerances in ounces per inch, as they are measured on an oil film and are only equivalent resultants measured in a rigid state. In this case, documented values should be used if there is a

check balance in another low-speed machine prior to operation, to retain the low vibration during testing.

4.8 Balance Tolerance

The degree of balance that must be obtained with a rotor is governed by the weight of the rotor and the speed at which it runs. The faster and lighter a rotor is, the more precise is the required balance. The usual tolerance is given by the formula adopted by API in 1987. It states that the maximum allowable residual imbalance per plane (journal) is to be calculated as follows:

$$U = \frac{4W}{N}$$

In SI units, this translates to:

$$U = \frac{6350W}{N}$$

where U = residual unbalance, oz-in (g-mm)
W = journal static weight load, lb (kg)
N = maximum continuous speed, r/min

Finally, during the shop test of the machine, assembled with the balanced rotor, operating at its maximum continuous speed or at any other speed within the specific operating speed range, the peak-to-peak amplitude of unfiltered vibration in any plane, measured on the shaft adjacent and relative to each radial bearing, may not exceed the following value or 2.0 mils (50 μm), whichever is less:

$$A = \sqrt{\frac{12,000}{N}}$$

In SI units, this translates to:

$$A = 25.4\sqrt{\frac{12,000}{N}}$$

where A = amplitude of unfiltered vibration, mils (μm) peak to peak
N = maximum continuous speed, r/min

Chapter

5

Rotors for Reaction Turbines

During the first two decades of this century, experienced manufacturers used solid forged rotors for small turbines. For larger units they preferred built-up rotors, consisting of a number of disks shrunk on to a central shaft. For small machines the solid forged rotor is still standard, but for the large machines the shrunk disk design was subsequently discarded because of its higher stress levels. Articles from a number of independent sources deal with the stress levels and quality of this type of rotor.

5.1 Solid Rotors

Solid rotors can be defined as those forged from a single piece (monoblock). Figure 5.1 shows a solid rotor forging being produced in a steel plant. Only after extensive testing, which is described later, is the rotor accepted for further machining. Figure 5.2 shows a solid rotor being machined in a lathe.

The bladed rotor of an 11.5-MW turbine is shown in Fig. 5.3, while Fig. 5.4 shows a section through a 15-MW mechanical drive turbine rotor.

The choice of whether a solid rotor can be used is dependent on the start-up procedure. The temperature and stress conditions arising because of the admission of hot steam onto the cold rotor must be exactly known and controlled. This is because the thermal stresses due to the temperature difference between the surface and the center line of the rotor determine whether it can be manufactured as a solid rotor.

The factors responsible for the stress levels in the rotor are basically the steam temperature, the geometry (diameter) of the rotor and, in particular, the time available to reach the full-load condition. The exact relationship between these criteria has been established following

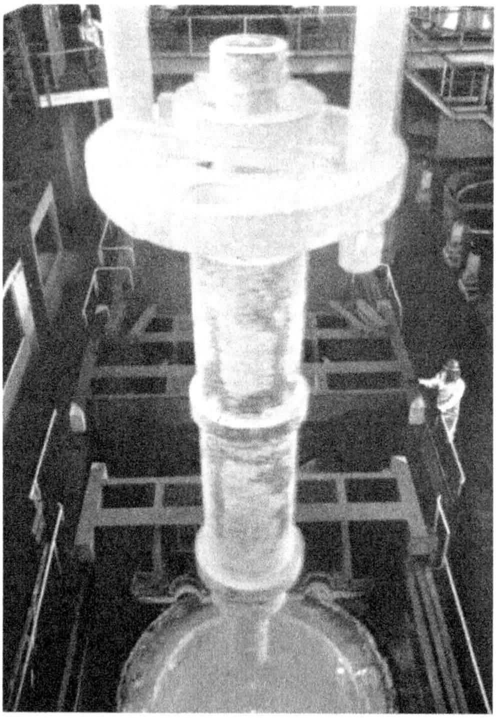

Figure 5.1 Solid rotor forging for reaction steam turbine being produced in a steel plant. *(Asea Brown-Boveri, Baden, Switzerland)*

extensive testing by such manufacturers as Brown-Boveri. If a particular operating condition exceeds the limits for a solid rotor, then welded rotors are employed (see Sec. 5.3). Figure 5.5 depicts an approximation of stresses set up in the rotor as a function of diameter and start-up procedure.

Figure 5.2 Solid rotor being machined in a lathe. *(Asea Brown-Boveri, Baden, Switzerland)*

Figure 5.3 Bladed solid rotor of an 11.5-MW steam turbine. *(Asea Brown-Boveri, Baden, Switzerland)*

Finite-element calculations allow the complete start-up procedure to be established and enable accurate predictions to be made about the temperature and stress conditions occurring at any time between the start-up and stationary condition.

Figure 5.6 shows the finite-element mesh of the high-temperature section of the solid rotor for a 45-MW condensing turbine. This mesh

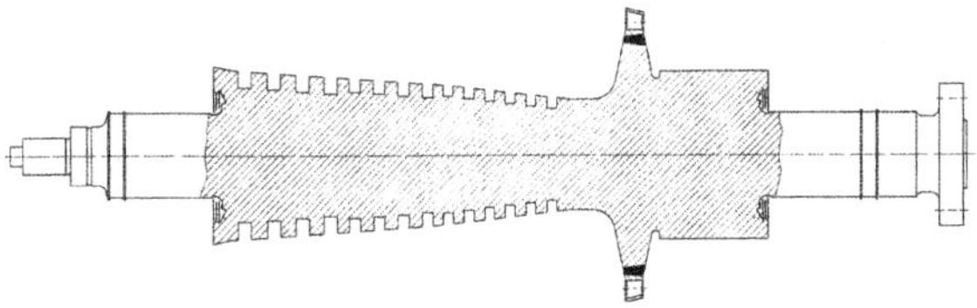

Figure 5.4 Section through the solid rotor of a 15-MW industrial steam turbine. *(Asea Brown-Boveri, Baden, Switzerland)*

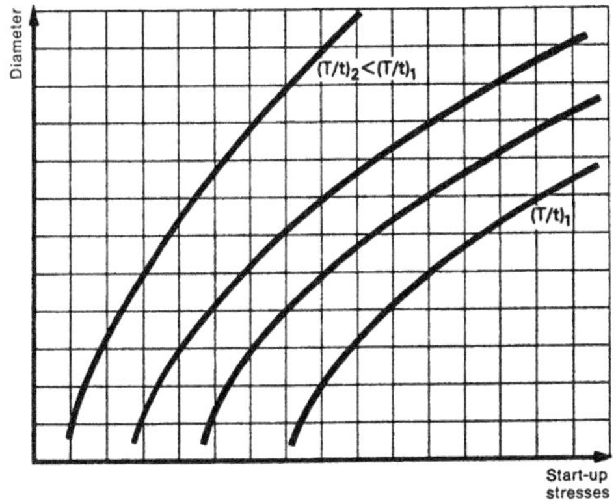

Figure 5.5 Relationship between stresses set up in a solid rotor, using diameter vs. start-up rate criteria (T, rotor temperature; t, time interval). *(Asea Brown-Boveri, Baden, Switzerland)*

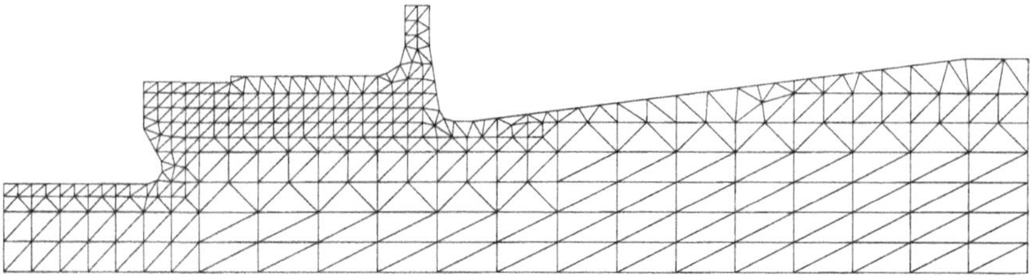

Figure 5.6 Finite-element mesh of a solid rotor for a 45-MW condensing steam turbine. *(Asea Brown-Boveri, Baden, Switzerland)*

enables the isothermal fields and the operating stresses to be established.

The isothermal meshes of a 23-MW turbine rotor are shown at various stages during start-up in Figs. 5.7 to 5.9. The lines shown in the mesh are lines of equal temperature; heat flux is perpendicular to the isotherms.

Figure 5.7 shows the temperature situation 10 min after start-up. The warming-up of the rotor from the outside to the inside can be clearly seen. Figure 5.9 shows the isothermal mesh 3 h after start-up (practically steady-state condition). The rotor is completely warmed up and the isotherms lie vertically to the axis. An intermediate condition 30 min after start-up is shown in Fig. 5.8.

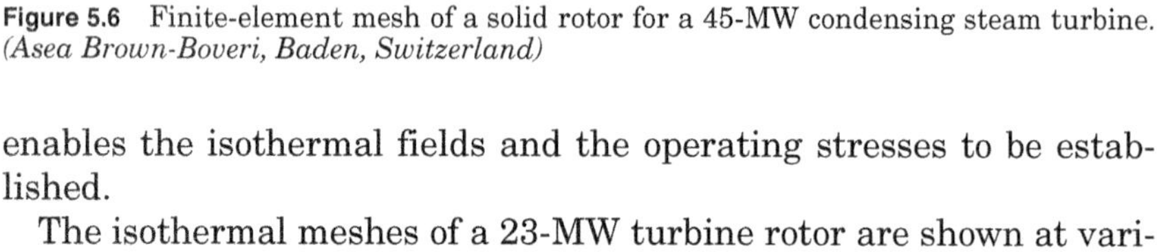

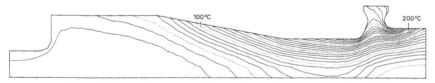

Figure 5.7 Isothermal mesh 10 min after start. *(Asea Brown-Boveri, Baden, Switzerland)*

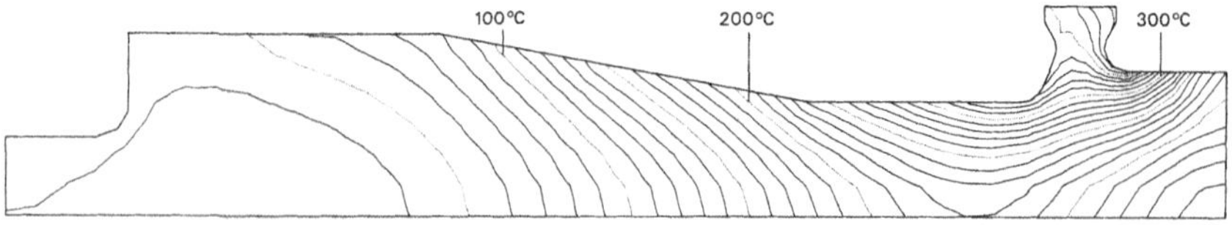

Figure 5.8 Isothermal mesh 30 min after start. *(Asea Brown-Boveri, Baden, Switzerland)*

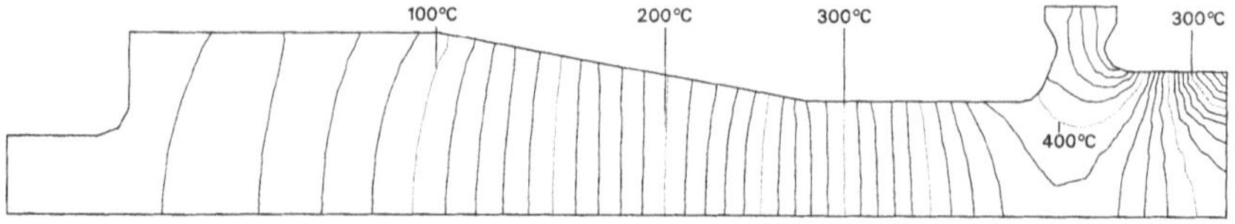

Figure 5.9 Isothermal mesh 3 h after start (stationary conditions). *(Asea Brown-Boveri, Baden, Switzerland)*

5.2 Materials for Solid Rotors

Experienced manufacturers often use St 460 TS and 461 TS steels for solid rotors. The mechanical properties of these steels and their behavior over longer periods have been under research for many years. Basic mechanical fracture investigations have given information on crack strength, crack propagation velocity, as well as the whole field of brittle fracture behavior. The chemical analysis and mechanical properties of *rotor materials* St 460 TS and St 461 TS are given in Table 5.1.

Figure 5.10 shows the stressing of the rotor of Figure 5.6 when the maximum temperature difference occurs between the rotor surface and the axis. The same comparative stress, resulting from the centrifugal and thermal stresses, prevails along a given line.

TABLE 5.1 Composition and Properties of Steels for Solid Rotors

	St 460 TS	St 461 TS
Chemical composition in %		
C	0.17 ÷ 0.25	
Mn	0.30 ÷ 0.50	
Cr	1.20 ÷ 1.50	
Ni	Max. 0.60/0.50 ÷ 0.80	
Mo	0.70 ÷ 1.20	
V	0.25 ÷ 0.35	
Mechanical properties, kgf/mm^2		
Fracture point at 20°C	σ_B	70 ÷ 85
Yield point at 20°C	σ_S	Min. 60
Resistance to fatigue		Min.
$\sigma_B/10^5$ h	350°C	44
	400°C	32
	450°C	22
	500°C	13
	550°C	7

These materials correspond to DIN specification 21 CrMoV 511.
SOURCE: Asea Brown-Boveri, Baden, Switzerland

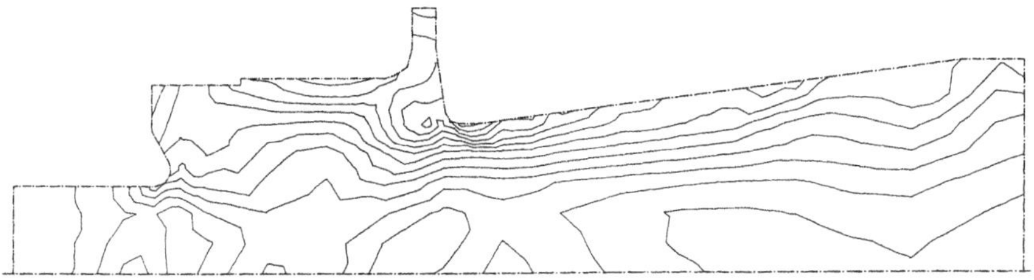

Figure 5.10 Stressing a solid rotor during start-up will yield the same comparative stresses along a given line. *(Asea Brown-Boveri, Baden, Switzerland)*

The relationship between this comparative stress and the allowable stress (dependent on the temperature) at any particular point represents the extent to which a rotor design is fully analyzed (Figure 5.11).

The rotors must be subjected to strict testing procedures. Every rotor should be premachined and ultrasonically tested in the steel plant. The mechanical properties are typically checked using test pieces before the forging is delivered. The test pieces are taken from the external section as well as the cone of the rotor (Figure 5.12). The values for tensile strength, yield point, and notch rupture strength must meet all stipulated requirements.

Where necessary, a warm run-out check is made on the rotor to establish its operating behavior. During this test the rotor will show a tendency to bend (unbalance) at high temperatures if its structure is not homogeneous and if irregularly distributed residual stresses are present because of heat treatment.

From the user's point of view, confidence in the integrity of solid steam turbine rotors should be made contingent upon favorable experience at the proposed point of manufacture. Calculation methods and design approaches must be verified for consistent accuracy and validity. Both steel supplier and turbine manufacturer must use extensive testing at the various stages of rotor production.

5.3 Welded Rotor Design

Welded rotor designs have been used in reaction turbines since the mid-1930s when a design was adopted using a number of disks welded together to form a solid rotor. Thus all the risks inherent in large one-piece forgings were avoided, and a high standard of fault detection was achieved, since the individual disks delivered by the steel plant are relatively small and can thus be very thoroughly tested.

Figure 5.13 depicts a cross section through the rotor of a 60-MW condensing turbine and clearly shows the forgings from which the rotor has been built up. The individual forgings are rough machined and ultra-

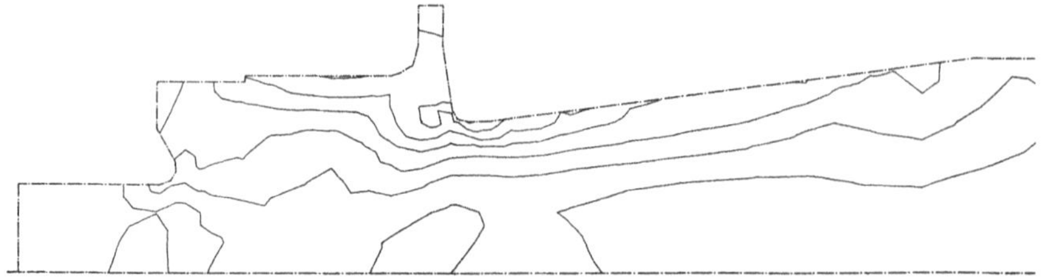

Figure 5.11 Final analysis of solid rotor shown in fig. 5.10, comparative stress/allowable stress. *(Asea Brown-Boveri, Baden, Switzerland)*

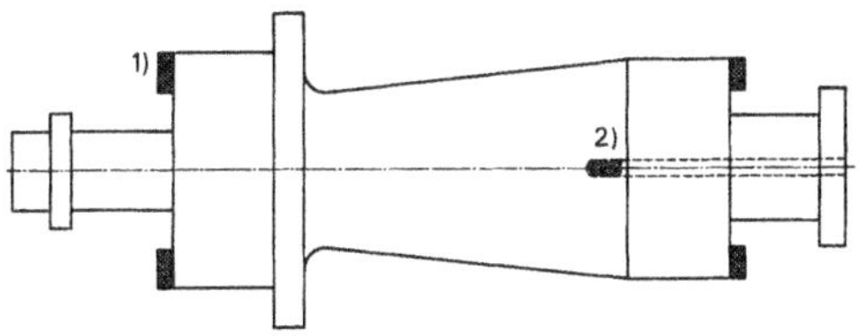

Figure 5.12 Location of test pieces, (1) and (2), for strength testing of solid rotors. *(Asea Brown-Boveri, Baden, Switzerland)*

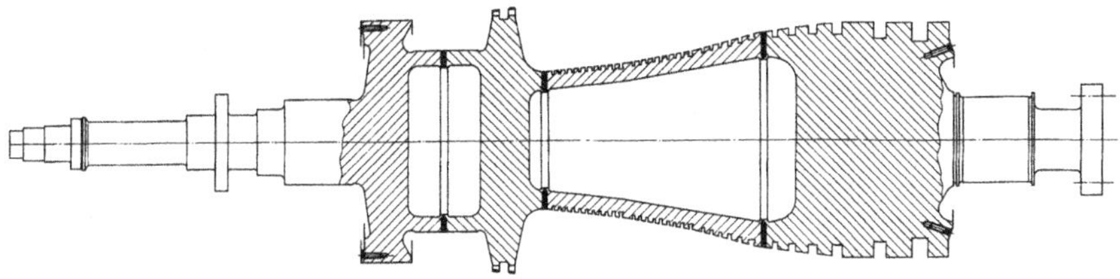

Figure 5.13 Section through the rotor of a 60-MW condensing turbine. This welded rotor will be fitted with reaction blading. *(Asea Brown-Boveri, Baden, Switzerland)*

sonically tested at the steel plant, and, in addition, mechanical tests are carried out on pieces from each disk before delivery is made to ensure that the mechanical characteristics are achieved. These tests determine the tensile strength, impact strength, and yield point of the material. All the disks are subjected to further standard inspection procedures in the shop, including chemical analysis, tensile and impact tests, before machining commences. In addition, the disks are ultrasonically examined for any internal flaws, such as cracks or inclusions. Only when confirmation has been received that all tests have been passed does premachining commence. This consists of turning the inner contour of the disk and the weld preparation contour. The condition of the rotor during this stage of manufacture can be seen in Fig. 5.14. After welding and machining, the rotor has the shape shown in Fig. 5.15.

Although no longer in the size range of mechanical drive steam turbines, large utility (power generation) turbines are of some interest.

Figure 5.14 Individual rotor disks are stacked prior to welding. *(Asea Brown-Boveri, Baden, Switzerland)*

Figure 5.15 Medium-sized condensing turbine rotor after machining. This welded rotor will receive reaction blading. *(Asea Brown-Boveri, Baden, Switzerland)*

The rotors for these large turbines consist of a number of disks welded together to form a single solid unit. The high-pressure rotor of a large turbine as shown in Fig. 5.16 contributes 440 MW to the total turbine-generator output of 1350 MW.

When building saturated steam turbines for nuclear power plants with an output of about 1000 MW, the large component weights of 1800 or 1500 r/min units become very important. In the low-pressure section for example, rotor weights between 150 and 200 tons and even

Figure 5.16 Reaction type utility turbine rotor after blade addition. *(Asea Brown-Boveri, Baden, Switzerland)*

higher are not unusual. With these large rotors the welded disk design has a very important advantage. The individual rotor components still have moderate weights, and the relatively small cross sections allow the material to be thoroughly forged. Thus high metallurgical quality is guaranteed with reduced risk of rejection. Figure 5.17 shows welding of the low-pressure rotor of a 1000 MW turbine for a nuclear power plant, and Fig. 5.18 depicts the same rotor after welding and annealing.

Figure 5.19 shows the various stages of development of the weld preparation during four decades. The deep weld technique adopted for today's rotors has been used by qualified manufacturers since about 1958 (Fig. 5.19*d*). Using this procedure and advanced welding methods, a rotor is produced where the stress values in the welded areas are similar to those in the base material of the forged disks. Regular tests on rotor welds provide a solid statistical background to the welded rotor design. Microsections through rotor welds (Fig. 5.20) are prepared to determine the quality of the weld, the extent to which the base material has been affected by the welding procedure, and the mechanical properties of the weld material.

Ultrasonic testing is now largely automated and enables flaws as small as 1 to 2 mm to be detected. If any weld is faulty, it is cut out, rewelded, and once again subjected to the heat treatment process.

Figure 5.17 Utility steam turbine rotor being welded. *(Asea Brown-Boveri, Baden, Switzerland)*

Figure 5.18 Welded reaction turbine rotor after heat treatment. *(Asea Brown-Boveri, Baden, Switzerland)*

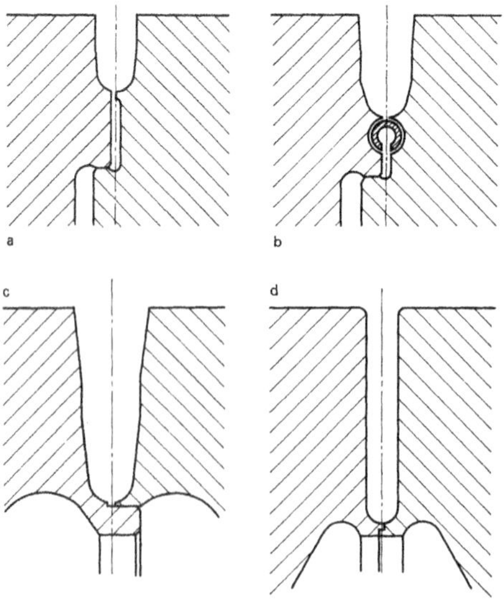

Figure 5.19 Evolution of weld preparation shapes, 1940 to 1980. *(Asea Brown-Boveri, Baden, Switzerland)*

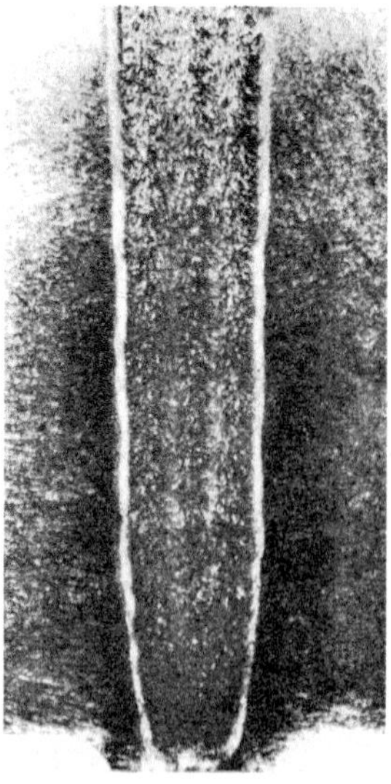

Figure 5.20 Microsection of a rotor weld. *(Asea Brown-Boveri, Baden, Switzerland)*

5.4 Welded Rotor Materials

Three different materials are used for the majority of welded rotors. For rotors subjected to the highest temperatures or to water droplet erosion, X20CrMoV121 steel can be used if necessary. Normally, 21CrMoV511 steel is used in the high-temperature region and a CrMo steel for units operating on lower steam conditions. These steels have been extensively tested during the past decades and have provided reliable statistical information on the creep properties. Fracture mechanics studies have also been carried out on these materials, providing information on crack resistance, crack growth rate, and the complete range of questions on brittle fracture. The use of the rotor welding technique presupposes a knowledge of the material properties of the weld material. The chemical analyses and mechanical properties of welded rotor steels used by one experienced manufacturer are given in Table 5.2. Note that welded rotors also employ the steels listed earlier in Table 5.1 for solid rotors.

TABLE 5.2 Chemical Properties and Mechanical Characteristics of the Materials Used in Welded Steam Turbine Rotors

BBC designation	ST 12 T	St 460 TS/461 TS	St 561 S
Chemical composition in %			
C	0.17–0.23	0.17–0.25	0.18–0.25
Mn	0.30–0.80	0.30–0.50	0.25–0.80
Cr	11.00–12.50	1.20–1.50	1.20–2.00
Ni	0.30–0.80	max. 0.60	0.90–1.10
Mo	0.80–1.20	0.70–1.20	0.50–0.80
V	0.25–0.35	0.25–0.35	max. 0.05
Ultimate strength and yield point at 20°C			
σ_B kgf/mm^2	80–100	70–85	75–90
σ_S kgf/mm^2	min. 70	min. 60	min. 60
Fatigue strength (min.) in 10^5 h kgf/mm^2 at different temperatures			
350°C	—	44	40
400°C	36	32	28
450°C	27	22	19
500°C	18	13	11
550°C	10	7	—
ASTM classification	A 565-66 Grade 616 HT	A 471-65 Class 7	
	AMS-USS 12 MoV Code 1406	A 470-65 Class 8	
		A 293-64 Class 6	
		AISI (SAE) 604	
DIN classification	~ X 20 Cr MoV 121 ~ X 22 Cr MoWV 121	~ 21 Cr MoV 511	

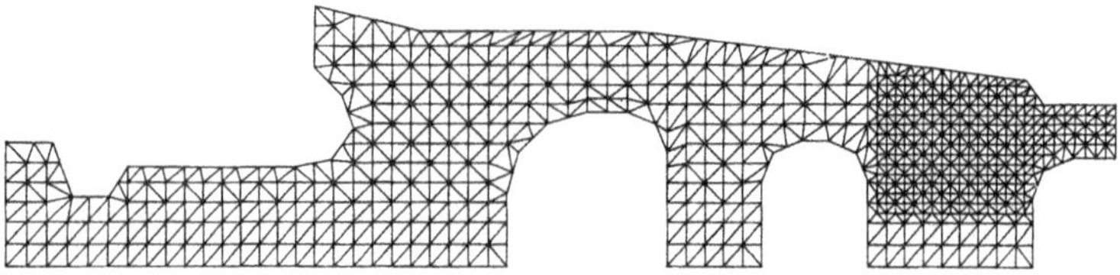

Figure 5.21 Finite-element mesh for determining isothermal fields and operating stresses of a welded rotor. *(Asea Brown-Boveri, Baden, Switzerland)*

Designing a rotor employing the latest technologies requires an appropriate computer study. Finite-element analysis makes it possible to calculate accurately the operating stresses in all parts of the rotor. Figure 5.21 shows the mesh used for the intermediate-pressure rotor of a 500-MW utility steam turbine to determine its isothermal field and operating stresses. For the same rotor the combined stresses due to the rotor speed and temperature (comparative stress) are shown under full-load conditions in Fig. 5.22. All points on any line in the figure have the same comparative stress level.

Extensive information is available on the stresses in steam turbine rotors during start-up.

Experienced manufacturers can make the claim that their welded steam turbine rotors are designed, manufactured, and inspected in such a way that a maximum of safety during operation can be guaranteed. The welded rotor has the following positive characteristics:

- Exceptionally large flexural rigidity, which favors smooth running operation
- Low stress levels since the solid disks have no central bore

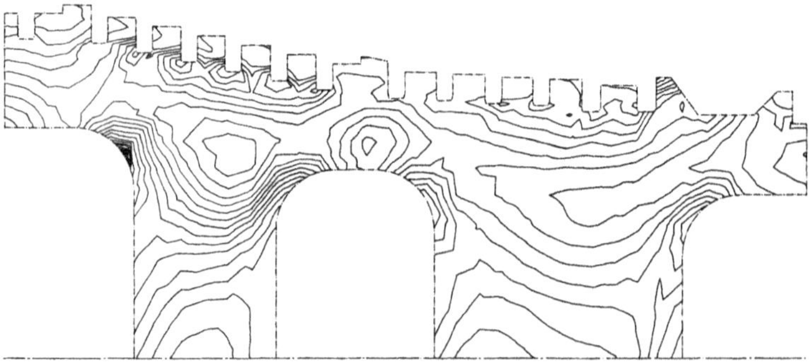

Figure 5.22 Stresses in the intermediate pressure rotor of a 500-MW utility steam turbine during steady-state operation. *(Asea Brown-Boveri, Baden, Switzerland)*

- Good quality of all highly stressed sections since the small disks can be more thoroughly forged
- Simple inspection of the individual pieces before welding and thus low risk of material defects
- Favorable heat flow during transient operation with no appreciable axial stresses at the center of the rotor since only a two-dimensional stress pattern exists.

Chapter

6

Turbine Blade Design Overview

The most critical aspect of steam turbine reliability centers on the bucket design. Since buckets, or rotating blades, are subjected to unsteady steam forces during operation, the phenomenon of vibration resonance must be considered. Resonance occurs when a stimulating frequency coincides with a natural frequency of the system. At resonance conditions, the amplitude of vibration is related primarily to the amount of stimulus and damping present in the system.

High bucket reliability requires designs with minimum resonant vibration. As will be seen later, the design process starts with accurate calculation of bucket natural frequencies in the tangential, axial, torsional, and complex modes, which are verified by test data. In addition, improved aerodynamic nozzle shapes and generous stage axial clearances are used to reduce bucket stimulus. Bucket covers are used on some or all stages to attenuate induced vibration. These design practices, together with advanced precision manufacturing techniques, ensure the necessary bucket reliability.

Almost all of the blading used in modern mechanical drive steam turbines is either of drawn or milled type construction. Drawn blades are machined from extruded airfoil shaped pieces of material stock. Refer to Fig. 6.1 for the steps in machining a drawn blade. Milled blades are machined from a rectangular piece of bar stock. Machining steps in the manufacture of milled blades are shown in Figs. 6.2 and 6.3. The cost of a drawn blade is much less than the cost of a milled blade, the reasons being obvious from Figs. 6.1 to 6.3 showing the number of steps needed to produce each blade.

As will be seen later, a certain percentage of steam turbine blades are neither drawn nor milled type construction. These blades are usually large, last-stage blades of steam turbines or jet gas expanders. They are either made by forging or a precision cast process.

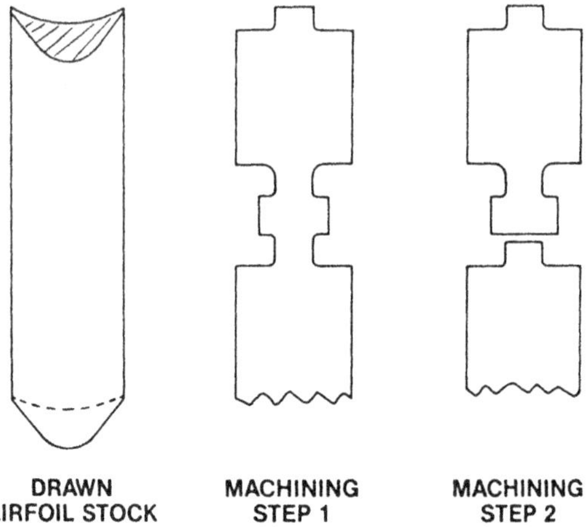

Figure 6.1 Machining steps in the manufacturing of drawn blades. *(Dresser-Rand Company, Wellsville, NY)*

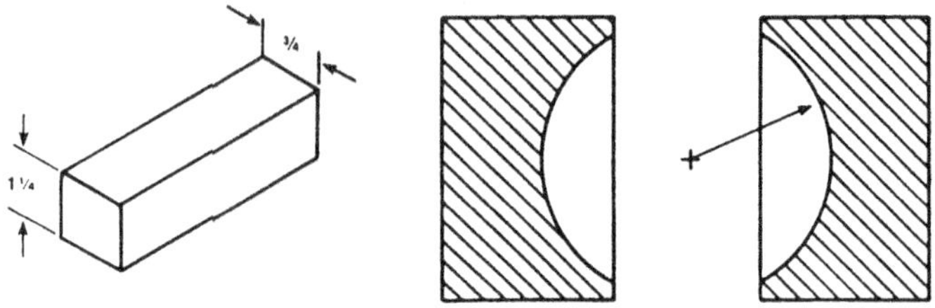

Figure 6.2 Machining steps in the manufacturing of milled blades. *(Dresser-Rand Company, Wellsville, NY)*

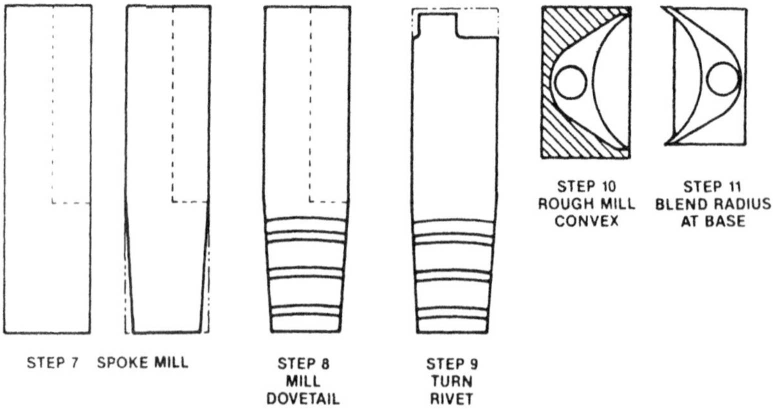

Figure 6.3 Machining steps in the manufacturing of milled blades, continued. *(Dresser-Rand Company, Wellsville, NY)*

6.1 Blade Materials

Among the different materials typically used for blading are 403 stainless steel, 422 stainless steel, A-286, and Haynes Stellite Alloy Number 31 and titanium alloy. The 403 stainless steel is essentially the industry's standard blade material and, on impulse steam turbines, it is probably found on over 90 percent of all the stages. It is used because of its high yield strength, endurance limit, ductility, toughness, erosion and corrosion resistance, and damping. It is used within a Brinell hardness range of 207 to 248 to maximize its damping and corrosion resistance. The 422 stainless steel material is applied only on high temperature stages (between 700 and 900°F or 371 and 482°C), where its higher yield, endurance, creep and rupture strengths are needed. The A-286 material is a nickel-based super alloy that is generally used in hot gas expanders with stage temperatures between 900 and 1150°F (482 and 621°C). The Haynes Stellite Alloy Number 31 is a cobalt-based super alloy and is used on jet expanders when precision cast blades are needed. The Haynes Stellite Number 31 is used at stage temperatures between 900 and 1200°F (482 and 649°C). Another blade material is titanium. Its high strength, low density, and good erosion resistance make it a good candidate for high-speed or long-last stage blading. For a typical materials overview, refer to Table 4.1.

6.2 Blade Root Attachments

The standard root attachment used on the majority of blade configurations is commonly referred to as the *dovetail root*. The drawn blading produced by Dresser-Rand uses a single tooth dovetail type root (Fig. 6.4) while the milled bladings have three variations of the dovetail root design. These variations are the *two tooth backside* (2TBS), *two tooth milled* (2TM), and *two tooth center milled* (2TCM) illustrated in Fig. 6.5. The 2TBS root is used for shorter, lower speed applications while the 2TM and 2TCM root is normally used for longer blades and higher speed applications. In addition to the dovetail type root designs manufacturers such as Dresser-Rand and others also use two additional special root attachments: the axial entry (fir tree) root and the finger type root. The axial entry root, Fig. 6.6 is primarily used where the root centrifugal stresses are high while the finger type root is used on large, last stage blades.

Some of the special design features of milled blading should be mentioned. The root teeth of milled blades are machined on a radius to match the groove diameter and to ensure full tooth contact along its pitch length. The concave (front) and convex (back) sides of the root are

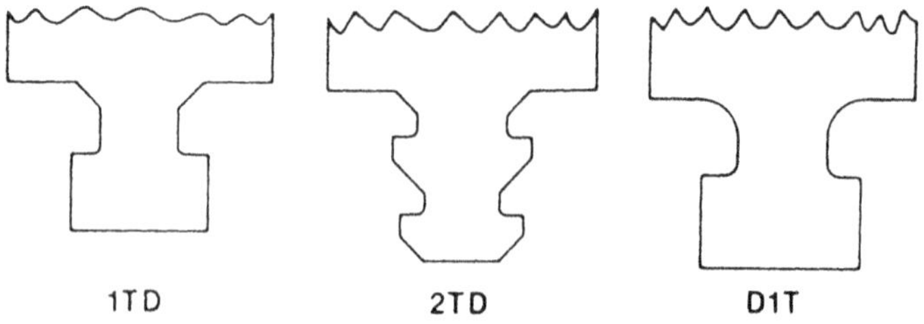

Figure 6.4 Drawn blade root attachments. *(Dresser-Rand Company, Wellsville, NY)*

spoke milled at an angle to give full contact along these faces when the blades are assembled. These features are illustrated in Fig. 6.7.

Also the dovetail teeth of milled blades are machined to fit the wheel groove in which they will be assembled. This is done by machining trial buckets until the dovetail cutters are properly adjusted to give four-tooth contact between the root and groove teeth.

Some manufacturers use a riveted shroud as standard construction for both drawn and milled blading. The purpose of the shroud is to help reduce steam flow leakage and to reduce blade vibratory stress. The shroud material used is the same as the blade material. Shrouds are typically assembled in five- or six-blade packets.

Riveted shrouds are used when shroud and rivet stresses permit. Shroud and rivet stresses become excessive when the turbine speed is high or the blade spacing is large, and for these cases an integral shroud or shroudless blade design is used. Other special types of shrouds are the interlocking (Z type) integral shroud, the butt type integral shroud, and the integral shroud with a lacing wire. A lacing wire, whether used in the shroud or in the airfoil, is used to provide additional damping to reduce blade vibratory stress.

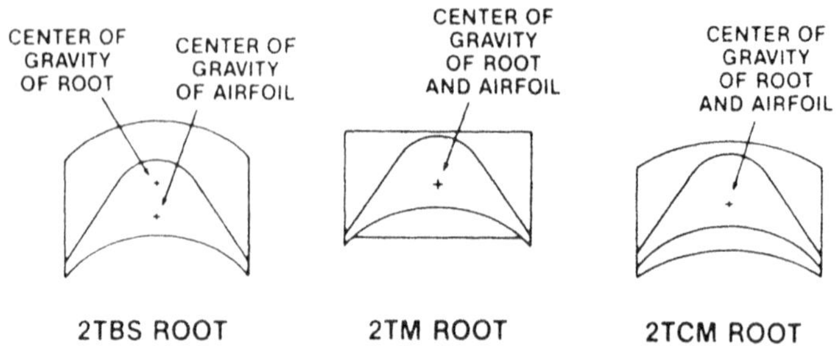

Figure 6.5 Milled blade root attachments. *(Dresser-Rand Company, Wellsville, NY)*

Figure 6.6 Axial entry root blading in control stage of a reaction turbine. *(Asea Brown-Boveri, Baden, Switzerland)*

6.3 Types of Airfoils and Blading Capabilities

Blade airfoils fall into three categories: constant area airfoils, tapered airfoils, and tapered twisted airfoils. The constant area airfoil is an impulse blade and it is generally used in short blades at the high pressure end of a turbine. The tapered airfoil is used on longer blades when a reduction in centrifugal stress is needed. The tapered twisted airfoil is basically a reaction blade. It is used when both a reduction in centrifugal stress and a change in blade angles, from hub to tip for thermodynamic efficiency, are required.

Drawn blading is used on both single-stage and multistage turbines when stresses permit. General guidelines for drawn blade limitations are 6000 r/min in speed, 3 to 3.5 in (75 to 89 mm) in blade height and about 1000 hp (746 kW) in stage loading. Milled blades are typically used when drawn blades are unacceptable because of high-stress action. Capabilities of existing blading include and often exceed 17.25-in (438 mm) length, tip speeds of 1386 ft/s (422.5 m/s) and horsepowers per blade of over 400 (300 kW).

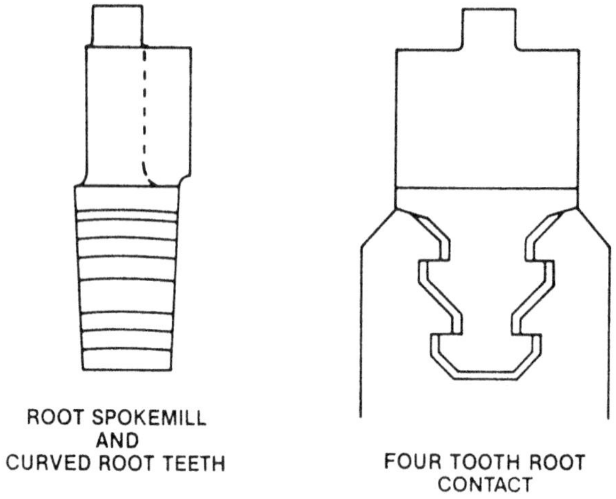

Figure 6.7 Design features of milled roots. *(Dresser-Rand Company, Wellsville, NY)*

6.4 Guide Blades for Reaction Turbines

It is reasonable to assume that impulse and reaction blading are not very different. In most reaction turbines, the *guide blades* (stationary blades) are constructed from drawn profiles with loose, milled spacers for maintaining the pitch (Fig. 6.8). As a rule, riveted shrouds are fitted.

The contours of the rotor blading with their roots and integral shrouds are milled from solid steel (Figs. 6.9 and 6.10).

Figure 6.8 Guide blade row constructed from drawn profile material. *(Siemens Power Corporation, Milwaukee, Wis. and Erlangen, Germany)*

The final blade of each row is identical with the normal blades except for the missing recesses in the T-root. The final blade is always secured with high-strength fasteners. Thus the spacing of the final blade does not differ from that of the normal blades.

The inserted, nonshrouded, freestanding blade has been largely discontinued in industrial turbine construction.

Guide blades and rotor blades typically have an identical blade profile. This profile is geometrically enlarged by a constant factor so that a range of geometrically similar blade profiles is created. For all profile sizes the optimal pitch ratio (pitch/profile chord length) is selected. The stagger angles of the blade profiles are graded and are equal for all profile sizes. In essence, geometrically similar blade cascades are thus created for all profile sizes.

The profile chord length remains constant for the full blade length. Only a few standardized stagger angles are generally used by a given manufacturer. The ratio of blade length to mean blade diameter is typically limited to 0.2. Therefore the pitch at hub and blade tip cannot deviate too much from the optimal value. At the same time the fan losses stay within a reasonable range.

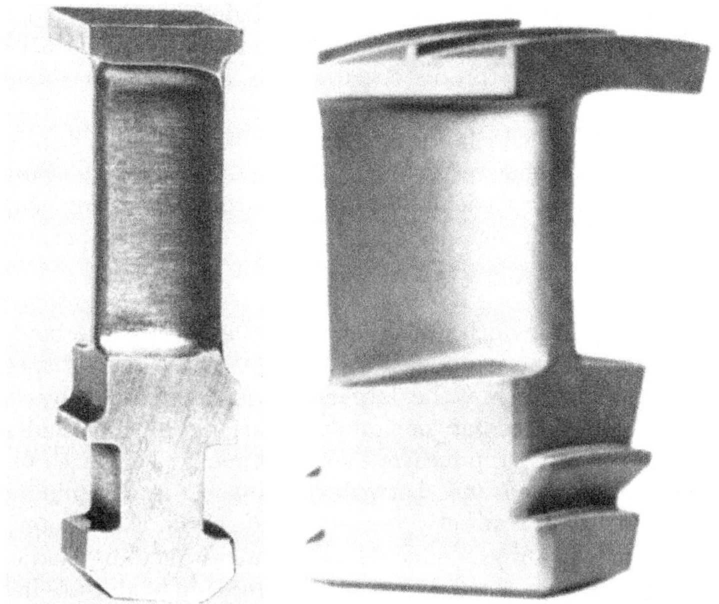

(a) (b)

Figure 6.9 Milled rotor blades with integral shrouds. *((a) Siemens Power Corporation, Milwaukee, Wis. and Erlangen, Germany (b) ASEA Brown-Boveri, Baden, Switzerland)*

Figure 6.10 Fully bladed reaction turbine rotor with integral shrouds. *(Siemens Power Corporation, Milwaukee, Wis. and Erlangen, Germany)*

To keep the lowest natural frequency of the blades principally above the sixth harmonic frequency of the turbine speed, the aspect ratio, i.e., the ratio of blade length to profile chord length, is limited to a value below 5.

In the transition zone, which is particularly endangered by vibration failures, this ratio is further reduced. *Transition zone* means the range of the turbine blading, which depending on the turbine operating point, alternately admits superheated steam or wet steam. The operating point is determined by the power generated by the turbine and the live steam conditions.

As a general rule the width of the axial gap between guide blades and moving blades is made at least 20 percent of the profile chord length. The actual value may be larger and is determined by the expected relative expansion between guide blades and moving blades.

Manufacturers usually standardize shroud dimensions for each profile chord length. The clearance between moving blade shrouds and guide blade carrier, as well as between guide blade shrouds and rotor is several millimeters. Sealing is effected by caulked-in sealing strips a few tenths of a millimeter thick (Fig. 6.11). The moving blades are held in the shaft groove by T-roots. Axial root dimensions typically equal the profile chord length. All sizes of T-roots produced by a given manufacturer are geometrically similar. For all the reaction blading only a single profile shape and a single root shape is necessary.

Blade roots and shrouds are sometimes designed in rhomboid shape. The rhomboid faces are ground and thus provide an optimal fit for the blade roots and blade shrouds.

Some notes on the stresses acting on the turbine blading will be of interest. The turbine blading is subject to dynamic forces because the steam flow entering the rotor blades in the circumferential direction is not homogeneous. Blades alternate with flow passages so that the rotating blades pass areas of differing flow velocities and directions. Since the forces affecting the rotor blading are generated by this flow, the blade stresses also vary. The magnitude of the stress variation depends very much on the quality of the blading. Poorly designed blading will often experience flow separation. This induces particularly high bending stresses on the blades. Dynamic blade stresses are also produced by ribs or other asymmetries in the flow area.

If the steam turbine is driving a compressor, surge events can induce high dynamic stresses in the rotor blades. These surges excite torsional vibrations of the turbine rotor which in turn excite bending oscillations in the blades. The severity of the alternating bending load in the blade due to the dynamic blade stresses depends on such parameters as magnitude of the dynamic blade force, frequency level of the blade, and the damping properties of the blade.

The frequency level is determined by the ratio of natural frequency to exciting frequency. With constant dynamic blade force the vibrational amplitude and thus the bending load increase with the decreasing difference of these two frequencies (resonance conditions).

With a given dynamic blade force and a given resonance condition the alternating bending stress is determined by the damping. Large

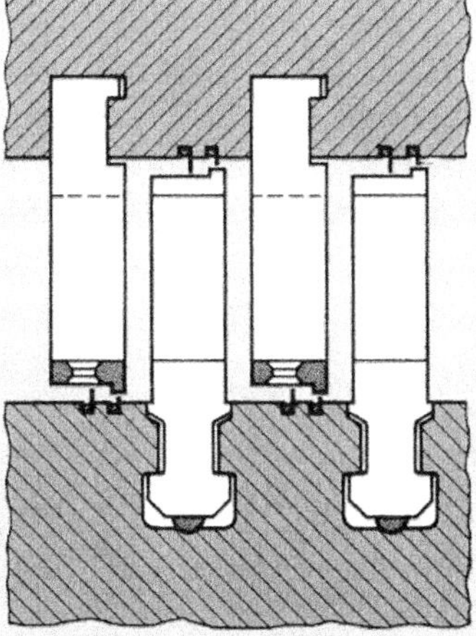

Figure 6.11 Cross section of reaction stages showing sealing strips, also called J strips or caulking strips inserted in shaft and blade carrier surfaces. *(Siemens Power Corporation, Milwaukee, Wis. and Erlangen, Germany)*

excitation forces and resonance conditions are not dangerous as long as the damping is high. So much of the vibration energy is transformed into heat that the vibration amplitude remains small.

The vibration of a blade is damped by the material-damping capacity, by the damping at the blade root and by the steam surrounding the blade. All cylindrical blades on drum rotors from such notable manufacturers as Siemens are machined with integral shrouds. When the blades of a row are assembled, these shrouds are pressed against each other and form a closed shroud ring (Fig. 6.10).

The complete shroudband links all blades of the stage to a coupled vibration system whose natural frequencies are substantially higher than those experienced by individual freestanding blades.

The transmitted energy of a vibration excitation into the linked blade system will be equally distributed to all blades within a row; the entire blade row has to be excited. For comparison, in an unlinked system (freestanding blades) the excitation energy will mainly be absorbed by the blade that has a natural frequency equal to the excitation frequency. This blade is then susceptible to breakage.

Some considerations of the effect of narrow gaps, which may form between the shrouds during operation, are given as follows (Fig. 6.12). Gaps could occur by:

1. Insertion of blades made from martensitic material (chrome steel) into a shaft made from ferritic material. The ferritic shaft material has a higher thermal expansion coefficient than the martensitic material. As shaft and blading heat up, there will be a proportionally larger expansion of the shroud in the radial direction than in the circumferential direction.
2. Expansion of shaft and lengthening of blades due to the centrifugal force at operating speed.

Gap formation will be eliminated through selection of suitable root and shroud geometry. Assembly-related forces on blade roots in the circumferential direction cause a small angular deflection in the blade profile/shroud section. In a completed blade row the counteracting tor-

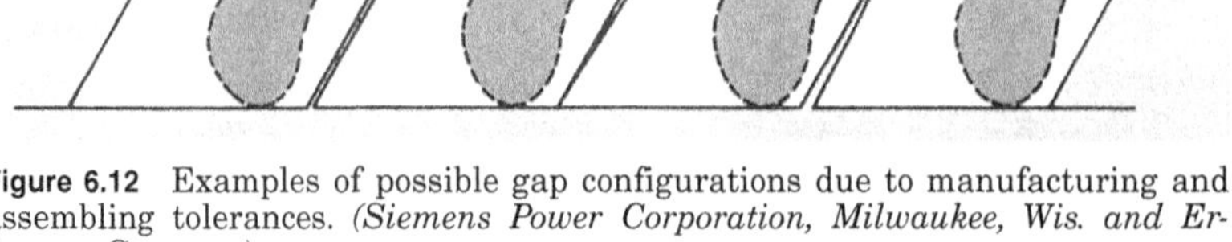

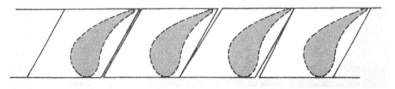

Figure 6.12 Examples of possible gap configurations due to manufacturing and assembling tolerances. *(Siemens Power Corporation, Milwaukee, Wis. and Erlangen, Germany)*

sional moment from each blade to its respective shroud prevents the formation of gaps as described by effects 1 and 2.

If the prestress in the shroud area is still not sufficient and gaps form because of extreme changes of the steam temperature in the blading, the vibration behavior of the circumferential unlinked shroudband is still substantially different from that of a row of freestanding blades.

All drum rotor blades have manufacturing and assembling tolerances, which cause the natural frequencies of the blades of a rotating row to be spread over a wide range. Therefore it is statistically impossible for all blades to get into resonance simultaneously.

The blade that is exactly in resonance is prevented from developing its maximum resonance amplitude by the neighboring blade, which is not in resonance. The shrouds of the neighboring blades act as amplitude limiters, and the vibration energy is transformed into heat by impact forces.

Energy is also dissipated from vibration amplitude by the following effect: Because of machining tolerances the existing gaps are not of uniform width, but wedge-shaped, crowned or another shape (see Fig. 6.12).

This, for instance, causes the energy of vibration about the axis of minimum inertia to be partly converted into a torsional vibration by impact against the neighboring shroud. The available vibration energy is thus distributed over several forms of vibration so that the maximum possible amplitude is decreased.

With existing gaps the shrouds act as amplitude limiters and vibration converters. The shrouds add further to the operational safety because there is a wide radial gap between shrouding and guide blade carrier. If because of a drop of the steam temperature the rotor or the casing should suffer distortion, the thin sealing strips are damaged without generation of excessive friction heat, but the radial clearance is never taken up so that the rotor cannot touch the casing.

As of 1992, more than 3000 reaction steam turbines had been fitted with integrally shrouded blades and their overall reliability was outstanding in every respect.

The operational reliability of turbines, which work in the transition zone of steam expansion depends decisively on the correct choice of permissible bending loads under these operating conditions.

First, a definition of *transition zone.* On admission of superheated steam, salts dissolved in the steam are deposited as a coating on the blade surface. Blading through which low-superheat steam normally passes can temporarily admit wet steam because of changes of the turbine operating conditions, i.e., an existing salt deposit is dissolved again by condensing steam.

On the other hand, it is possible for superheated steam to pass through stages normally operating in the wet steam zone and for salts to be then deposited.

By a systematic investigation of the possible operating conditions of a turbine, the range of stages can be determined that may be subject to this change of steam conditions. The upper limit of the design condition is defined by minimum superheat, the lower limit by maximum steam content.

Upstream of the transition zone the blading admits superheated steam only, downstream only wet steam. The first and last drum rotor stage of the transition zone can be determined by plotting the transition zone on the design expansion curve. What is special about the stages situated in the transition zone is that temporarily, during the dissolving of the salt deposits on the blading, highly concentrated sodium chloride solutions can act on the blade material.

Fatigue strength tests have proved that the fatigue strength under vibration stress measured in a concentrated sodium chloride solution is much lower than that measured in air. Consequently, the material is likely to fail because of stress corrosion cracking. For blades employed in the transition zone the allowable bending stress is thus only a fraction of that for blades in the superheated steam zone.

Extensive use is made of computerized calculation and design techniques for both mechanical and thermodynamic stage characteristics. Standardization is applied as far as practical. Among the design variables are the distribution of enthalpy drop, blade height, profile chord length, and stagger angles of the profiles.

6.5 Low-Pressure Final Stage Blading

With increasing turbine ratings there is an obvious technical requirement for increasing exhaust area of the last stage of condensing turbines. One way of satisfying this requirement is to arrange two or more last stages in parallel. Particularly with single-casing turbines this leads to a large bearing span and, hence, to sensitive rotors.

A second, more obvious means, is to increase the tip to hub ratio of the last stage, i.e., to make the last-stage blade longer. Slender blading can still conform to the obvious reliability requirements as long as high blade stiffness or a high first mode natural frequency is being achieved.

Experienced steam turbine manufacturers are therefore opting for small blade-to-chord ratios. This design results in low-steam bending stresses and thus low dynamic blade stresses. The blades are attached to the rotor by wide straddle roots and taper pins. This type of root can be made with such close tolerances, even in small sizes, that all the prongs are loaded uniformly.

The peripheral speed varies considerably between the hub and the blade tip so that the blade foil profile and the stagger angle along the blade height must be matched to the always varying flow directions.

Flow velocities are high in slender last-stage blades because circumferential velocities are high. The steam in the blade passages is accelerated from subsonic to supersonic flow and, thus, transonic flow is obtained.

Because of the large centrifugal forces acting in long last-stage blades, their profile thickness is reduced from the hub toward the blade tip.

Because the blades are designed for variable speed operation, they must be capable of running safely at resonance points. The resonance frequencies should, therefore, be as high as possible, and the blade row should have sufficient damping.

It is not possible on last-stage freestanding blades to provide integral shrouding because of the blade taper. The last stage blades are therefore interconnected with loosely fitted lacing bars. The blind bores required for these lacing bars are machined into integrally forged bosses to minimize weakening and prevent excessive stress concentration around the bore (Fig. 6.13).

When the blades are rotated, centrifugal force presses the lacing bars against the upper outer wall of the bores. All blades are thus essentially coupled to a single vibrational system. This coupling raises the resonance frequencies and makes the resonance criterion more unlikely: in a coupled system resonance only occurs when the excitation frequency

Figure 6.13 Integrally forged bosses at low-pressure blading of reaction turbine. *(Siemens Power Corporation, Milwaukee, Wis. and Erlangen, Germany)*

is equal to the natural frequency and when the accompanying excitation forces at the circumference are simultaneously equal to the characteristic mode.

Alternating bending stresses occurring in laced blades are lower than stresses experienced in freestanding blades under resonant speed conditions. This is due to the higher resonance frequencies and out-of-phase vibration of laced blading.

If large vibration amplitudes occur, friction between the lacing bar and the inside of the bore further dampens vibration amplitudes.

Since the lacing bars are well supported at both ends, coupling is independent of typical fabrication tolerances of the bore; coupling of all blades is thus ensured even at low speeds.

Damping coefficients and the influence of lacing bars on the natural frequency of a row of blading are difficult to calculate. They are, therefore, generally determined by measurements (Figs. 6.14 and 6.15). This is why strain gauge measurements of alternating bending stresses are more typically made in actual blades under operating conditions. Figure 6.16 shows one set of results: no excessively high values over the entire mass flow and speed range.

Slender case-stage blades are used in steam turbines ranging in speed from 15,000 r/min to 3000 r/min. The blades are generally made as a series of geometrically similar blades. If the same peripheral speed is

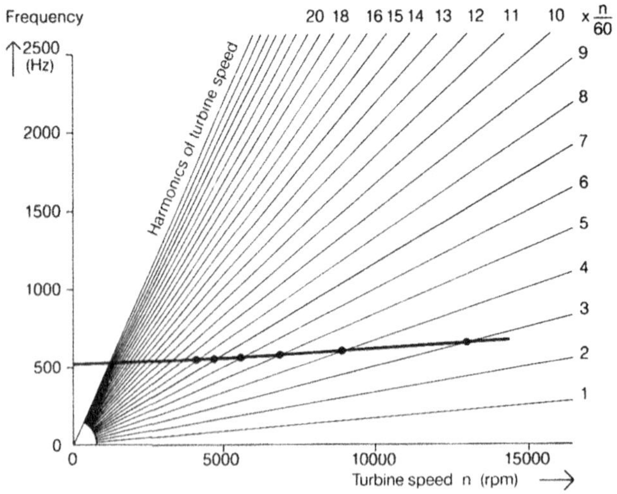

Figure 6.14 Campbell diagram of a freestanding blade. The points of resonance were measured during turbine operation. *(Siemens Power Corporation, Milwaukee, Wis. and Erlangen, Germany)*

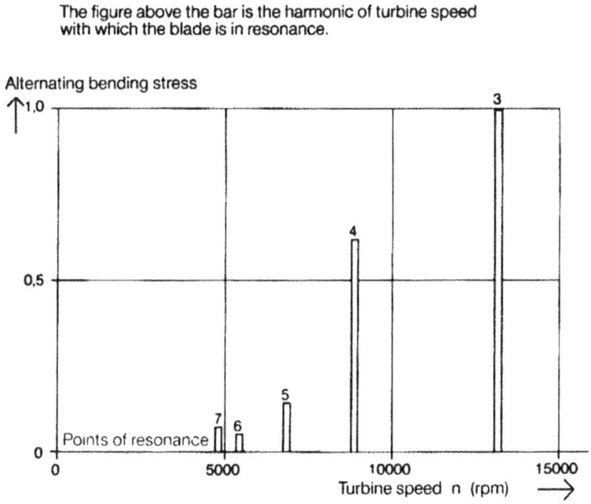

Figure 6.15 Measurements of alternating bending stresses of loaded freestanding blade at points of resonance. Maximum alternating bending stress of blade with lacing bar is only 10 percent of stress of freestanding blade. *(Siemens Power Corporation, Milwaukee, Wis. and Erlangen, Germany)*

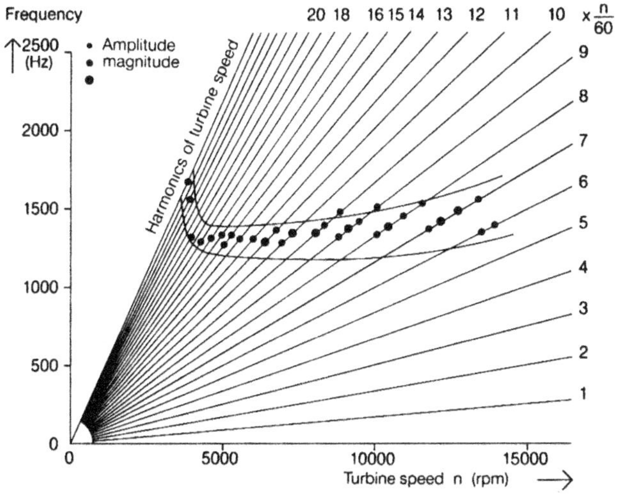

Figure 6.16 Campbell diagram of the same blade damped by loosely fitted lacing bar. *(Siemens Power Corporation, Milwaukee, Wis. and Erlangen, Germany)*

selected for every blade size, the mechanical and aerodynamic characteristics will also be the same for every blade size. Typical designs strive to accomplish a number of objectives:

- All stresses produced by centrifugal forces and steam bending moments in the blade and the root have the same magnitude for all blade sizes.
- The ratio of the blade natural frequency to the maximum rotational frequency is identical for all blade sizes.
- Steam conditions being equal before and after the blade annulus, the flow velocities for every point of the flow area are the same for every blade size. Hence, the Mach numbers are also the same.
- If steam conditions upstream and downstream of the blade annulus and peripheral speeds of various blade sizes are maintained constant, the blade efficiency does not depend on the blade size. Moreover, the exit loss as a function of volume flow and peripheral speed is the same for all blade sizes. Consequently, mechanical and aerodynamic measurements obtained on a particular blade size can be transferred to all the other sizes of a blade series. This also applies to the operating experience as a whole, i.e., the experience gained with certain blades of this series will apply to all blade sizes of this series.

Chapter

7

Turbine Auxiliaries

Among steam turbine auxiliary systems we find lubricating oil supply consoles, barring or turning gear units, trip-throttle or similar emergency stop valves, gland sealing arrangements, and lube oil reclaimers or purifiers.

7.1 Lube Systems

Lube systems for steam turbines are not significantly different from oil supplies for centrifugal compressors or other turbomachines. Although lube oil supply systems come in numerous layouts and configurations, they would essentially embody the principal components shown in Fig. 7.1.

The oil supply system is designed to cope with the requirements of the turboset or equipment train. Figure 7.1 shows the lubricating system for a turbine used to drive a generator direct, i.e., without gearing.

When a process machine is driven through gearing, the main oil pump (1) is often driven by a gear shaft. In high-reliability units, three independent oil pumps ensure that the bearings receive sufficient lubricating oil at all operating conditions.

- The main oil pump (1) driven either off the turbine shaft or through a gearing shaft supplies oil for lubrication while the turboset is running.
- The auxiliary oil pump (2) designed for 100 percent capacity supplies lubricating oil while the set is being started up or running down. It is usually a gear oil pump (like the main pump) and is driven by an AC motor.
- The battery-operated emergency oil pump (3) is generally a centrifugal pump and designed to supply 40 percent of the required amount

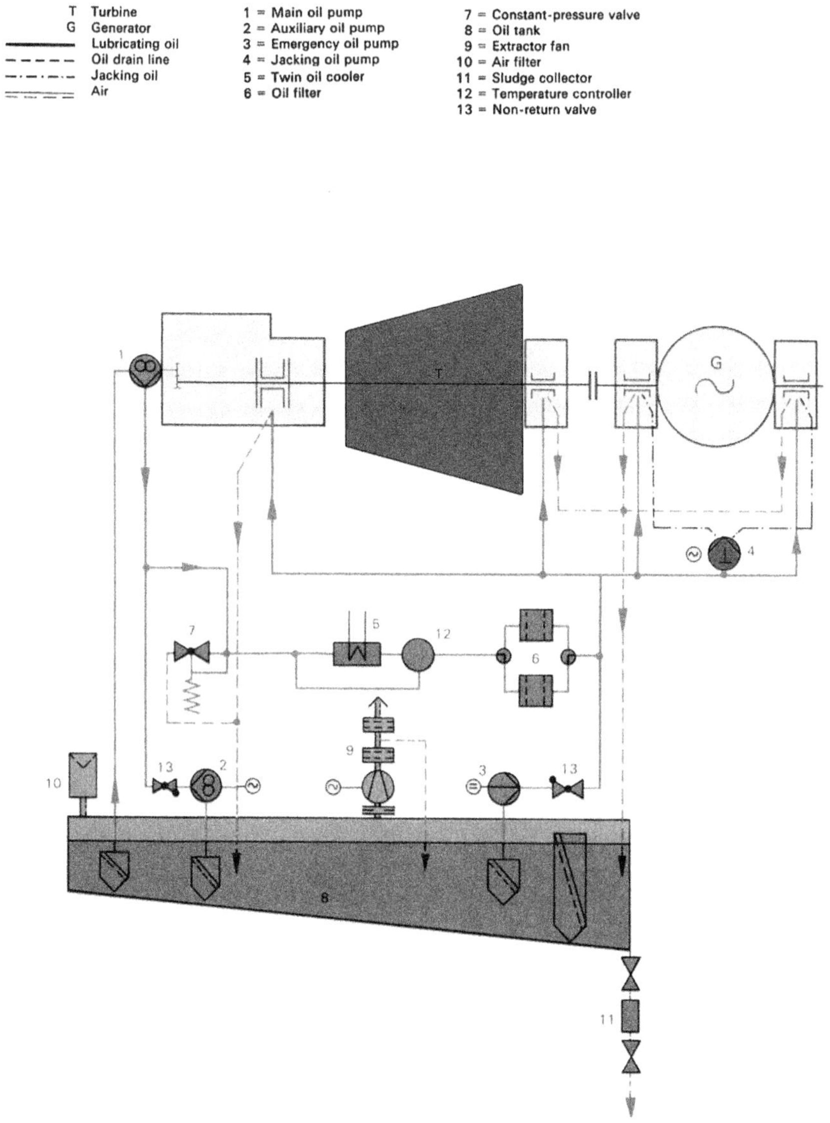

Figure 7.1 Lubricating oil system for an industrial steam turbine. *(Asea Brown-Boveri, Baden, Switzerland)*

of oil. In the event of the auxiliary pump failing, it ensures that the turboset receives sufficient oil while running down.

To lessen the breakaway torque in the bearings of the driven machine, a reciprocating high-pressure pump is sometimes provided. On large turbines the bearings are also fed with jacking oil. A jacking

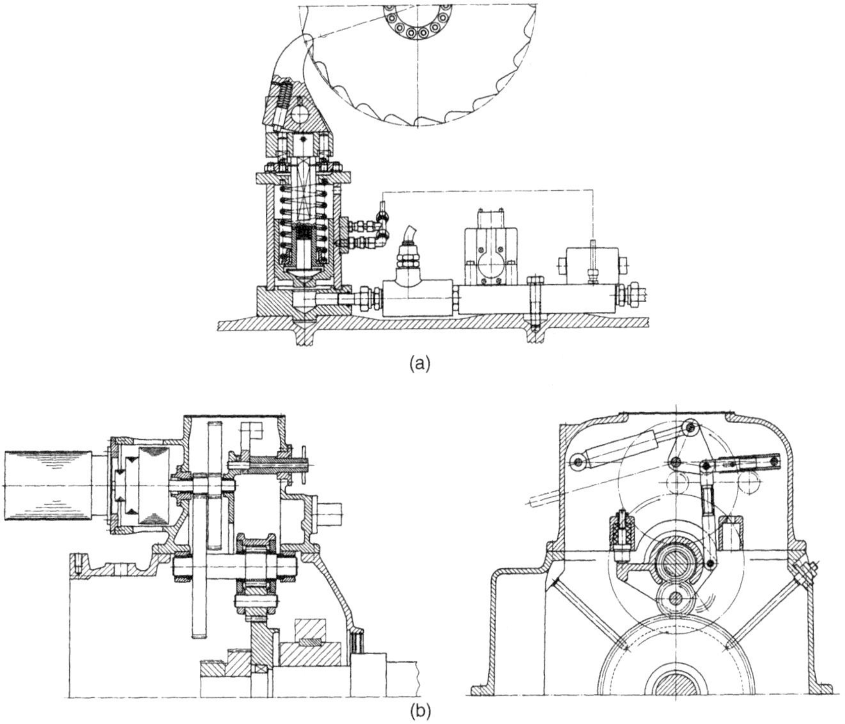

Figure 7.2 Barring (slow turning) gear arrangements for steam turbine rotors—Hydraulic (*a*) and Mechanical Versions (*b*). *(Asea Brown-Boveri, Baden, Switzerland)*

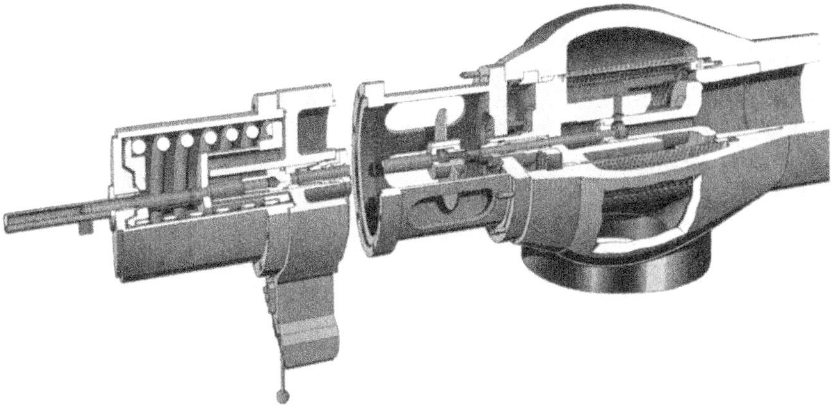

Figure 7.3 Emergency stop valve (trip valve). *(Asea Brown-Boveri, Baden, Switzerland)*

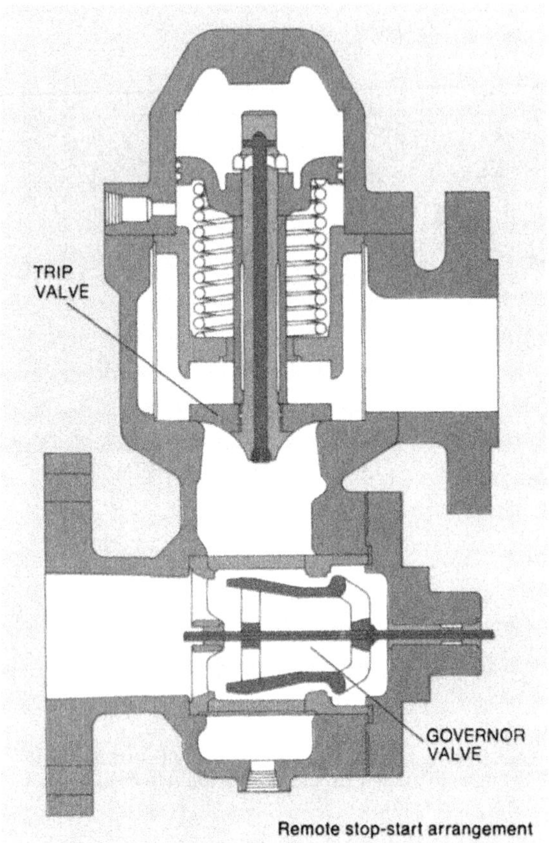

Figure 7.4 Trip valve arrangement used on some small steam turbines. *(Dresser-Rand Company, Wellsville, NY)*

oil pump (4) makes sure that an oil film exists between bearing and journal (shaft) surfaces from the very instant rotation commences.

The oil cooler (5) and filter (6) are duplicated so that they can be maintained while the set is running. The oil filter has a mesh width of 10 to 15 μm.

Many turbines obtain their lubrication requirements from a combined lube system that also serves the driven machinery and that would customarily be furnished by the driven equipment vendor. Our companion text, *A Practical Guide to Compressor Technology,* deals with these supply systems in considerably greater detail.

7.2 Barring or Turning Gears

Barring or turning gear units are depicted in Fig. 7.2. Large steam turbines are customarily equipped with these devices. Turning gear units

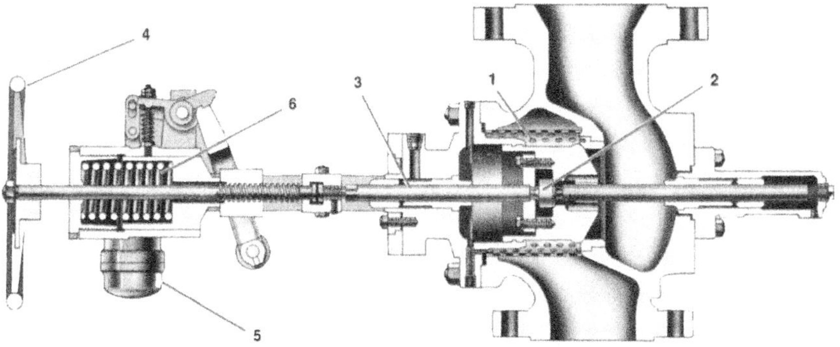

1. Protection of trip-valve and downstream parts from foreign particles provided by *perforated sleeve strainer* of stainless steel or monel metal.
2. Reduced operating force required by *inner valve* which opens first to reduce effort required to open main valve.
3. Inadvertent tightening of packing guarded against by use of *metallic labyrinth* packing with leak off.
4. *Handwheel operated* reset and valve operating mechanism.
5. Elimination of mechanical trip linkage with *latch tripping cylinder* actuated by dump valve.
6. Positive closing action due to heavy duty *spring operator closing mechanism*.
7. Patented *trip latch mechanism* for increased load capacity and tripping reliability.

Figure 7.5 Oil actuated trip-throttle valve. This balanced valve is easy to operate and permits hand throttling of the turbine against full-line pressure. The valve acts as a powerful, quick-closing valve to shut off all steam to the turbine when actuated by the turbine's tripping devices or on loss of oil pressure. *(General Electric Company, Fitchburg, Mass.)*

are typically put in operation before the turbine is started up and just after it is shut down. Turning at approximately 20 r/min prevents temporary sag of the rotor at standstill while still hot, or while it is being heated up. With the hydraulic barring gear the turbine rotor (1) is intermittently turned by a hydraulically actuated ram (5) through a sprocket (2). Mechanically driven turning gears operate on a similar principle. If failure of the auxiliary power supply occurs, the turning gear can also be operated manually.

Occasionally, steam turbines are executed with provisions to automatically engage the turning gear whenever the rotor has slowed down to a preset speed. Upon start-up and steam admission, the turning gear will automatically disengage whenever the rotor again reaches a preset speed of 40 to 60 r/min.

7.3 Trip-Throttle or Main Stop Valves

Trip, or main stop valves, are provided upstream of the steam chest. These valves are opened either under oil pressure or by manually turning a hand wheel connected to a valve spindle.

When one of the safety devices trips, the pressure is relieved in the oil system, causing the valves to close within a fraction of a second under

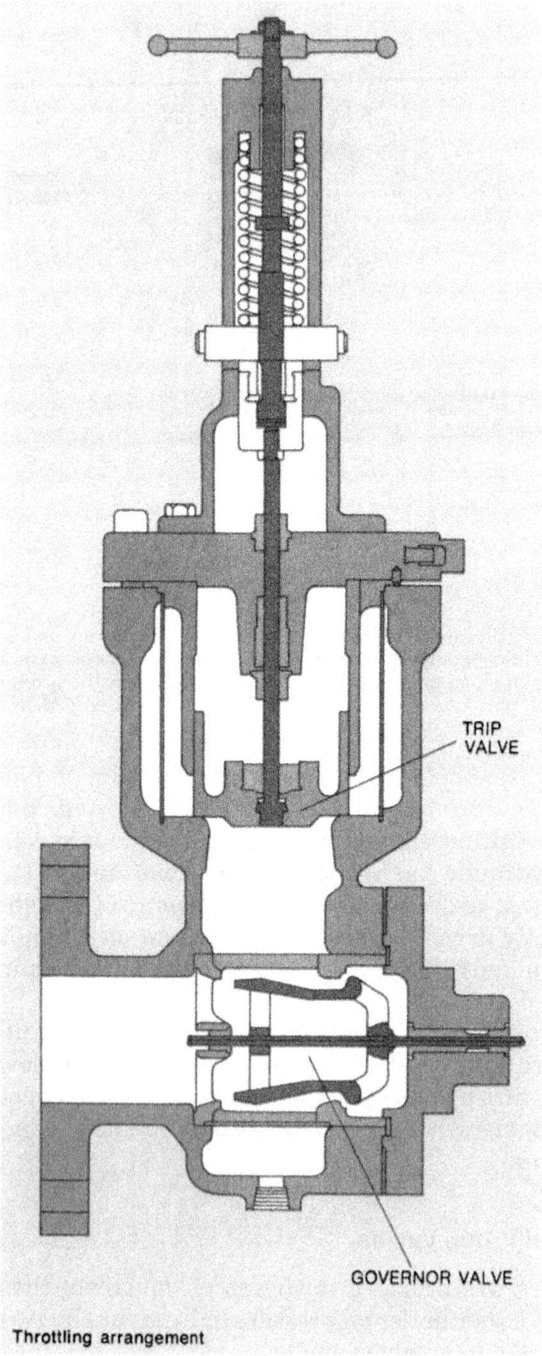

Figure 7.6 Trip-throttle valve used on some small steam turbines. *(Dresser-Rand Company, Wellsville, NY)*

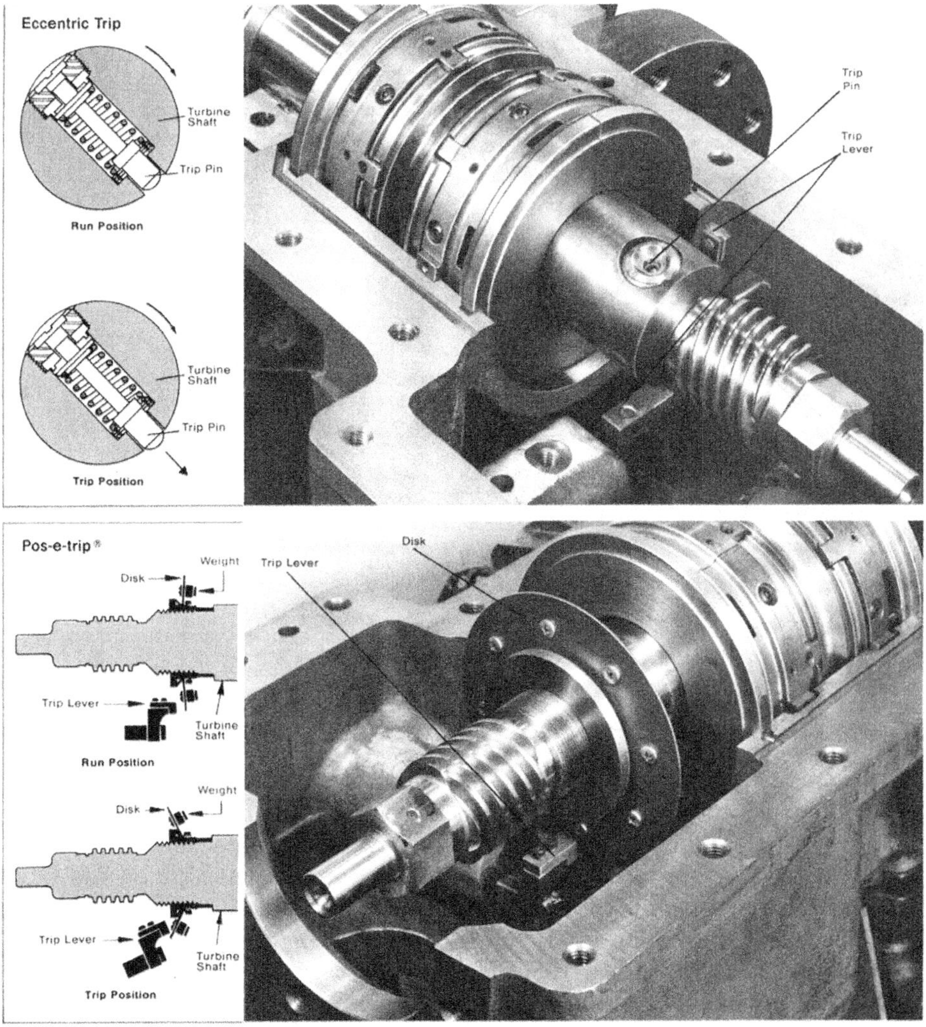

Figure 7.7 Eccentric trip pin (top) and disk type overspeed trip device (bottom). *(Elliott Company, Jeannette, Pa.)*

the action of springs and the steam pressure. These valves can be "exercised" or checked for correct operation while in service. A built-in steam strainer is usually provided to protect the blading against solid particles from the boiler and pipework and also to distribute the flow of steam more evenly.

Trip, or main stop valves, which incorporate the ability to manually or hydraulically effect a progressive and precise valve opening for starting up a turbine are called *trip-throttle valves.* Figures 7.3 and 7.4 show trip valves, and Figs. 7.5 and 7.6 depict trip-throttle valves.

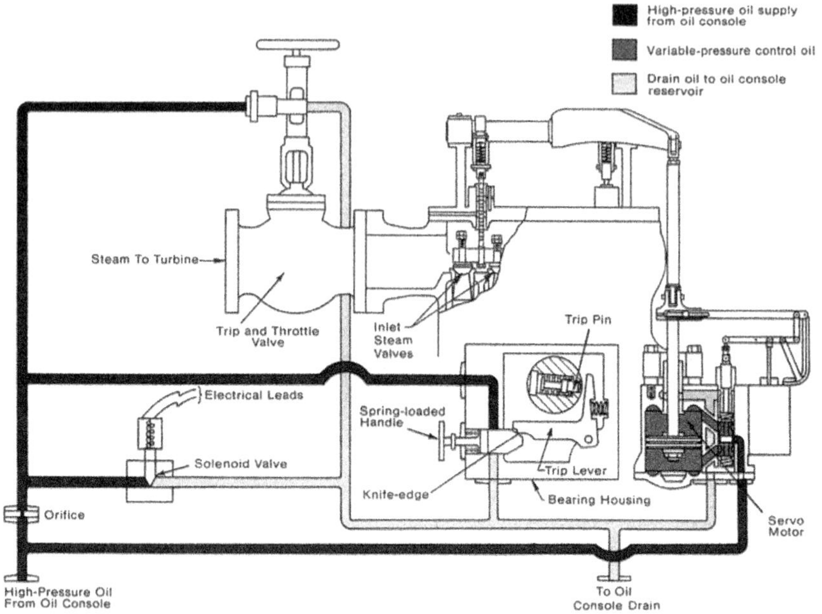

Figure 7.8 Steam turbine trip system showing an eccentric trip pin mounted in turbine shaft. (*Elliott Company, Jeannette, Pa.*)

7.4 Overspeed Trip Devices

Overspeed trip devices, Fig. 7.7, are activated whenever the turbine or driven, connected machinery reach a predetermined maximum, or "trip" setting. Although electronically activated overspeed monitors are both feasible and relatively common, adjustable mechanical overspeed trip devices continue to be used extensively.

Regardless of whether electronic or mechanical systems are used, they should be separate from the speed control system. This way a failure in one system does not prevent the remaining system from functioning, thus limiting the overspeed condition.

When trip speed is reached, centrifugal force on the pin causes it to strike the trip lever. This relieves the oil pressure on the trip cylinder of the trip and throttle valve, instantly stopping the flow of steam to the turbine.

Trip speed determines which of the two overspeed devices shown in Fig. 7.7 is used.

Turbines tripping at speeds below 8000 r/min are generally fitted with a conventional eccentric trip pin mounted in the turbine shaft.

Units tripping at speeds over 8000 r/min are often provided with a disk type overspeed control which consistently trips within $\frac{1}{10}$ of 1 per-

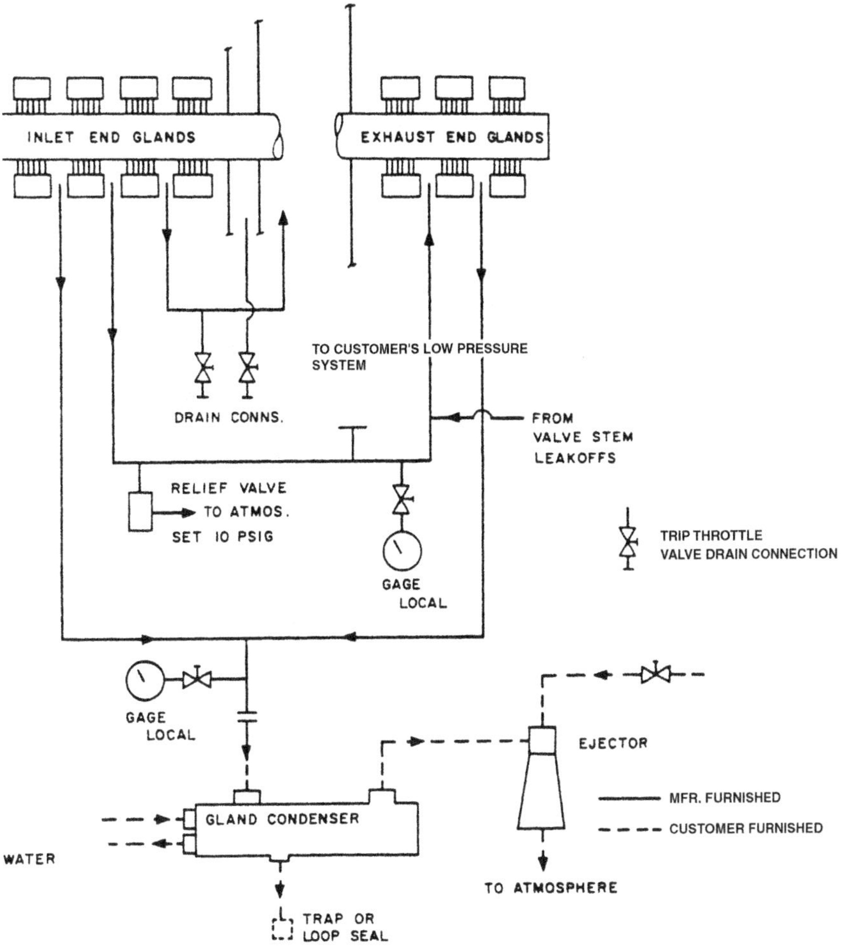

Figure 7.9 Steam leakoff and sealing schematic.

cent of trip speed. Instead of the conventional trip pin, a weighted disk is mounted on the shaft. At trip speed, centrifugal force causes the disk to snap to the tripped position and strike the trip lever, stopping the steam flow.

A trip system incorporating an eccentric trip pin mounted in the turbine shaft is shown in Fig. 7.8. At trip speed the trip pin strikes the trip lever, causing the spring-loaded handle to uncover the oil drain. Trip circuit oil pressure falls, causing the trip and throttle valve to stop the steam flow to the turbine. The solenoid valve opens to shut down the turbine in response to low oil pressure, remote pushbuttons, or abnormal process conditions.

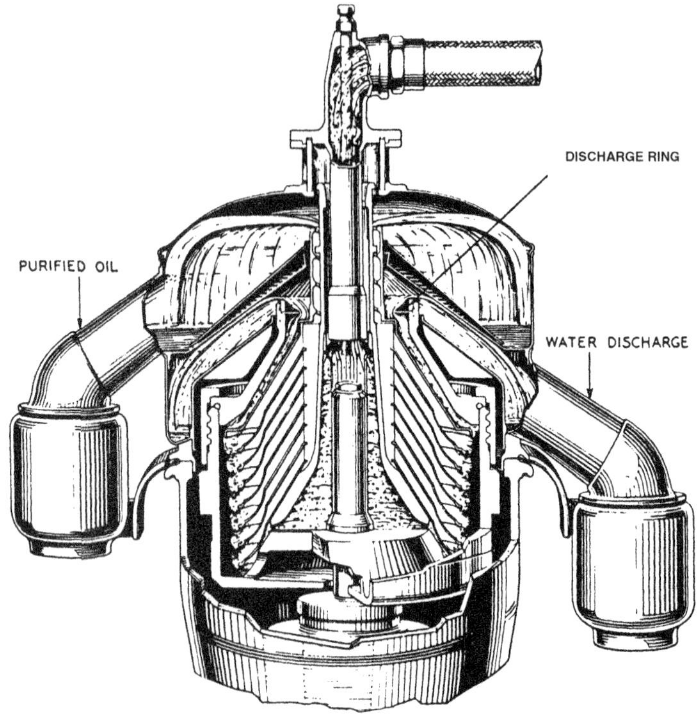

(a)

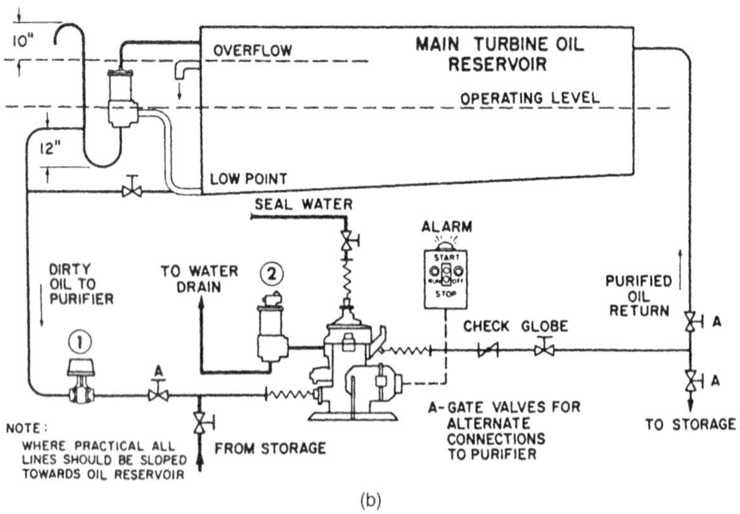

(b)

Figure 7.10 Typical centrifugal separator cross section (top) and installation schematic (bottom). *(DeLaval Separator Company, Poughkeepsie, NY)*

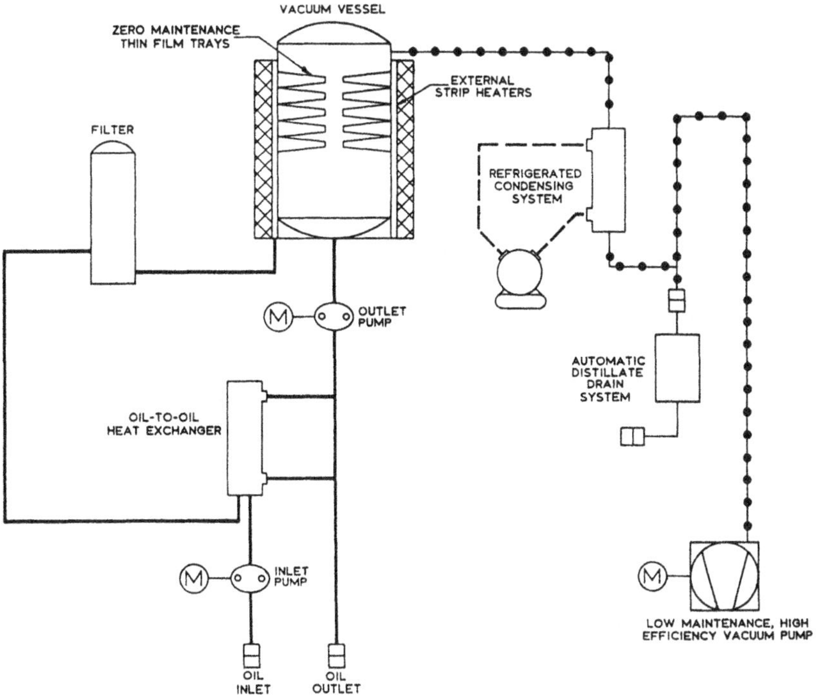

Figure 7.11 Typical schematic of a vacuum dehydrator system used for purifying steam turbine lubricants. *(Allen Filters Inc., Springfield, Mo.)*

7.5 Gland Seal Systems

Gland seal systems are piping arrangements that take steam from high-pressure labyrinth or carbon seal ports to locations operating at lesser pressure. It can be noted from Fig. 7.9 that leakoff steam exiting from a higher pressure labyrinth port becomes injected steam at a lower pressure labyrinth port. Gland condensers are also part of the sealing system for large steam turbines. (See also Fig. 2.27.)

7.6 Lube Oil Purifiers

Lube oil reclaimers or purifiers are needed if long-term reliable operation of steam turbines is desired. While it is generally well known that particulate or dirt contamination can be addressed by proper filtration, water contamination concerns are sometimes overlooked.

Water contamination causes the physical and chemical properties of oil to deteriorate. Viscosity, oxidation stability, lubricity, additives, and other oil properties could be affected. The oil must be replaced or reconditioned to avoid causing performance degradation and damage to

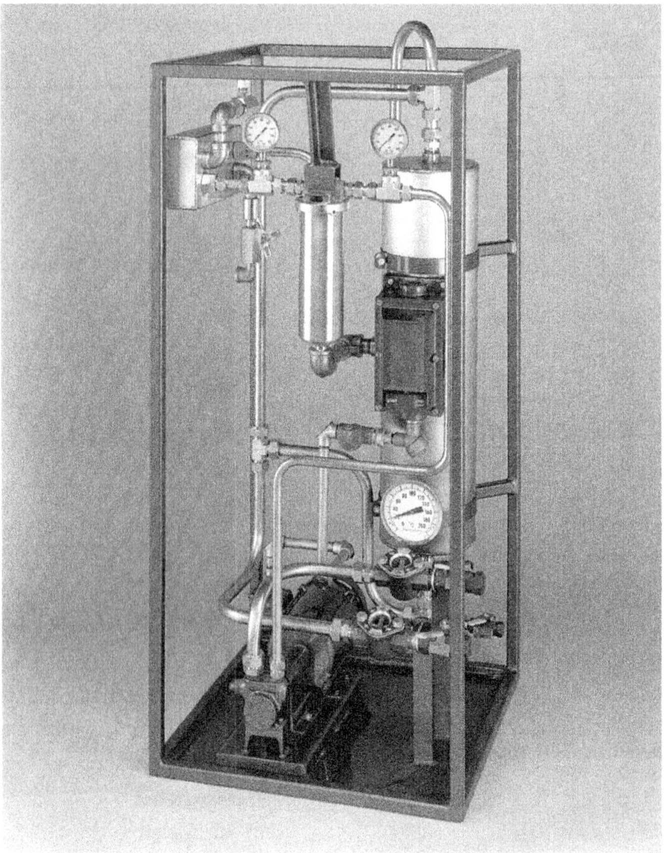

Figure 7.12 Elementary air stripping apparatus. *(Ausdel Pty, Ltd, Springvale South, Victoria, Australia)*

machinery. Oil life can be extended indefinitely by continuous removal of particles and free water, oil/water emulsions, plus dissolved water.

Centrifuges (Fig. 7.10) are capable of removing all except trace amounts of free water from turbine lube oils. They must be operated continuously if the accumulation of free water is to be positively avoided. Vacuum dehydrators, Fig. 7.11, will effectively remove not only free water but also dissolved water and gases. Used in conjunction with a conscientiously followed lube oil analysis schedule, vacuum dehydrators could be operated intermittently. Although they consume more energy than either centrifuges or coalescers, the yearly energy costs for the three methods can be considered a standoff.

Reclaimers based on gas stripping technology represent a fourth and even more efficient water removal method. Figure 7.12 depicts one of these lowest cost units.

Chapter

8

Governors and Control Systems

8.1 General

Mechanical-hydraulic governor systems as illustrated in Figs. 8.1 and 8.2 are suitable for most turbine-driven pumps, fans, compressors, small generators, etc. Other systems with NEMA D and C performance are available for installations requiring better steady-state speed regulation or remote adjustment features. The separation of governor and trip systems is evident from Fig. 8.2.

Motion of the flyweights depicted in Fig. 8.2 causes the pilot valve plunger to change the power piston oil pressures, producing movement of the terminal shaft that is connected to the balanced, double-seated valve through linkage and a single lever.

The overspeed trip pin, lever, and trip valve interaction can also be visualized from Fig. 8.2. A spring-opposed pin in the turbine shaft is thrown outward by excessive speed and strikes a nonsparking aluminum silicon bronze-tipped trip plunger. Plunger movement rotates the latch to disengage the knife edges. The double-seated trip valve is snapped shut by a closing spring. A torsion spring normally keeps the knife edges engaged.

Although the foregoing explanation demonstrates simplicity, steam turbine control systems can, at first, appear complex and mysterious. In reality, however, they can be described as a combination of simple elements. For any system, each element must be considered and analyzed independently. Basically these elements are sensing, transmitting, and correcting.

Consider first the *sensing* element. Generally, this is speed. The primary function of a speed governor is to maintain the speed of a turbine within set limits for the total load range of the unit. Speed sensing

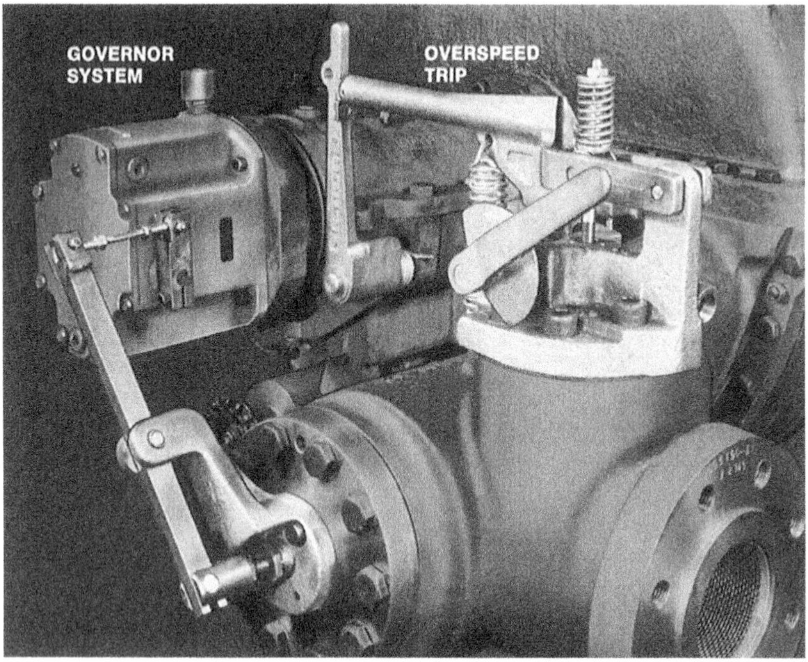

Figure 8.1 Type T sealed, mechanical-hydraulic oil relay governor (NEMA Class A). The overspeed trip device is a separate subassembly on this single-valve, single-stage turbine. *(Elliott Company, Jeannette, Pa.)*

devices use the flyball weight to sense changes to centrifugal force as a function of speed change, an electric generator or pickup whose signal can be modified and corrected for control purposes, or the pressure from a positive displacement pump where a flow (and pressure) change reflects speed.

The *transmitting* elements can be mechanical linkage, hydraulic or pneumatic pressures, electric signals, or a combination of these elements. Some systems require amplifying elements to put strength into a weak signal. For example, an ounce in force at the speed governor may be felt as several hundred pounds at a steam admission valve. Such amplifiers are called prepilots and servomotors. The tail end of the control system (*correcting* element) is the steam admission valve. These valves are found in all manner of sizes and shapes. And, although they are officially an integral part of the turbine proper, their function is clearly control system related and merits being mentioned again at this time. (The reader will recall our overview discussion earlier under the heading Steam Admission Section [Sec. 2.2].)

Most single-valve turbines utilize a double-seated valve. This type of valve has a characteristically high flow at relatively low lifts with small

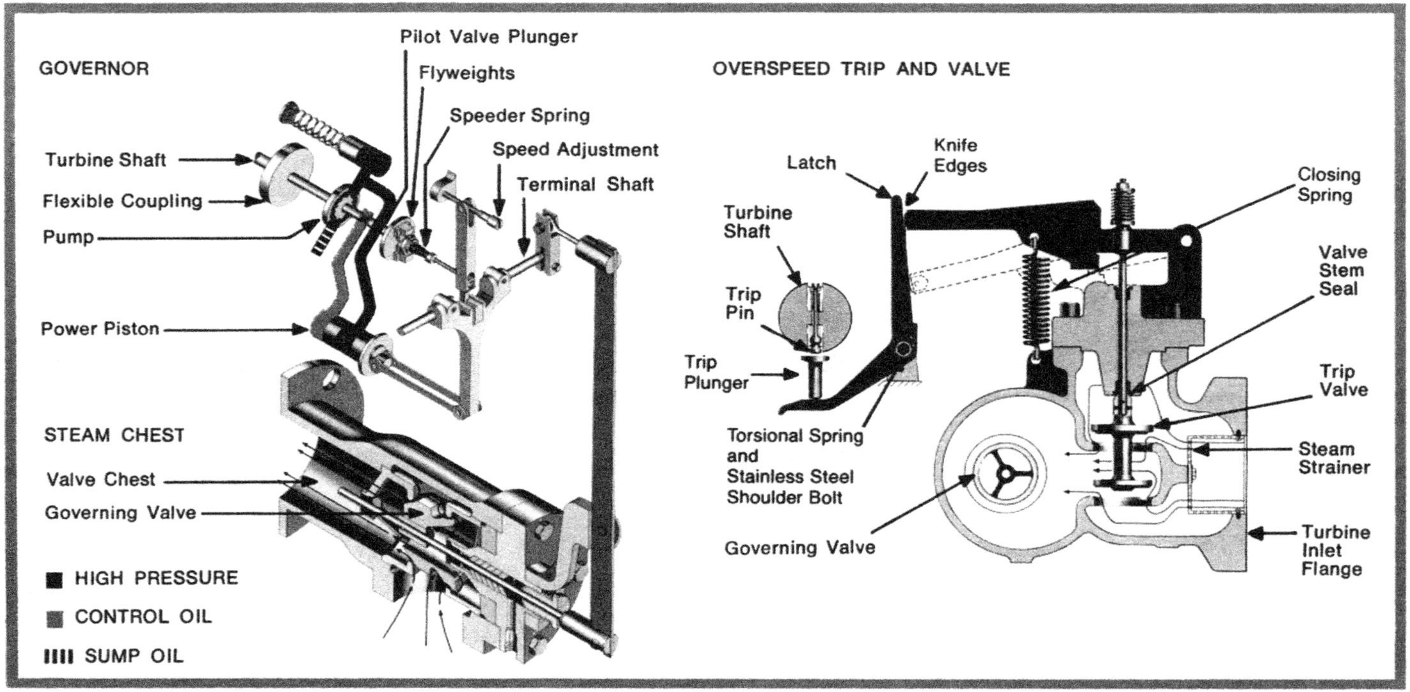

Figure 8.2 Functional arrangement of governor and overspeed trip assemblies on single-stage steam turbine shown in Fig. 8.1. *(Elliott Company, Jeannette, Pa.)*

unbalanced forces over the normal lift range. A properly designed double-seated valve will exhibit the same magnitude of unbalance at any valve lift point. For high flows and pressures a venturi type valve is used. This valve has superior flow lift characteristics over the double-seated type valve.

If controls are combinations of simple elements, why is there trouble with them and about them? Like any other mechanical or hydraulic device, governors are prone to certain deterrents. Principally, these are:

- Friction
- Lost motion
- Misunderstanding

Friction cannot be eliminated entirely, as we well realize, but it can be reduced to a low acceptable value (i.e., proper alignment, lubrication, proper valve stem design and adjustment). Most steam chest covers are designed with a valve steam leakoff arrangement and connection to reduce valve steam leakage without excessively increasing the pressure of packing on the valve stem. Frictional binding sensed by the packing can upset an otherwise reliable control system.

Similarly, lost motion can be controlled. For example, excessive tolerance in connector pins will cause lost motion, which is nothing more than the governor or sensing element responding to a small change in speed although not great enough to effect a change in the steam admission valve due to the linkage pin or connector tolerance. This problem can usually be traced and corrected.

Principally, an understanding of a governor and/or system is most necessary for its proper application and installation. Even an excellently designed control system with proper linkage and adjustment can offer poor or unacceptable performance if the selection of the system was not properly made. An attempt will be made to detail various types of governors and controls along with a classification, specification, variation, performance, and recommended application of each type.

8.2 Governor System Terminology

As a means of rating the performance of a governor or control system, the requirements are usually specified in terms of percent speed regulation.

8.2.1 Speed regulation

Speed regulation, expressed as a percentage of rated speed, is defined as the change in speed when the power output of the turbine is gradually changed from rated output to zero output, under the following conditions:

1. Steam conditions constant
2. Speed changer set at rated conditions
3. Turbine under no external control device

$$\% \text{ Speed regulation} = \frac{(\text{r/min @ zero output}) - (\text{r/min @ rated power output})}{(0.01)\,(\text{r/min at rated output})}$$

Most steam turbines have some speed droop—i.e., there occurs a drop in speed as the load is applied to the turbine. A governor that has no droop, i.e., constant speed, with load changes, is termed an *isochronous* governor. A NEMA, Class D governor is basically an isochronous governor. Its speed vs. load relationship is shown in Fig. 8.3*a*.

8.2.2 Speed variation

Speed variation, expressed as a percentage, is the total magnitude of speed change, or fluctuations from the set speed. This characteristic includes dead band and sustained oscillation.

$$\% \text{ Speed variation} = \frac{(\text{change in r/min above set speed}) + (\text{change in r/min below set speed})}{(0.01)\,(\text{rated speed})}$$

8.2.3 Dead band

Dead band is a characteristic of the speed-governing system that is commonly known as *wander*. It is the insensitivity of the speed governing system that is defined as the total speed change during which there is no resultant change in the position of the governing valves to compensate for the speed change.

8.2.4 Stability

Stability is the measure of the ability of the speed-governing system to position the governor-controlled valves so that sustained oscillations of speed are not produced during a sustained load demand, or following a change to a new load demand. A stable governor would conform to the speed vs. time relationship depicted in Fig. 8.3*b*.

Extreme oscillation about the set point is commonly called *hunting*.

8.2.5 Speed rise

The governor must be capable of catching the speed increase when the load is dropped instantaneously and return the system to the set point. *Speed rise* is defined as the maximum momentary increase in speed

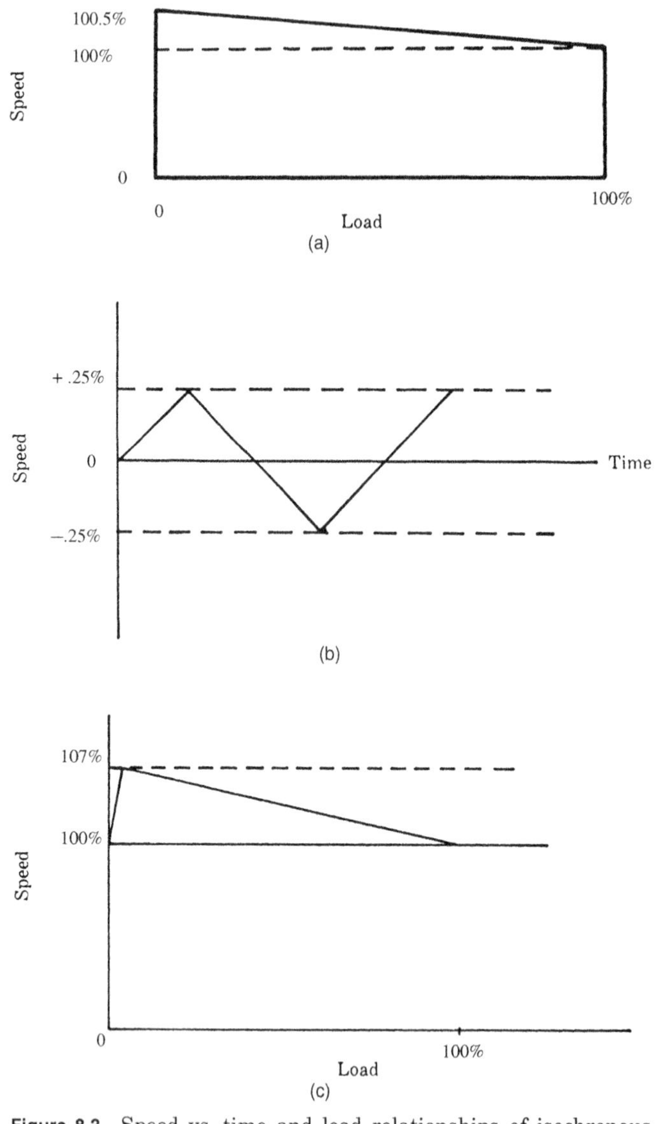

Figure 8.3 Speed vs. time and load relationships of isochronous governors. (*a*) speed vs. load; (*b*) speed vs. time of a stable system; (*c*) allowable maximum speed rise vs. load. *(Elliott Company, Jeannette, Pa.)*

obtained when the turbine is developing rated power output at rated speed, and the load is suddenly dropped to zero. An industrially allowable maximum speed rise vs. load relationship is shown in Fig. 8.3*c*.

$$\%\ \text{Speed rise} = \frac{(\text{max. speed @ zero power output}) - \text{rated speed}}{(0.01)\,(\text{rated speed})}$$

To restate and define, we should note the following:

- *Set point* is the value of the controlled variable that the governor or the regulator is set to maintain under fixed conditions.
- *Speed compensation* is a device that responds to movement in the governing system and loads the governor a predetermined amount. This loading compensates for the inherent regulation of the governor and results in a settled regulation of lower value than the inherent. It is often referred to as *reset device, droop correction, proportional speed,* or *floating action.*
- *Isochronous governor* is a governor that controls for constant speed or, theoretically, has no regulation.
- *Direct-acting* governor systems allow only the speed sensitive portion of the governor to supply the force to position the governor valve.
- A *relay governor system* provides for a servo or a power supply element to amplify the force positioning the governor valve.

8.3 NEMA Classifications

In the United States, governors in general are rated by NEMA (National Electrical Manufacturer's Association) classifications as shown in Table 8.1 and are used by industry in specifying governors. However, there are some modifications to these NEMA standards. For example, a governor may meet the requirements of NEMA Class A but not be capable of a 65 percent speed range. Further, a NEMA Class D

TABLE 8.1 NEMA Classifications for Speed-Governing Systems

NEMA class	Adjustable speed range, %	Maximum steady-state speed regulation, % no load to full load	Maximum speed variation plus or minus, % at rated cond.	Maximum speed rise, %
A	10–65	10	0.75	13
B	10–80	6	0.50	10
C	10–80	4	0.25	7
D	10–90	0.50	0.25	7

Overspeed Trip System Settings

NEMA class	Trip speed, % above rated speed
A	15
B	10
C	10
D	10

recommends 10 percent trip speed, but for a compressor drive, 5 percent could be sufficient because a maximum speed rise (instantaneous loss of load) would never occur unless a coupling failed. In some cases, a 5 percent trip has been used on compressor drives.

8.4 Valves

8.4.1 Single-valve turbines

The correcting element of a control system is the steam admission valve. Most ¾- and 1-in valves are single seated. Differently tapered plugs permit proper selection of desired valve lift or travel at full-load flow. These small valves are used for relatively low flow-rated turbines. Valve sizes of 2, 3, 4, 5-½, and 7 in are double seated and are available in steam chests of the same size or larger than the valve size. Figure 8.4 shows such a valve.

In a single-valve turbine, all the steam flows through a single governing or throttling valve to the turbine nozzles. Since the quantity of steam passing through the turbine nozzles depends on the pressure of the steam (density), the quantity of steam flowing through these nozzles will depend directly on the pressure in the turbine steam chest. Hence, variations in steam flow are accomplished by changing (throttling) the steam pressure.

At part loads, wherein much throttling may be required, the turbine will be inefficient because of the energy dissipated by throttling. To reduce the part load steam consumption, one or more hand valves are usually provided, which can reduce some of the nozzles, thereby reduc-

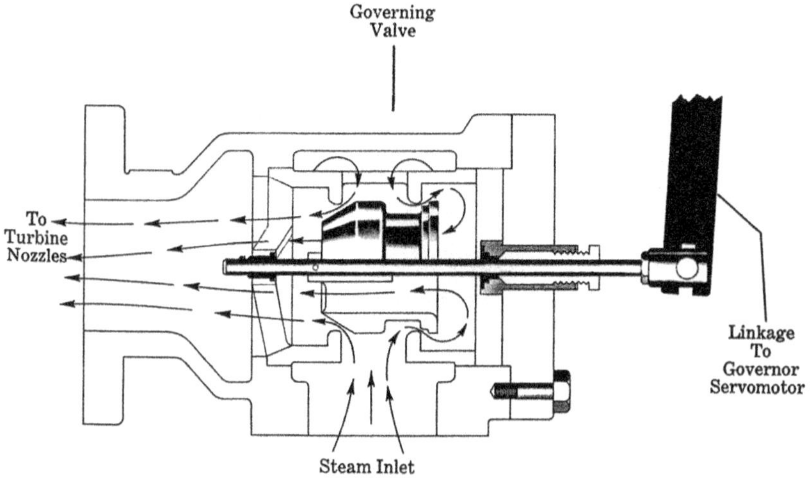

Figure 8.4 Double-seated steam admission valve. *(Elliott Company, Jeannette, Pa.)*

ing the nozzle area and, consequently, the steam flow. The resulting relationship was shown earlier in Fig. 2.8.

8.4.2 Multivalve turbines

Multivalve turbines usually use a type of valve entirely different from single-valve machines. Such a valve is commonly called a *venturi valve.* It is used for higher pressures and flows because of its superior flow lift characteristics. A series of such valves are then used (that is, 4, 5, 6, or 7) for finer control and better part load economy. Valve sizes vary up to a nominal 5-in diameter.

Grid type valves are also used, usually for extraction service. This type of valve offers very high flow areas required for extraction pressures of less than 50 psig. Occasionally, grid valves have also been used for inlet service. Venturi valves are used for extraction pressures greater than 50 psig (3.5 bar).

As was discussed earlier, on multivalve machines venturi valves are located in the top half of the steam chest casing and are lifted by a bar or cam mechanism. In bar mechanisms the valves and stems are set in the bar with different stem heights to allow them to be lifted from their seats in the proper sequence and move positively in the opening direction. Motive force to move the valves in the closing direction comes from the weight of the valve itself and the unbalanced steam pressures across it. This is again illustrated in the bar lift arrangement of Fig. 8.5. In cam mechanisms, Figs. 2.4 and 2.6, each valve is individually opened or closed through the direct contact action of a contoured cam.

In a multivalve turbine, the steam flow is divided and directed to several nozzle groups. Each nozzle group is controlled by an individual valve. An increase in steam flow through the turbine is obtained by opening the valve on successive nozzle groups, thereby increasing the nozzle flow area. The lifting beam has the valve stem length so graduated that the minimum number of nozzle groups are passing steam to satisfy the load requirements. When these individual valves are open, there is no throttling.

In essence, the multivalve turbine changes steam flow by cutting in or cutting out small increments of nozzle flow area. Since little, if any, throttling is done, the multivalve turbine is more efficient, particularly at part loads, or under varying load conditions.

8.5 PG Governors

Automation and process requirements are expanding and demanding more and finer control. Thus, users have more need for a governor sys-

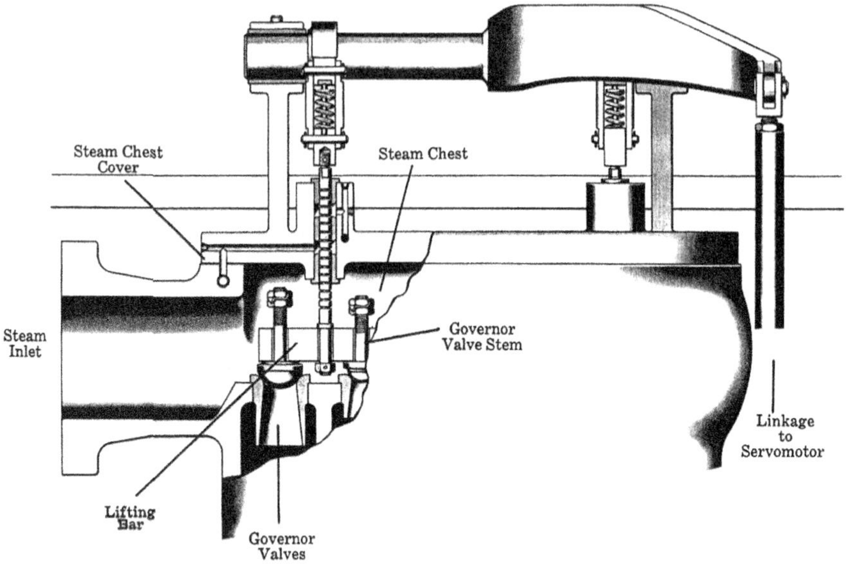

Figure 8.5 Venturi valves and bar lift mechanism. *(Elliott Company, Jeannette, Pa.)*

tem with strength and precision control with specific optional modifications. The Woodward PG governor with its power cylinder (sometimes referred to as a servomotor) and a compensating system meets these additional governor system requirements and has been the industry standard in the decades leading up to 1980. The governor is a NEMA Class C or D governor recommended for centrifugal process compressor, generator drive, reciprocating process compressor, decoking and debarking pumps, and sugar mill and paper mill applications.

All PG governors have similar basic elements regardless of how simple or complex the control may be. These basic elements include:

1. Oil pump with complete built-in governor oil system including the relief valve to limit maximum oil pressure.
2. Centrifugal flyweight head pilot valve assembly to control flow of oil to and from the governor power cylinder assembly.
3. Power cylinder assembly to operate turbine control valve.
4. Compensating system to give stability to the governor.
5. Means of adjusting the governor speed setting.

This governor including the governor oil pump is driven by a worm wheel from the turbine shaft. The standard drive gear ratios are typically selected to provide governor speed of from 150 to 1500 r/min. It is desirable to limit the maximum governor speed to 1200 r/min to eliminate the need for a governor oil cooler. The governor drive arrangement is generally applied in conjunction with turbine lube oil systems supplying approximately 0.5 g/min (2 l/min) of oil for drive gear lubrication.

The Woodward PG governor (Fig. 8.6) is a mechanical oil relay governor. The mechanical force developed by the governor weights is transmitted by oil pressure to a power piston moving the governor valve. The advantage of this system is that only a small amount of force is required to move the pilot valve, as opposed to the large amount of mechanical force necessary in a direct-acting system. Consequently, smaller, more accurate elements may be used. Friction losses are thus reduced to a minimum. The result is a better governor control. The governor system is not truly isochronous because of governor valve(s) unbalance and system friction. The speed is adjusted manually by means of an external knob that is connected to the governor spring. The speed is, therefore, varied by changing the spring tension, which changes the force on the governor pilot. This governor is also available with an electrical speed changer that uses an electric motor connected to the governor spring.

The PG-PL governor includes an integral air speed changer; otherwise, it is similar in operation to the standard PG governor. The air signal positions a secondary pilot valve. This pilot valve directs oil to the speed-setting piston mounted on the speeder spring to change the tension of the speeder spring, similar to action on the hand speed changer in the PG governor. With a given air signal applied, the governor speed

Figure 8.6 Single-valve steam turbine with Woodward PG-PL governor. *(Elliott Company, Jeannette, Pa.)*

may be changed, but governor speed should remain within the following limits: Normal speed between 250 and 1050 r/min, but a low speed of 150 r/min and a high speed of 1500 r/min can be obtained if required. Governor speed changes of from 8 to 120 r/min per psi (8 to 120 r/min per 0.07 bar) control air pressure change can be effected for PG-PL governors.

The PG-PH governor is a modified PG governor used for turbine-generator units where the regulation is required to permit the generator to parallel a utility system or another generator.

The output of the power piston is fed back into the speeder spring. The governor valve movement is, therefore, proportional to spring force. This provides the governor with regulation. The governor also has a built-in regulation adjustment. The regulation can be adjusted from 0 to 6 percent. The regulation adjustment control is mounted outside the governor permitting the regulation to be readily set for each application. This governor also has a load limit feature.

8.6 Electronic Governors

The Woodward electric type governor is used on large numbers of mechanical drive steam turbines and is described here because it is rather typical of modern devices. It is designated by EG and is intended for speed control of steam turbines driving generators, pumps, fans, compressors, and paper mill machinery. The EG governor consists basically of three separate assemblies: a control box, a speed-adjusting potentiometer, and a hydraulic actuator.

There are two basic types of EG control boxes. The EGA model is used primarily for generator drive and receives its power supply and speed signal from the generator system. The EGM model is used primarily for mechanical drive applications and requires a separate power source. The speed signal originates from the electrical impulses generated by a magnetic pickup.

The operation of the EGM governor system is as follows (see Fig. 8.7).

1. A magnetic proximity speed pickup is mounted on the turbine to sense speed from a gear mounted on the turbine shaft. The pickup is mounted as close to the bearing as possible to reduce speed errors caused by shaft vibration and runout.
2. The speed reference signal from the pickup is fed into the EGM control box. These speed pulses are then converted into a dc voltage proportional to speed. This voltage signal is then fed from the control box to the electrohydraulic transducer portion of the governor system. The EGM consists of a chassis, cover, and two encapsulated

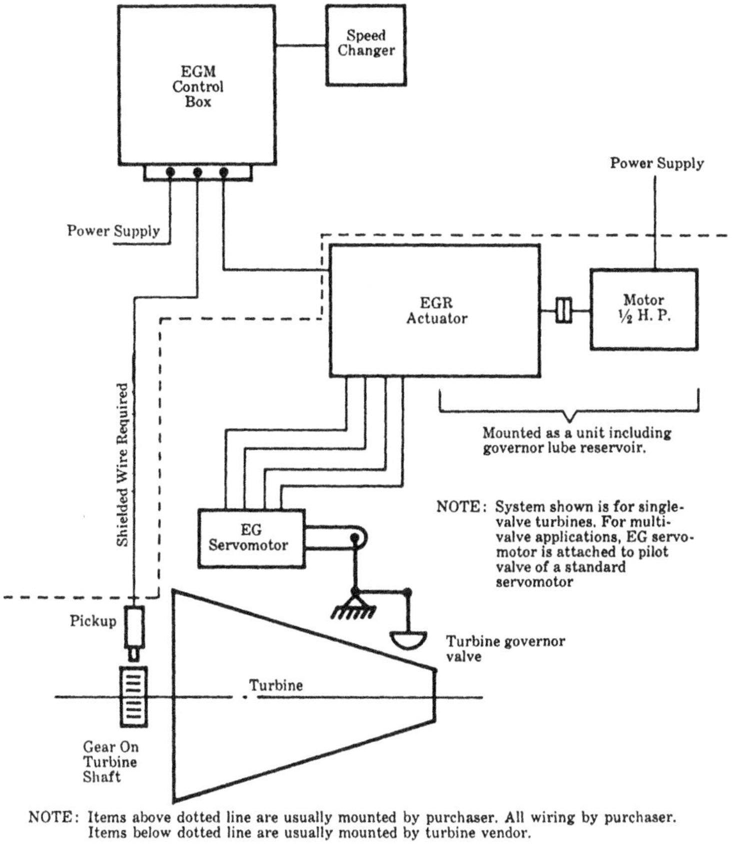

Figure 8.7 Woodward electronic governor. *(Elliott Company, Jeannette, Pa.)*

printed circuit boards. The chassis internally carries the terminal strips for the input/output connections, transformer, diodes, transistors, potentiometers, etc.

The speed-setting potentiometer establishes the reference voltage at the input to the EGM box and thus establishes the set point. This speed-setting potentiometer can be motor operated if remote control manual speed setting is required. A pneumatic speed changer is also available.

3. The EGM signal feeds the EGR hydraulic actuator of the system. The hydraulic actuator adjusts turbine speed as dictated by the signal produced by the EGM control box. The essential element of the EGR actuator is the electrohydraulic transducer that directs oil to and from the power piston or servomotor. This servo is remotely mounted on the turbine similar to PG governors on multistage turbines.

Since the rotating pilot of the EGR is similar to the standard PG series, the EGR requires a motor driver and a lube system. This system is usually offered as a complete package with the only customer connection being an electrical power supply connection to the motor.

8.7 Governor Systems

8.7.1 General

One sensing element for the multivalve turbine control is the Woodward governor. This governor contains its own servo with given power outputs. For single valve applications this power output is generally sufficient to move a double-seated valve throughout its design range. However, for multivalve applications the forces required to operate the valves can approach 20,000 lb or 9070 kg. The Woodward servo cannot generate forces of this magnitude. To achieve these force levels, the Woodward servo is remotely mounted to serve as a prepilot (or slave) to a master pilot that controls the flow of high pressure oil to a large piston. This assembly of Woodward servo (prepilot), pilot valve, and piston is called a *servomotor.* Such a control system is shown in Fig. 8.8. Servomotors are available in various diameters. For example, using 75 psig (approximately 5 bar) oil pressure and an 8-in (203-mm)-diameter servomotor piston, the force that can be generated is 3760 lb (1700 kg) and with a 10-in (254-mm)-diameter servomotor piston utilizing 125 psig (8.6 bar) oil pressure the force generated is in the neighborhood of 9800 lb (4440 kg). Thus, it is readily seen how a few ounces in governor force can be multiplied through a hydraulic mechanical advantage to generate forces required to operate multivalve turbine governor valves.

8.7.2 Extraction control

Steam at constant pressure for process use can be supplied from a steam turbine by adding a pressure-regulating system to the speed control system. This combined speed-pressure control system is called *extraction control.*

The simplest manner in which to explain extraction principles and the action of the turbine is to think of the turbine as having two steam flows, exhaust and extraction, as indicated in Fig. 8.9. The power generated by the steam depends on the available energy and the quantity, in the conventional manner. Thus, if exhaust flows and extraction flows are equal, the exhaust flow will generate more horsepower than extraction flow, because it is subjected to a greater pressure drop.

These simple rationalizations are shown in the extraction diagram, Fig. 8.10, in terms of tangible pounds of steam and effective horse-

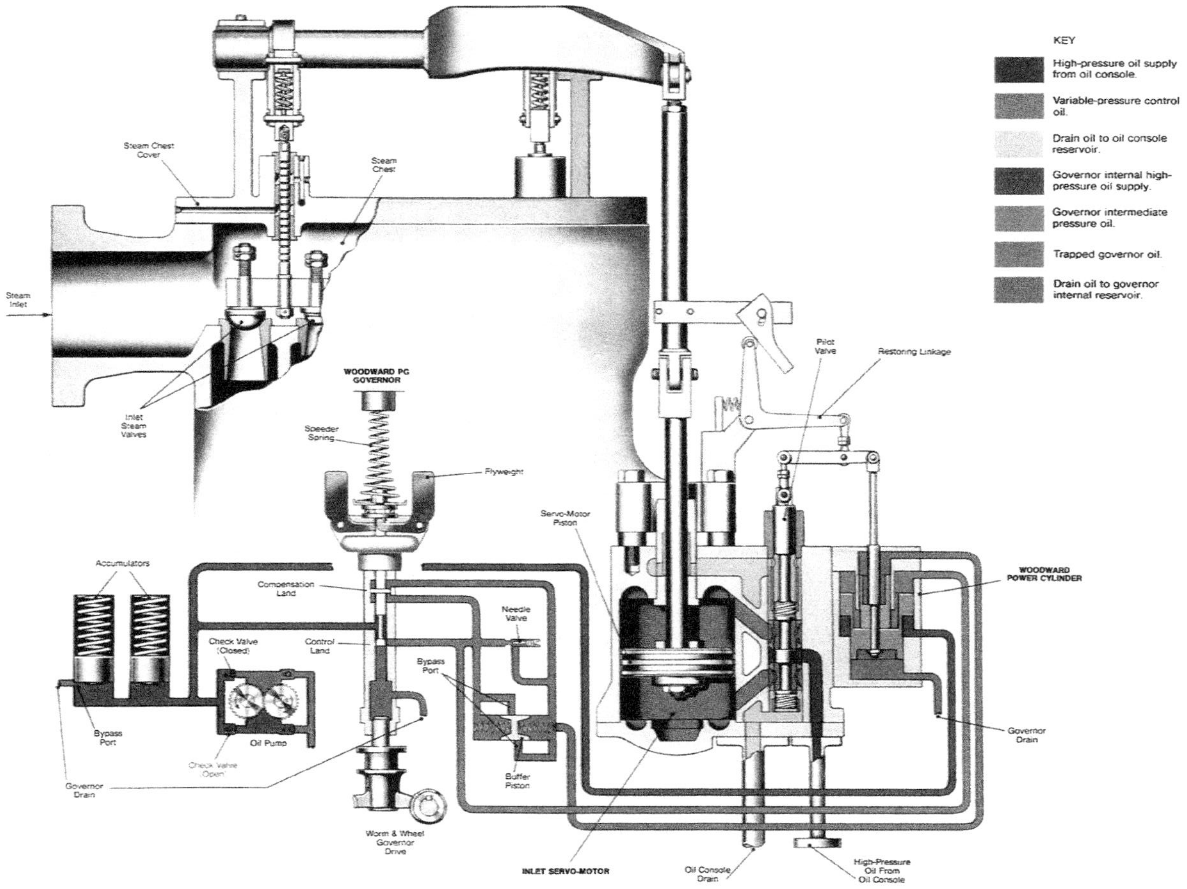

Figure 8.8 Typical steam turbine control system using Woodward PG type governor. *(Elliott Company, Jeannette, Pa.)*

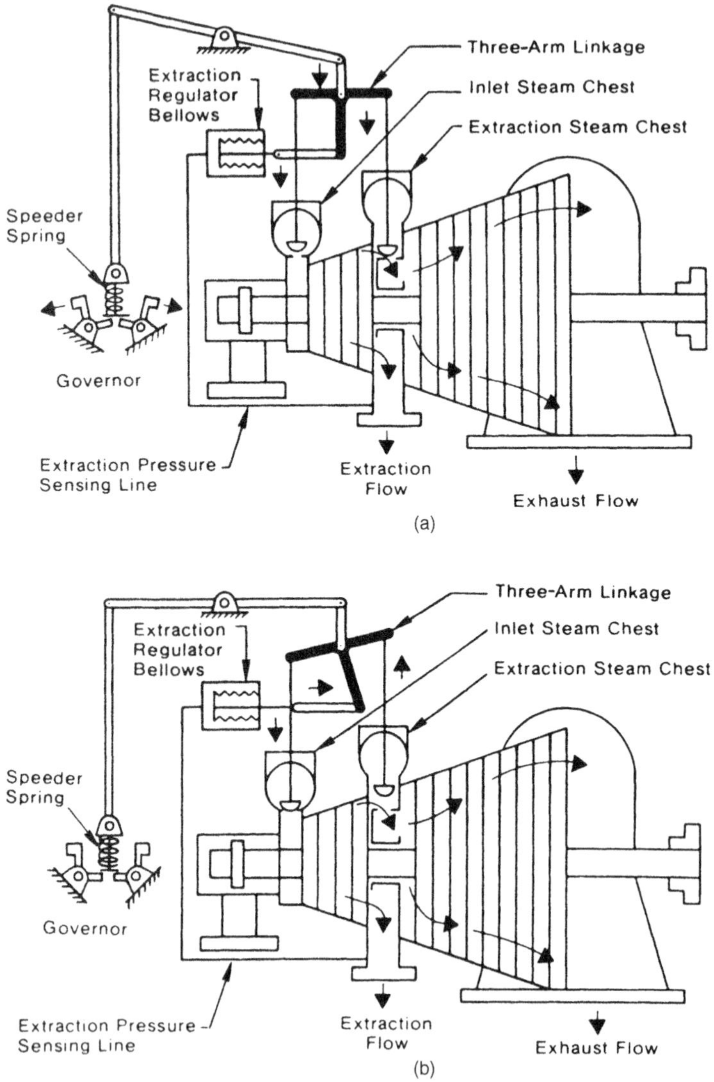

Figure 8.9 The steam extraction principle: Turbine speed changes; extraction steam demand is constant (*a*). Extraction steam demand reduces; turbine speed is constant (*b*). *(Elliott Company, Jeannette, Pa.)*

power. The vertical coordinate represents the total steam flow through the turbine inlet, the sum of exhaust and extraction flows, and the horizontal coordinate shows the total horsepower developed. The zero extraction line represents the steam flows required for various loads when extraction flow is zero. The total extraction line represents the steam flow required when exhaust flow is zero (except for cooling

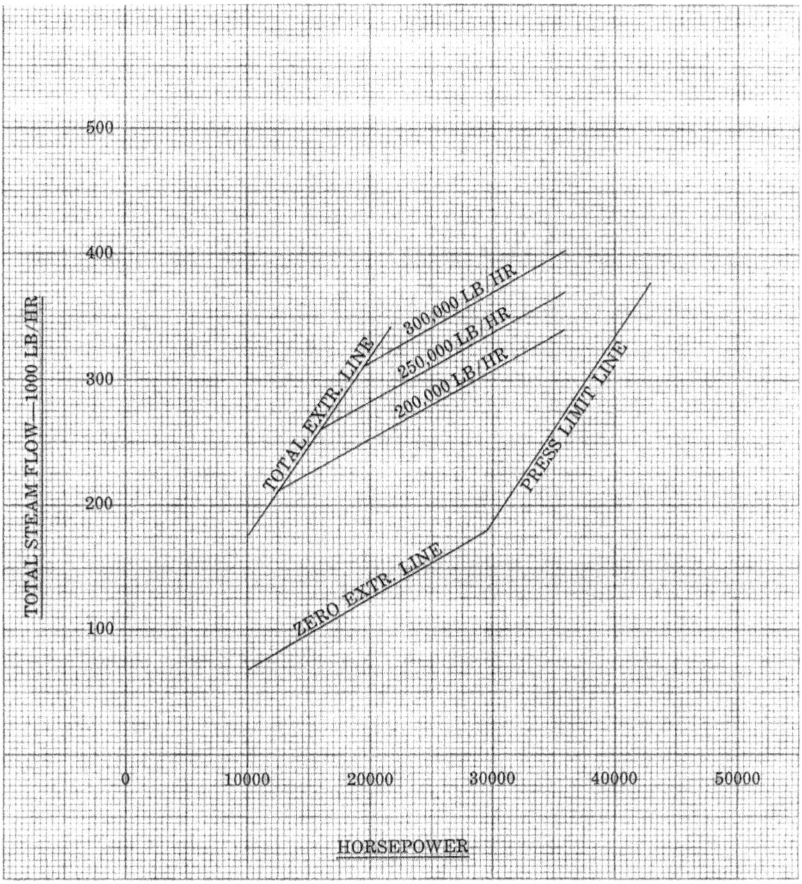

Figure 8.10 Extraction diagram. *(Elliott Company, Jeannette, Pa.)*

steam). The extraction lines show the total inlet steam flow for various horsepower loads when the flow indicated is extracted. Thus, 32,500 hp with an extraction flow to the extraction line of 250,000 lb/h would require a total flow of 350,000 lb/h. By indulging in some "kitchen arithmetic," exhaust flow is obtained as the difference between total flow and extraction flow, that is, 100,000 lb/h.

At 16,000 hp the 250,000-lb/h line terminates in the total extraction line, and at 12,500 hp there is no 250,000-lb/h extraction line. The story as told by the chart is that if the process needs 250,000 lb/h of steam, the turbine must be loaded to at least 16,000 hp. At 12,500 hp the turbine will give 200,000 lb/h for extraction flow, and if 250,000 lb/h is required, an additional 50,000 lb/h must be supplied by a reducing valve in the steam header system or by another source.

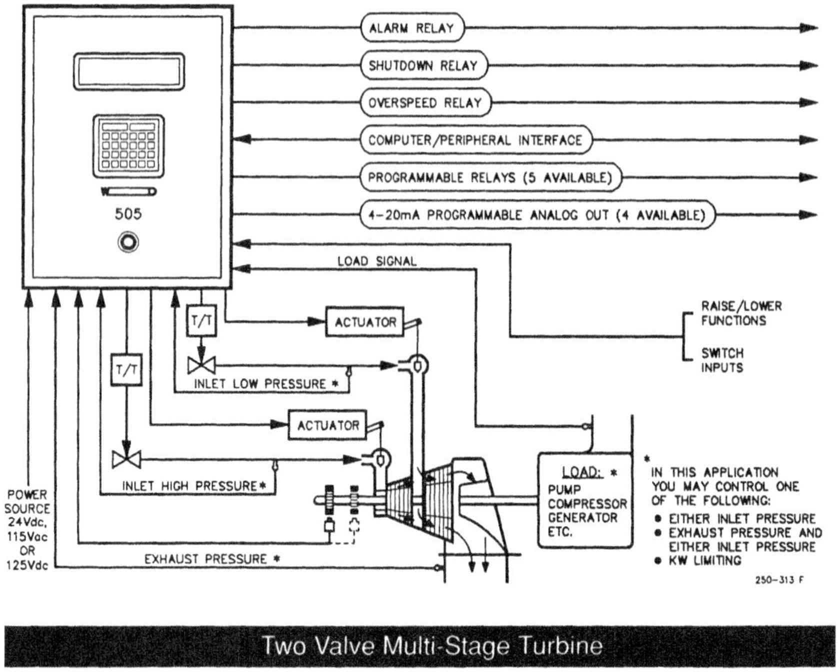

Figure 8.11 Electronic extraction governor system, load limiting. *(Woodward Governor Company, Loveland and Fort Collins, Colo.)*

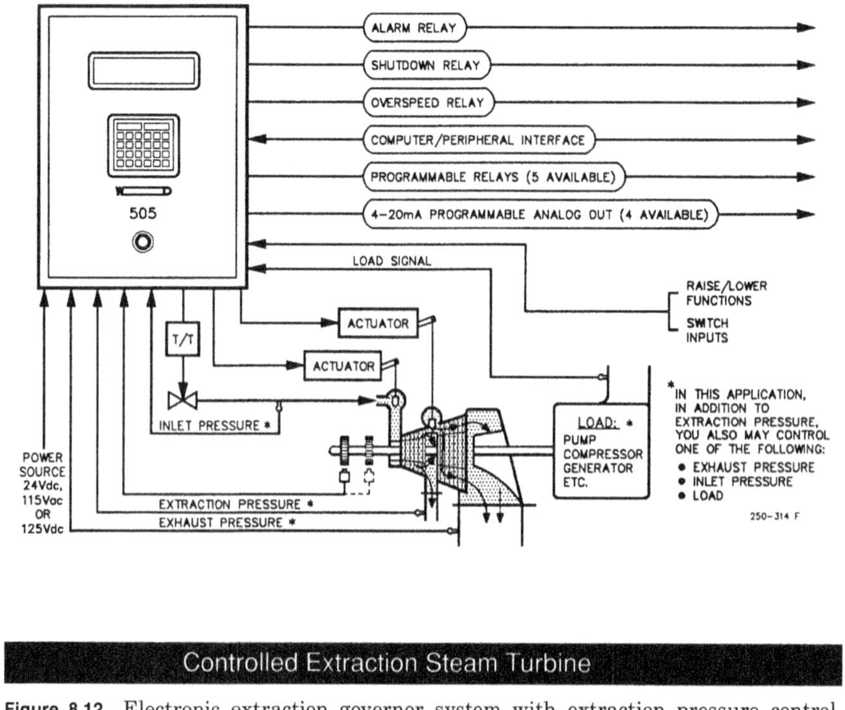

Figure 8.12 Electronic extraction governor system with extraction pressure control. *(Woodward Governor Company, Loveland and Fort Collins, Colo.)*

The pressure limit line is the limit line for exhaust flow and is the upper terminus for the extraction lines. To the right of this line, exhaust flow exceeds the capacity of the turbine exhaust, and no more steam can make its exit by that route without an extraction stage pressure increase.

The operating limits of the extraction turbine are between the total extraction line and the pressure limit line, with the size of the turbine inlet as the upper limit, and the zero extraction line as the lower limit (exhaust size). Inside of these limits the turbine is free to perform and satisfy demands for extracted steam. And if these limiting conditions are fully understood and considered in the original design, a most satisfactory system, capable of the necessary flexibility, will result. (For additional information on extraction diagrams, see sec. 15.3.)

Modern extraction control systems consist of electronic devices and interfaces, as depicted in Figs. 8.11 and 8.12. Note that the schematic of Fig. 8.11 does not provide extraction pressure control; Fig. 8.12 incorporates this feature.

Recall that the extraction valves are an integral part of the turbine and are not valves in the process line. These valves limit turbine exhaust end flow. When closed, maximum extraction flow occurs and when open, maximum turbine exhaust end flow results.

Chapter

9

Couplings and Coupling Considerations

There are numerous coupling types and configurations available in the marketplace. However, from practical and reliability-oriented points of view the overall selection narrows down quickly to just a few types. These are described below; for additional details, please refer to our companion text, *A Practical Guide to Compressor Technology*.

In principle there are two possibilities: the flexible coupling and the rigid coupling. Flexible couplings for steam turbines are further subdivided into lubricated and nonlube types. Finally, nonlube types could incorporate one or more flexing metallic disks or, in the case of small turbines, elastomeric flexing elements. The three principal coupling types of interest are shown in Fig. 9.1.

9.1 Power Transmission

The diaphragm coupling (Fig. 9.1*a*) is a worthy replacement of the older gear coupling. This is indicated by comparing the selection criteria of Table 9.1.

The gear coupling transmits the torque through the gear teeth of the flange halves and the coupling sleeve (Fig. 9.1*b*). This coupling design allows a certain degree of misalignment between the shafts and also some eccentricity of the turbine and compressor shafts. In addition, gear couplings can accommodate a certain amount of axial displacement of both shafts relative to one another. These compensating characteristics are the only reason why gear couplings are in use because, on the other hand, this type is more complicated than the rigid coupling and is limited in its application.

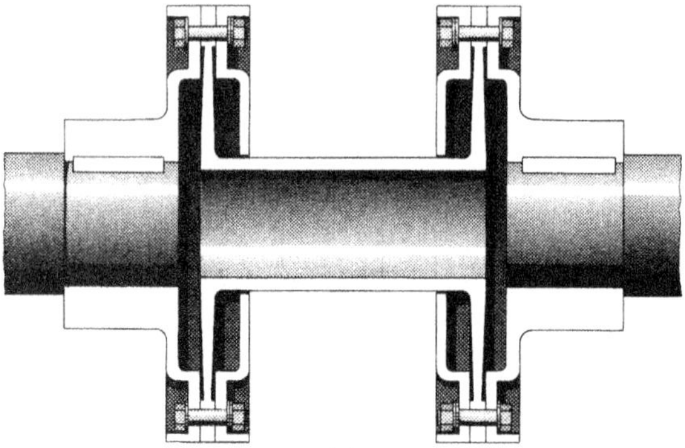

(a)

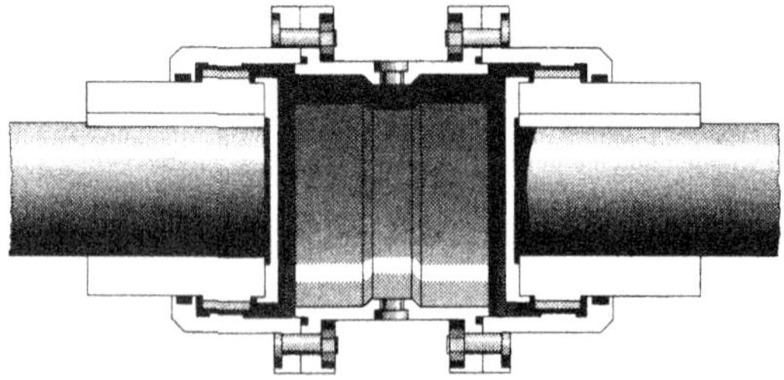

(b)

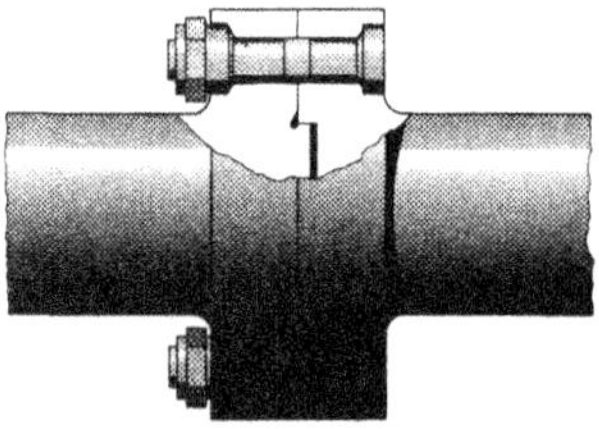

(c)

Figure 9.1 Principal coupling types used in major turbomachinery: (*a*) diaphragm coupling; (*b*) gear type coupling; (*c*) solid coupling. *(Asea Brown-Boveri, Baden, Switzerland)*

The flanges of a rigid or solid coupling are normally forged on as shown in Figs. 9.1*c* and 9.2, although they may occasionally be shrunk and keyed on. They are connected together by means of bolts. The friction between the two coupling halves is sufficient to transmit the entire torque. Only a flanged coupling can be used in the case of large turbogenerator units due to large short-circuit torques.

Occasionally, an intermediate shaft, flexible enough to allow for misalignment of the same order of magnitude as other types of couplings, connects the two shaft ends of the machines to be coupled together.

TABLE 9.1 Coupling Selection Criteria for Major Turbomachinery

Feature	Gear type coupling	Diaphragm coupling	Solid coupling
Power transmission	Line contact	Friction	
Limitation of power and speed	Limited	Limited, depending on conditions	Unlimited
Erection	Great care required		Normal care
Lubrication	Very clean oil. Problems with high peripheral speeds	Not required	
Wear	Possible if not assembled correctly	None	
Number of thrust bearings	Separate thrust bearings for turbine and driven machine		Only one thrust bearing
Loading of thrust bearing	Influenced by friction in the coupling	Influence of diaphragm deflection possible	Well-defined
Balancing of axial thrust	Not possible		Possible
Transmission of differential expansion	No		Yes
Mutual effect upon critical speeds	Slight		Normal
Windage	Small	Large	Small
Unbalance	Possible if not correctly machined and/or assembled		Negligible

SOURCE: Asea Brown-Boveri, Baden, Switzerland

Axial alignment of the train requires the same conventional procedure and expertise.

With direct turbine drive, normally the turbine is equipped with its own thrust bearing and coupled to the compressor train with a gear coupling. Compressors and expanders are solid coupled, with the thrust bearing usually located in the low-pressure compressor.

If an intermediate gear is necessary, one manufacturer (Sulzer) often uses single helical gears provided with thrust collars on the pinion shaft as shown in Figs. 9.3 and 9.4. Here, the thrust collars not only neutralize the axial thrust created by the meshing of the teeth cut at an angle to the axis of the shaft but also transmit the unbalanced axial thrust of the high-speed rotor train to the thrust bearing on the low-speed section.

Because of the oil film, the pressure zone is spread over an enlarged surface with a pressure distribution very similar to that of a standard oil-lubricated journal bearing. The thrust transmission is therefore

Figure 9.2 Rigid (integral flange) coupling on reaction stream turbine rotor. *(Asea Brown-Boveri, Baden, Switzerland)*

effected with almost no mechanical losses. The considerably reduced losses of the single thrust bearing as compared with the high losses of individual thrust bearings lead to a substantial power saving.

With motor drive the normal practice is the same. The main gear is equipped with a simple shoulder bearing, does not have thrust collars, and is connected to the compressor train with a solid coupling.

9.2 Shaft Alignment

During operation the shaft alignment of a turbomachinery unit may be adversely affected for several reasons, e.g., a bearing pedestal may grow because of temperature influences; different bearing loading and bearing diameters may cause an unequal shaft lift; or foundation settling may disturb the original shaft alignment and affect the critical speed.

In the design of a rigid coupling the temperature influences and hydrodynamic effects must be incorporated in the instructions for aligning the shaft.

Well-designed bearing pedestals and widely spaced bearing center distances improve the insensitivity of the shafts against various types of misalignment.

It is not feasible to fully compensate for the possible effects of incorrect *shaft alignment* with a gear coupling, regardless of whether these imperfections originate from site assembly or from actual operation. In other words, a gear coupling is not necessarily the answer to all problems. The larger the output and the higher the speed, the smaller the permissible imperfections in shaft alignment become. Coupled sets of high-speed machinery require a high degree of alignment accuracy. At every revolution of a gear coupling slight axial movements occur, resulting in friction and heat rise.

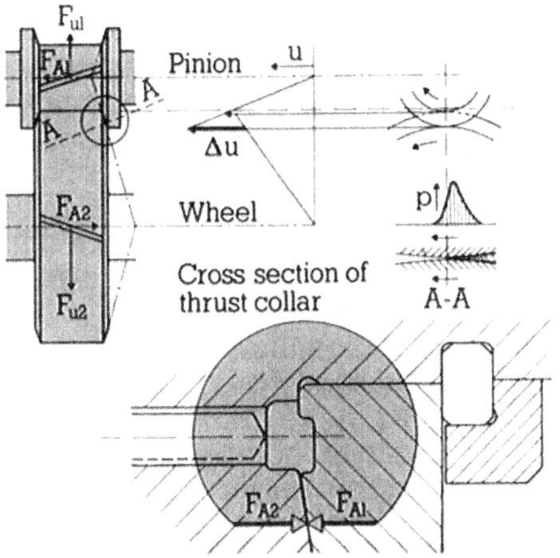

Figure 9.3 Method of axial thrust transfer in a single helical gear with thrust collar. *(Sulzer Brothers, Ltd., Winterthur, Switzerland)*

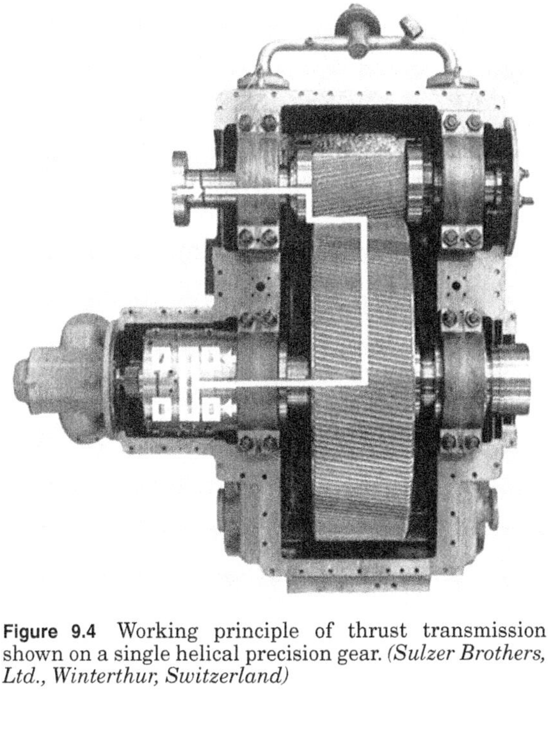

Figure 9.4 Working principle of thrust transmission shown on a single helical precision gear. *(Sulzer Brothers, Ltd., Winterthur, Switzerland)*

9.3 Maintenance

Lubrication with oil or grease reduces friction in a *gear tooth coupling.* At higher speeds and outputs it is difficult to maintain a reasonable degree of balance with a grease-filled coupling, and the surrounding air does not provide adequate cooling to remove the heat generated by friction.

Much better results are obtained with continuous oil flow lubrication. However, oil impurities may be centrifuged in the sleeve and for that reason only very clean, dehydrated oil ensures safe and reliable operation. At very high speeds there is again a balancing problem, and shaft alignment to a high degree of accuracy is required.

A *rigid coupling* has no maintenance requirements. Neither does a *diaphragm coupling,* although it requires careful protection against overstretching and/or accidental scratching of the relatively thin diaphragms.

9.4 Influence on the Critical Speeds

The *rigid coupling* has a marked influence on the critical speed of the individual shafts; in fact, it increases the critical speeds. Calculating the critical speed of shaft strings with several bearings is easily accomplished with a routine computer program. However, it is most important to know such important parameters as stiffness and damping constants of bearing pedestals and various other factors.

9.5 Differential Expansions

Expansion and contraction of steam turbine components must be taken into account whenever rigid couplings are applied. This is easily accomplished at the design stage but could prove difficult if rigid couplings are chosen as an afterthought.

With a gear type coupling the cumulative effect of differential expansions is avoided, since the shafts can slide axially within the sleeve of the coupling. However, they slide under the force effect of the torque acting on the connected shafts. This axial force can be expressed as:

$$F = \frac{T\mu}{r}$$

where T = torque
r = distance from shaft center to gear pitch line
μ = coefficient of sliding friction

On inadequately lubricated gear couplings, μ has been observed to approach values as high as 0.25.

9.6 Axial Thrusts

In the turbine as well as in the compressor or other connected machine relatively high axial thrust forces must be absorbed by suitably designed tilting pad axial thrust bearings.

The *rigid connection* between turbine and driven equipment shafts allows the axial forces of both turbine and driven machine to be absorbed by only one axial thrust bearing. The axial thrusts of both machines are in opposite directions, so that the thrust bearing only has to absorb the resulting differential thrust.

Gear type couplings are expected to slide when axial forces are applied. For this reason the driven machine and the turbine must each be equipped with an axial thrust bearing.

9.7 Limits of Application

Flexible couplings at higher torques and speeds soon reach a technical limit for application. Beyond that there is only the rigid coupling. There is no other acceptable solution. However, the operational results of multishaft machines with rigid couplings up to 1350 MW show that properly designed equipment will perform flawlessly for decades.

Chapter

10

Rotor Dynamics Technology

Rotor dynamics analysis began in the 1870s when efforts were made to calculate the fundamental natural frequencies of shafts. Fifty years later the effects of an unbalanced force on the response of a single rotating disk were considered. For 25 more years, until 1944, analyses were limited to simple rotors, or to graphical solutions, for the determination of fundamental frequencies. At that time, M. A. Prohl of the General Electric Company developed a general calculation method that could be applied to any rotor geometry for the determination of many frequencies. This was a key breakthrough for rotor analysis and constitutes the basis for the present-day representation of rotors.

10.1 Rotor Model

Prohl's method enabled the analyst to represent the rotor by a mathematical model that closely resembled the actual geometry. There was no longer any need for the simplifying assumptions required in previous calculated and graphical solutions. In addition, the second and higher mode frequencies and mode shapes could be obtained as readily as the first.

The method divided the rotor into a series of finite elements consisting of concentrated masses connected by massless springs. In current practice, mass stations are established at virtually every rotor location where there is a change in shaft diameter and at the wheel, thrust collar, coupling, and bearing locations as shown in Fig. 10.1. This is a high-speed rotor that has a total of 48 mass stations. The bearing centerlines are at stations 5 and 45.

The resulting mathematical model has the same mass and stiffness distribution as the actual rotor. Refinements are included such as the effective stiffnesses of the shaft sections through the integral wheels

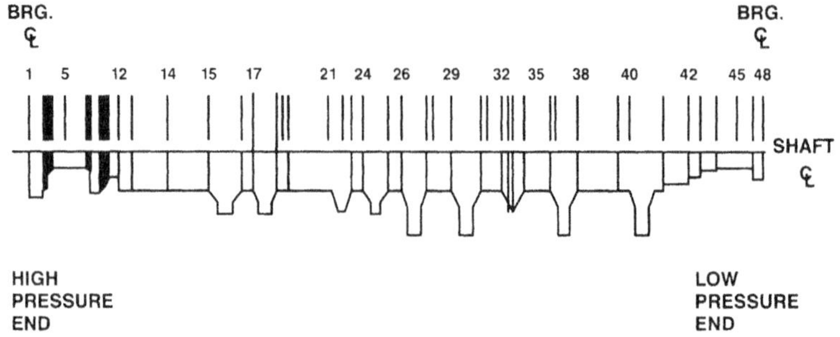

Figure 10.1 Impulse type steam turbine rotor model with mass stations identified. *(General Electric Company, Fitchburg, Mass.)*

and the coupling flanges and the gyroscopic effects. The calculation proceeds, station by station, from one end of the rotor to the other. At each station, the moment, the shear force, the slope, and the deflection are calculated for an assumed frequency of vibration as functions of the initial boundary conditions. The assumed frequency is varied until the boundary conditions at the last station are satisfied. The assumed frequency then equals the natural frequency.

External stiffness and damping are introduced at the journal locations. Unbalanced weights can then be added at any location along the rotor, and a vibration response calculation can be made.

For a number of years following the development of Prohl's method, the knowledge of bearing oil film and support characteristics was uncertain. The capability of making bearing analyses lagged behind the rotor analyses. Because of this, efforts were made to evaluate the rotor support characteristics and their effect on natural frequencies. In the mid-1950s, the dynamic stiffness concept was developed for this purpose. It considered the combined bearing and support system.

10.2 Dynamic Stiffness

The dynamic stiffness concept was developed to emphasize the crucial role that rotor support characteristics have in determining the system vibration behavior. It is still in use today and is commonly referred to as the *critical speed map.*

The need for this approach was apparent because of the circumstances that existed. Prohl's method made it possible to calculate the dynamic characteristics of the rotor, but it did not have the capability to analytically predict the dynamic characteristics of the rotor supports. Using an impedance-matching technique, it was possible to derive the support characteristics for a line of turbine designs that had the same type of bearings and casing supports. The method combined the analytical and experimental results. Rotor critical speeds were calculated and plotted as a function of assumed dynamic stiffness at the journal surfaces. Special response tests were then run on a number of production turbines by installing unbalanced weights in the rotors and identifying the critical speeds by the shaft response amplitudes. Mode shape indications were inferred from the relative phase relationships of the measured shaft amplitudes. Having obtained the actual critical speeds, the effective dynamic stiffnesses of the total support systems could be derived from the rotor dynamic stiffness curves.

A typical plot of the original dynamic stiffness concept is shown in Fig. 10.2. The total effective bearing support stiffness was derived from actual critical speeds observed during factory tests.

In this way, a design range of support dynamic stiffness was established, as shown in Fig. 10.3, and used for many years in the dynamic analysis of new rotor designs. Eventually, the capability of accurately predicting bearing oil film stiffness and damping characteristics was developed to replace the empirical support stiffness design curves.

Today, the dynamic stiffness concept is useful in the basic understanding and effective preliminary evaluation that can be made when various combinations of rotor geometries and types of bearings are being considered. The separate effects of rotor changes and the use of different bearings on the placement of critical speeds is readily apparent. However, the concept must not be used as the sole rotor design criterion, but only for the initial consideration of rotor geometry and bearing selection because they affect critical speed locations relative to operating speeds. The capability of predicting the vibra-

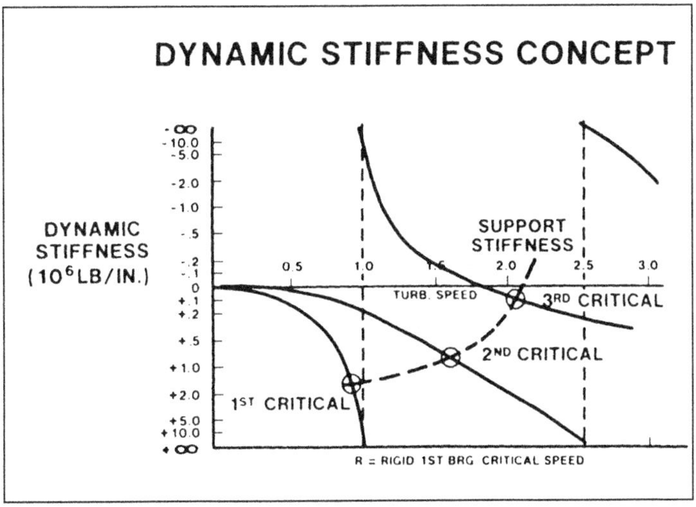

Figure 10.2 The original dynamic stiffness concept. *(General Electric Company, Fitchburg, Mass.)*

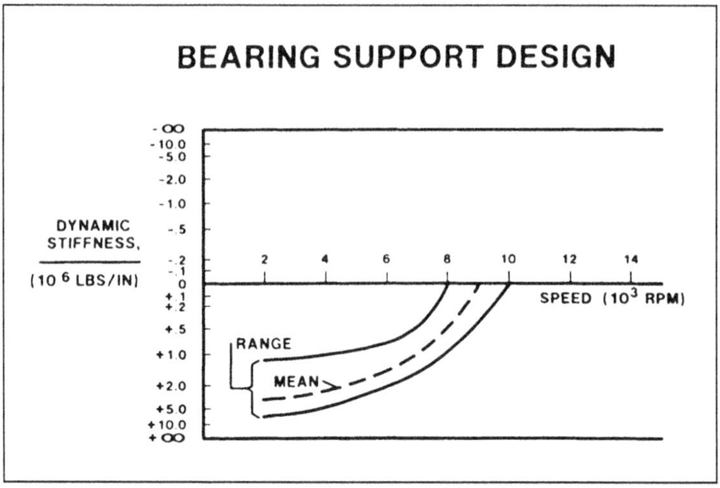

Figure 10.3 Bearing support design data. *(General Electric Company, Fitchburg, Mass.)*

tion response of the rotor due to assumed unbalanced forces has made the response calculation the ultimate criterion for successful turbine operation.

In its basic format, the dynamic stiffness concept assumes that the rotor has two bearings with equal support stiffness characteristics. In practice, the two bearings are usually the same type and do not differ greatly in diameter, length, or reaction loading. A good approximation of effective dynamic stiffness is an average of the two. Different vertical and horizontal bearing stiffnesses are considered separately, which theoretically gives rise to pairs of critical speeds. Each pair involves the same mode shape, but they occur at different speeds corresponding to the vertical or horizontal orientations. The effects of bearing damping on critical speeds are usually neglected as are the cross-coupling terms. The effects of the casing support stiffness are included in series with the bearing oil film stiffness.

Figure 10.4 shows what happens when both the vertical and horizontal bearing support stiffness are different. Each rotor mode curve is intersected twice; therefore, there will be pairs of critical speeds for each mode. The horizontal bearing stiffness is usually lower, and the lower critical of the pair will exhibit an elliptical shaft whirl orbit that is horizontally oriented. As speed increases, the shaft orbit will rotate to a vertically oriented ellipse at the speed where the vertical support stiffness curve intersects the rotor mode curve.

In actuality, many of the discrete theoretical critical speeds are not evident in an operating turbine because of damping. In the case of sep-

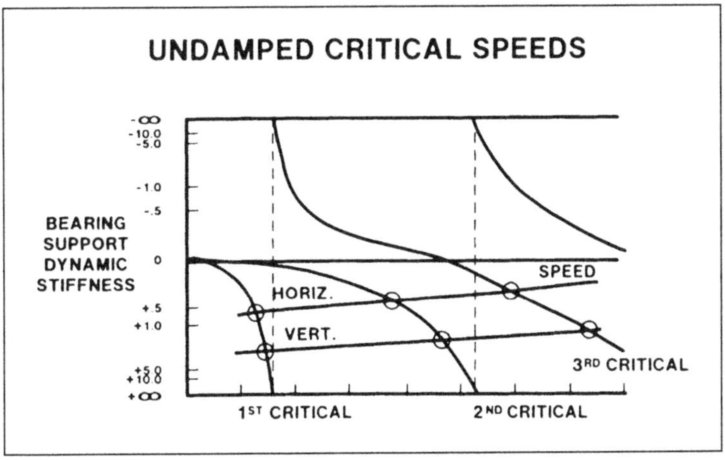

Figure 10.4 Pairs of critical speeds exist for each mode whenever both the vertical and horizontal bearing support stiffness are different. *(General Electric Company, Fitchburg, Mass.)*

arate vertical and horizontal criticals that are reasonably close in speed, the rotor will whirl in an orbit that exhibits one broad response peak instead of two sharp peaks. Depending on the mode shape, particularly the second critical, a heavily damped response peak may not be discernible at all. This often leads to a misinterpretation of the test data and needless efforts to establish operating speed margins for theoretical, undamped criticals.

10.3 Effects of Damping on Critical Speed Prediction

The damping forces within the bearing oil film are functions of vibration velocity and are normal to the journal deflection. The effect of dashpot damping on a simple spring-mass system is to depress the natural frequency as well as to reduce the peak amplitude. However, bearing oil film damping increases the equivalent stiffness and tends to increase the resonant speeds. Since the journal whirls in an orbit, the effect of damping on the equivalent stiffness will also depend on the shape of the orbit.

The prediction of critical speeds, using the dynamic stiffness curves, can be improved if an equivalent stiffness includes the approximate effects of damping. Two examples are presented in Table 10.1 to show the critical speeds predicted by the dynamic stiffness map as compared to the response peaks calculated by the rotor response analysis. Example A is a large, medium-speed rotor with four-pad, tilting-pad bear-

TABLE 10.1 Critical Speed Predictions

Example A: Medium-Speed Rotor		
	First critical	Second critical
Case 1. Stiffness only	1200 r/min (−17%)	3000 r/min (−40%)
Case 2. Stiffness and damping	1300 r/min (−8%)	4200 r/min (0%)
Response calculation	1400 r/min (0%)	4200 r/min (0%)

Example B: High-Speed Rotor			
	First critical	Second critical	Third critical
Case 1. Stiffness only	2800 r/min (−11%)	6000 r/min (−22%)	11,700 r/min (−11%)
Case 2. Stiffness and damping	2900 r/min (−7%)	7000 r/min (−4%)	12,700 r/min (−2%)
Response calculation	3100 r/min (0%)	7300 r/min (0%)	13,000 r/min (0%)

SOURCE: General Electric Company, Fitchburg, Mass.

ings. Example B is a high-speed rotor with five-pad, tilting-pad bearings. Critical speed predictions are presented for the following cases, in which it is assumed that the dynamic stiffness includes:

- Bearing and support stiffnesses only
- Combined bearing and support stiffness with damping effects

The critical speeds are listed for the case when stiffnesses alone is considered in comparison to predicted critical speeds when damping effects are included. The peak speeds, as calculated by the vibration response to unbalance program, are also listed for comparison with the simple dynamic stiffness predictions. Using stiffnesses only, the first criticals are 11 percent to 17 percent low, the second criticals are 22 percent to 40 percent low, and the third critical is 11 percent low. With damping effects, the dynamic stiffness predictions are much closer to the peaks calculated with the complete rotor response program, particularly for the second and third criticals.

10.4 Bearing-Related Developments

In the rotor support system, the bearing oil film stiffness and damping characteristics have, by far, the most significant effects on the vibration response of the rotor. The key breakthrough in this area was the result of three major developments. The first was the advent of high-speed, high-capacity computers, which made it practical to perform the theoretical analyses of bearings without the need for many simplifying assumptions that were required previously. The second was the availability of new instrumentation, which made it possible to obtain accurate measurements of vibration amplitudes and phase angles. The noncontacting vibration probe and the tracking filter were significant component developments.

The third major development, which was essential to the evolution of rotor dynamics technology, was the establishment of laboratory bearing test facilities to make dynamic measurements of the journal orbit motion while running in different types of full-size bearings.

Correlations were obtained between calculated and experimentally observed bearing-induced rotor behavior. As a result of these developments, a new design philosophy emerged in which bearings were considered as more than just hydraulic supports for the rotor. The dynamic characteristics of the bearings were matched to specific rotor characteristics to evaluate their effect on the vibration response of the rotor-bearing system. For example, tilting-pad bearings were originally selected because of half-frequency, oil-whip problems. Their use was limited to high-speed, light-load applications. Currently, tilting-pad bearings are being applied to all rotors regardless of speed or load because of their stiffness and damping characteristics as they affect synchronous rotor response.

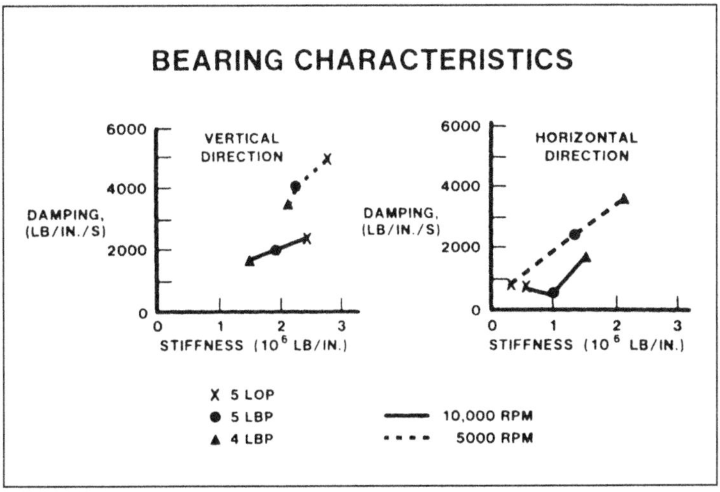

Figure 10.5 Bearing characteristics affected by damping and stiffness (LOP = load on pad; LBP = load between pads). *(General Electric Company, Fitchburg, Mass.)*

A variety of characteristics can be obtained for different numbers of pads and different load orientations (Fig. 10.5). It should be noted that there is no single *best* bearing for all rotors that can be assumed from such comparisons. The complex interaction between stiffness and damping effects in the vertical and horizontal directions, plus significant cross-coupling effects in fixed-arc bearings, must be evaluated by rotor response calculations. In the design process, such calculations are made for the rotor with different bearings and clearances. The final bearing selection is then based on the results of the rotor dynamics analysis.

10.5 Refinements

As a consequence of laboratory bearing tests, several mathematical options have been added to the basic analytical bearing model. A modified computer program is now capable of evaluating their effects. In practice, they do not result in appreciable changes in calculated rotor response for rotor-bearing support systems that are relatively insensitive to unbalanced forces. In the case of more sensitive systems, however, the analysis may show significant changes. One of these effects is associated with local deformation of the pad contact surfaces where the concentrated compressive stresses are high. This hertzian effect results in a decrease in the equivalent bearing stiffness values.

Another effect is due to slight thermal distortions of the pads, which tend to produce the effects of preload. The oil temperature at the babbitt surface of the pad is greater than at the back of the pad, which effectively increases the bearing radius while the minimum bearing clearance decreases. This introduces a small amount of preload that can change the bearing characteristics.

Deliberate preload by design can cause significant changes in the stiffness and damping coefficients. These changes must be evaluated for specific applications to determine whether or not they are beneficial. However, the general design philosophy is to produce a rotor-bearing system that is not dependent on preload for success. One reason for this is that it may be difficult to maintain the desired preload. For example, slight bearing surface wear or wipes, or manual rework, may inadvertently remove the preload and result in a vibration problem if the design is dependent on preload.

Other effects such as pad inertia and fluid inertia can also be considered. Pad inertia is not usually of practical significance. The pad size is such that it readily tracks the shaft orbit and is not subject to flutter or the introduction of cross-coupling effects. Fluid inertia is the carry-over of momentum from one pad to the next. Its major effect is to sustain an oil film wedge in the upper pads, which normally experience little or no loading. Without this effect, the upper pads may theoretically nose-dive because of the moment caused by shear forces along the babbitt surface.

10.6 Bearing Support Considerations

The dynamic characteristics of the structure that supports the bearings can have an appreciable effect on rotor response, depending primarily on the interaction between the oil film coefficients and the support stiffnesses. The support theoretically includes all of the structures installed between the oil film and the earth. Part of this is a function of the turbine design; the other major part is in the foundation design.

The bearing itself must be aligned to the journal and held firmly. For fixed-arc bearings, this is accomplished by the use of spherical bearing liner seats and an interference fit between the bearing liner and the bearing bracket. The purpose of the spherical seat is to obtain initial alignment. Once this is done, the bearing is clamped in place by the interference fit and cannot change its position. Tilting-pad bearings that have pads with spherical pivots do not require spherical liner seats for alignment, since the pads may tilt in any direction. An interference fit is still provided between the bearing liner and the bearing bracket to prevent looseness that may result in vibration at multiples of the running speed.

The turbine casing support at the thrust bearing end of the turbine is usually a flex-plate design as shown in Figs. 1.5 and 8.6. It accommodates axial thermal expansions by bending but is stiff in the vertical and transverse directions. Another type of support occasionally used in special applications is the sliding support. It is free to slide axially and is keyed to restrict transverse motion. Its stiffness is somewhat variable, depending on the static loading and the key fit. Impedance test measurements indicate that it is comparable to the flex-plate design in stiffness.

The bearing support at the drive end of condensing turbines includes the exhaust casing wall and the casing support feet. The support feet are free to move transversely to accommodate casing thermal expansions, and they are keyed to resist axial movement. Considerable design effort has been made in the modern exhaust casings to achieve increased stiffness in the transmission path from the drive end bearing, through the back wall, and into the casing support feet. Modern noncondensing turbine casings are supported at the drive end by trunnions. The bearing itself is directly supported by a pedestal that provides a stiff connection to the foundation.

One of the important reasons for maximizing stiffness between the bearings and the foundation is that it optimizes the inherent damping characteristics within the bearing oil film due to the relative motion between the journal and the bearing.

10.7 Foundations

The most common foundation designs are relatively massive, reinforced concrete structures that present a high impedance to the turbine at the connection points. In this respect, they are dynamically similar to the special bases that support the turbine during vibration response tests run in the factory. The experience has been that the factory test results have been predictive of subsequent operation in the field.

From this standpoint, concrete foundations are preferable to fabricated steel structures that tend to exhibit resonant frequency behavior that may be difficult to predict and more difficult to modify after the structure is in place. Local foundation resonances will affect turbine response, because they introduce low dynamic stiffnesses into the total bearing support system.

10.8 Impedance

The total bearing support structure is complex. Thus, its dynamic characteristics are difficult to represent with a mathematical model. On the other hand, experience indicates that good correlation is obtained between rotor dynamic analyses and unbalanced vibration response tests when a simple approximation of support stiffness is used.

In the past, support stiffness was based on static deflection, either measured or calculated. More recently, impedance tests have been made to determine the dynamic stiffness for various turbine designs. In the case of new designs, impedance tests have been made on scale models and later verified by tests on the full-scale prototype. A typical impedance plot is shown in Fig. 10.6 for the drive end bearing support of a high-speed turbine. The exciting force was applied at the bearing location, and the motion was measured at the same location and in the same direction. The ratio of driving force to vibration velocity is the direct or driving point impedance.

The ordinate scale is the *mechanical impedance ratio,* which is pounds of force divided by velocity in inches per second. The abscissa is frequency, in Hz. The diagonals that slope downward to the right are *dynamic stiffness ratios,* which are pounds of force divided by amplitude in inches. The diagonals that slope upward to the right are effective mass lines in pounds of weight. Each valley in the impedance plot is a resonant frequency, and each peak is an antiresonant frequency. At the resonant frequencies, the supporting structure presents a lower impedance (dynamic stiffness) to the rotor-bearing, oil film system at

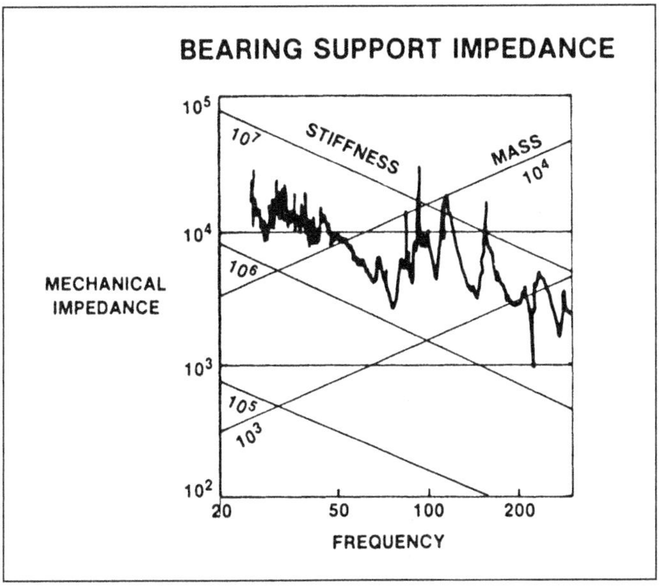

Figure 10.6 Typical impedance plot for drive end bearing support of a high-speed turbine. *(General Electric Company, Fitchburg, Mass.)*

the interface between them. This means that the unbalanced force transmitted from the rotor will cause an increase in bearing bracket amplitude at that frequency. The opposite is true at the antiresonant frequencies. In this case, the dynamic stiffness at the bearing bracket is high, which results in a decrease in bracket amplitude due to rotor unbalanced forces. In terms of modal analysis, the mode shape of the supporting structure has a node at the bracket, which is a point of low amplitude. (Theoretically, it would be zero.) The mechanical impedance at the bottom of each valley is the system damping at that frequency. The nearly constant dynamic stiffness at low frequencies extends to the static stiffness value.

It may be possible to curve fit the dynamic stiffness plot in piecewise fashion with simple spring-mass-damper combinations or to reproduce the variable dynamic stiffness itself as a function of frequency. It is not necessary, however, to specify the exact value of the impedance and its variation within the turbine's operating speed range, since this refinement does not significantly change the general rotor response to unbalanced curve. The effect of the support impedance variations is to cause the minor peaks and valleys that appear in the response test data. The support can be effectively approximated within an operating speed range by a spring with a single value of dynamic stiffness.

A curve of dynamic stiffness vs. operating speed is used for design purposes (Fig. 10.7). The irregular lines are a simplified approximation derived from the impedance plot in Fig. 10.6 for the 6000 to 12,000 r/min range (100 to 200 Hz). The straight lines are average design values of dynamic stiffness. For preliminary design calculations, the lowest stiffness value within the specified operating range would be used to represent a spring in series with the bearing oil film springs and dampers. This gives a conservative assessment of rotor response to unbalance. A more realistic model may be calculated for a specific speed of interest, or for narrow speed range, by using the corresponding stiffness value from the design curve.

Impedance tests on scale models made of steel can be used to predict natural frequencies and dynamic stiffness by directly applying the scale factor. However, damping does not scale proportionately. Thus, the peaks and valleys in the impedance plots are usually sharper and have more dynamic range. A typical plot for a one-sixth scale model of a large, low-speed turbine exhaust casing is shown in Fig. 10.8*a,* and the corresponding plot for the full-size turbine is shown in Fig. 10.8*b.* Applying the scale factor of one-sixth to the scale model resonant frequency of 400 Hz gives 67 Hz, which is the same as the first natural frequency of the full-size support. Similarly, the scale model dynamic stiffness of 1×10^6 lb/in in the 100- to 300-Hz frequency range would be 6×10^6 lb/in for comparison with the full-size value of 10×10^6 lb/in for the 20- to 40-Hz range.

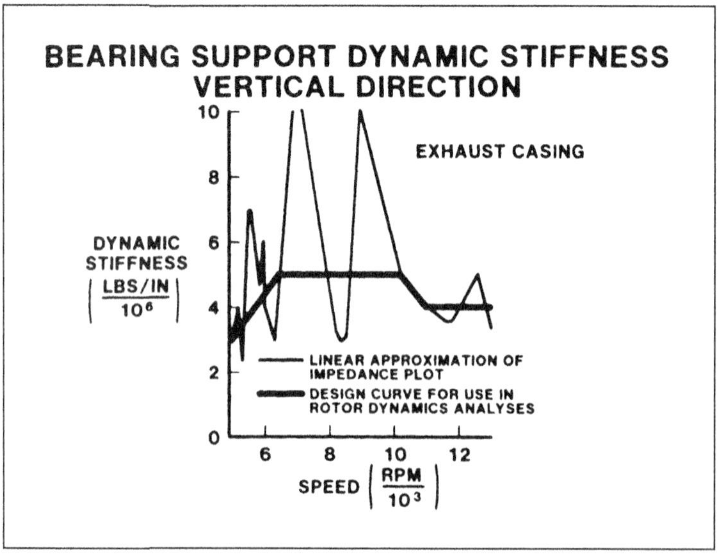

Figure 10.7 Bearing support dynamic stiffness values vs. rotor speed, experimental values, and approximations used in turbine design. *(General Electric Company, Fitchburg, Mass.)*

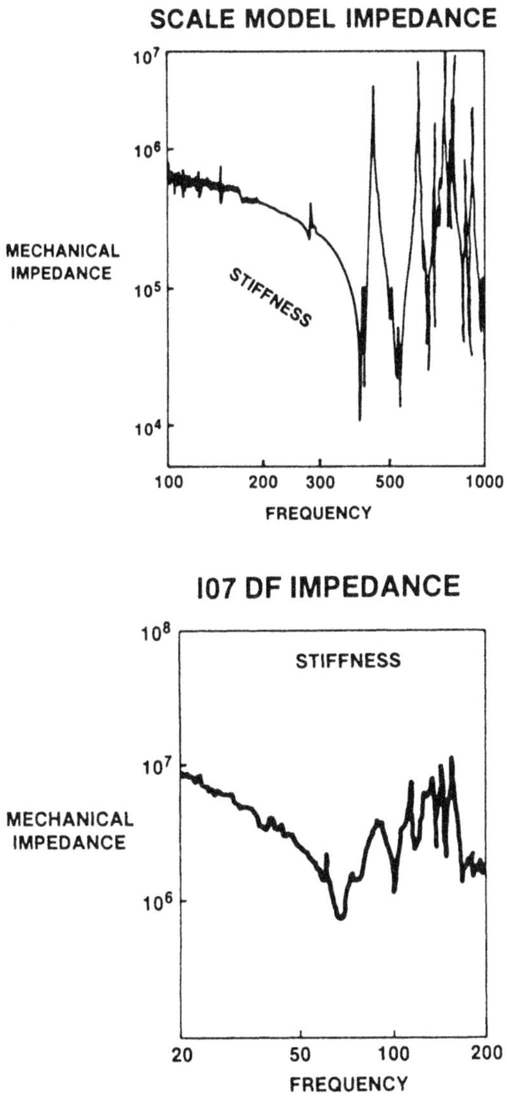

Figure 10.8 Scale model and full-size turbine impedance plots. *(General Electric Company, Fitchburg, Mass.)*

10.9 Partial Arc Forces

For many years, it has been the practice of turbine designers to arrange the steam control valve-opening sequence so that the first valve to open will result in a downward steam force on the rotor. The reason for this was to avoid unloading the bearings during partial arc operation, which might cause undesirable rotor vibration.

More recently, in the case of a small, high-power turbine in service, it was discovered that the entire valve-opening sequence must be considered, because the partial arc steam forces were large relative to the rotor weight. This could push the rotor into a sector of the bearing where the dynamic characteristics of the bearing would be different than what was expected. The vertical gravity load was directed between the pads of a four-pad bearing. But during partial arc operation, the combined gravity and steam force vector rotated as much as 60° so that the resultant load could be directly in line with the adjacent pad when two control valves

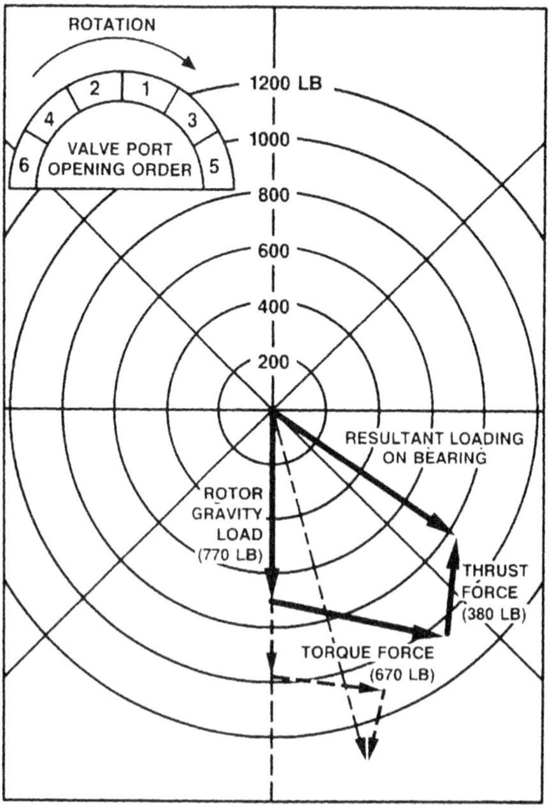

Figure 10.9 Force vector plot shows load orientation on bearing changing with steam admission sequence. *(General Electric Company, Fitchburg, Mass.)*

were open. This created a four-pad load on the pad configuration with its different bearing characteristics, which changed the shaft orbit and increased the rotor response. This phenomenon is illustrated in Fig. 10.9, which is a force vector plot for one of the bearings.

The vertical force is due to rotor weight. The sketch shows a 360° nozzle box with the valve-opening sequence and the direction of rotation, as indicated. The directions of the resultant forces within each nozzle arc are shown.

The first valve supplies 90° of nozzle arc in the lower half. The steam flow is limited to about a quarter of full flow, but a large proportion of the total turbine power is developed in the first-stage partial arc. As the second valve opens, the resultant force vector rotates toward the pivot point angle of the adjacent tilting-pad. When it has rotated 45°, it is in line with the pad, and the bearing characteristics are those of a four-pad load on the pad configuration.

When the rotor force vector is directed between the pads of a four-pad bearing, the oil film stiffness and damping coefficients are nearly symmetrical. The shaft orbits are reasonably circular and well damped. When the force vector is directed toward a pad, the coefficients in line with the pad are vastly different from those that are perpendicular to the force vector. This produces a narrow, elliptical (line) orbit of much greater amplitude. In addition, the critical speed responses change and become more complicated. This condition will not occur during a no-load factory test, but only during operation with load.

10.10 Design Procedure

To avoid large changes in bearing characteristics due to force vector changes, a design procedure has been established to evaluate the first-stage, control valve, partial arc forces due to the valve-opening sequence. Extraction valve forces are analyzed in the same way. Worst-case combinations of the forces from both control stages are then identified. The effect of these forces relative to the rotor weight, the bearing loads, and the bearing type is then determined.

Another aspect of the same problem occurs in conjunction with turbines that have partial arc diaphragms in the steam path, which are required for efficiency reasons. The design practice is to circumferentially index the partial arcs of the succeeding stages in the direction of rotation. This diaphragm partial arc effect exists whether or not the first stage is a partial arc. Even in some cases of longer, heavier rotors, this effect results in significant changes in the bearing loads and can rotate the resultant rotor force vectors around to the adjacent pad pivot points.

In modern steam turbine design, the combined effects of the gravity forces and the partial arc forces (both control stages and the diaphragm stages) are evaluated, and bearings are selected so that the rotation of the force vector does not cause the journal attitude angle to approach the adjacent pad pivot point. In some cases, the partial arc diaphragms are repositioned to avoid this condition.

10.11 Rotor Response

If it has been determined that the combination of partial arc forces and rotor weight can produce a resultant force vector that rotates by a significant amount, the bearing stiffness and damping coefficients are calculated for this condition. With the new coefficients, rotor response to unbalanced calculations are made to ensure that the vibration response is acceptable with the partial arc conditions.

10.12 Instability Mechanisms

Over the years, many causes of unusually high, nonsynchronous vibration due to instability have been experienced. The most common cause has probably been bearing oil film whirl, which has been associated with high-speed, lightly loaded, fixed-arc bearings. The use of tilting-pad bearings, which have extremely small oil film cross-coupling effects, has been the solution to this type of problem.

Instability due to hysteretic damping is associated with rotors having parts, such as wheels, that are shrunk-on to the shafts. This problem is avoided by designing rotors with integral wheels, thrust collars, and coupling flanges.

Rotors that have bore holes or internal cavities may be subject to instability due to trapped fluid. With modern forging practices, process controls, and ultrasonic inspection results, mechanical drive turbine rotors have not required bore holes for the removal of centerline indications or for bore sonic inspections.

Bore holes are usually required for larger rotors, in which case the hole is sealed at each end to prevent the entrapment of fluid.

Other types of instability have been caused by close, nonuniform, seal or blade tip clearances. Contact or near-contact friction has also caused problems.

Stability criteria that have been developed for bearing oil film forces and for steam forces are successful in providing assurance that the classic violent vibration associated with instability will not occur.

10.13 Subsynchronous Vibration

Many of the known instability mechanisms cause vibration at frequencies lower than those corresponding to running speed. Most often, the vibration frequency coincides with the fundamental rotor natural frequency at slightly less than one-half of the running speed frequency. When unusually high-vibration amplitudes have been encountered in the past, the vibration signal has been analyzed for its frequency content to determine if the high vibration was subsynchronous and, therefore, related to an instability mechanism. This type of problem has been identified as a bearing oil film instability caused by improper loading of fixed-arc bearings. The problem has always been resolved by redesigning the bearing to increase its eccentricity or, in some cases, by installing tilting-pad bearings. With the extensive application of tilting-pad bearings over the past several years, this type of problem has not occurred in well-designed equipment.

In the context of normal, acceptable vibration behavior, the total vibration signal includes many frequency components, both synchronous and nonsynchronous. Generally, the synchronous component related to rotor speed has the largest amplitude, depending on the degree of rotor balance. In the past there has been little concern regarding the frequency spectrum and the identification of the various components. This has resulted in very little documentation, for turbines in service, in the form of spectrum analyses that define exactly the subsynchronous components. The total, unfiltered, vibration signal is always somewhat greater in amplitude than the filtered, synchronous vibration amplitude, and subsynchronous vibration may account for some of the difference. When a classic instability occurs, the total vibration is several times larger than the synchronous vibration amplitude.

Vibration limits are usually specified as the total amplitude of the unfiltered vibration signal. This is compatible with present control room vibration readouts that are unfiltered. In factory tests, more emphasis is placed on the synchronous vibration component that can be reduced to very low levels by precision balancing while other frequency components are unaffected. Since the final vibration limits are unfiltered, any sizeable subsynchronous vibration components that occurred during factory tests would be apparent regardless of the degree of balance. This is also true of field installations, where the total vibration signal is monitored.

The presence of subsynchronous vibration components can be detected by vibration meters, which display only the total unfiltered amplitude, because of the fluctuation in the reading. The synchronous amplitude and its multiples are usually fairly steady in amplitude. But when subsynchronous vibration is superposed, it causes periodic reinforcement and cancellation of the total steady signal that produces the amplitude fluctuation, or bounce, in the amplitude reading. This type of behavior has been observed on many turbines in the past.

In recent years, signature analysis has evolved as a possible indicator of the condition of rotating machinery. Frequency spectra of vibration signals are obtained, usually with a real-time analyzer. This instrument performs a Fast Fourier Transform analysis of the total signal from which plots of vibration amplitude vs. frequency can be generated. By periodic monitoring and comparisons with baseline or reference plots, changes in the frequency spectra can be identified, and vibration trends can be defined. Changes in vibration at frequency multiples of the rotor speed may be attributed to misalignment, loose connections, or rotating dissymmetries such as buckets, gear teeth, and impeller vanes.

Changes in vibration at frequencies not associated with rotor speed may be attributed to changes in the external excitation or to responses of the stationary and rotating structures at any of their natural frequencies.

Quite often, real-time analysis spectra will show low-amplitude vibration at partial frequencies of rotor speed. Because of bad experiences in the past with classic instability, which also occurs at partial frequencies, many operators and consultants become concerned when they observe the presence of partial frequency vibration components. These vibration indications tend to be erratic and fluctuating (sometimes disappearing), but they remain limited and low in amplitude in contrast to classic instability vibration, which rapidly increases to very high amplitudes.

Many different types of rotating machinery (steam turbines, gas turbines, compressors, gear sets, and generators) have exhibited limited amplitude, subsynchronous vibration with no deleterious effects in operation or reliability. This vibration has occurred in machines with different types of bearings, fixed-arc and tilting-pad. Many mechanical drive steam turbines in successful operation for many years have been involved in signature analysis programs. The analyses revealed that their frequency spectra exhibited limited amplitude, subsynchronous, vibration components. It can be inferred that such vibrations existed from the time the turbines went into service but were not documented at the time because the necessary sensitive instrumentation and spectrum analyses were not available.

On the basis of such experiences and other observations, it was concluded that limited amplitude, subsynchronous vibration is fairly common and does not affect reliable operation in any way.

An important distinction must be drawn between an unstable system and a stable system. Unstable systems are self-excited and exhibit the characteristics of negative damping. The exciting force is created by shaft motion and is normal to the direction of shaft deflection. When the shaft is deflected, the destabilizing force increases proportionately. This causes the rotor to whirl at its natural frequency with rapidly increasing amplitude until physical contact occurs in close clearance areas or until some other nonlinear restraint changes the system and prevents further amplitude increases. In a stable system, the exciting force does not increase with increasing amplitude. Therefore, the amplitude is limited to a low level by the restoring damping force.

Again, the presence of limited amplitude, subsynchronous vibration in operating turbines has probably always existed but has not been documented by frequency spectrum analyses.

10.14 Service Examples

The amount of spectrum data available for turbines in service is limited to relatively few installations. This is particularly true for the older turbines. However, some examples of subsynchronous vibration in turbines having long service times are shown in Figs. 10.10 through 10.12. This type of vibration has existed for many years, and no problems because of it have been reported to date.

The first example (Fig. 10.10) is a six-stage extraction turbine rated 14,000 hp at 12,915 r/min. The unit went into service in 1967. In 1981, a frequency spectrum analysis was made while the turbine was operating at 98 percent of rated output at 12,225 r/min. The synchronous shaft vibration amplitude is labelled 1X. The subsynchronous vibration is at slightly less than 0.25X running speed. The first critical speed had previously been observed at 80 r/s, but no indication of this appears in the spectrum plots. The high-pressure and low-pressure shaft ends show 1X amplitudes that are much different relative to one another, but the other vibration components are about the same. This turbine has tilting-pad journal bearings.

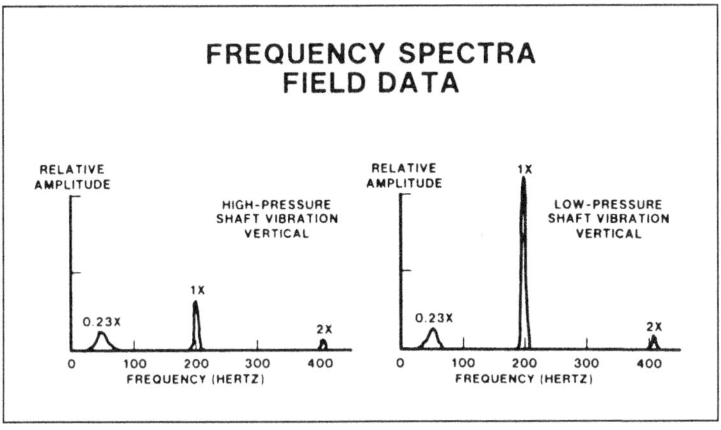

Figure 10.10 Frequency spectra for large steam turbine show relative amplitudes of vibration at 23 percent of running frequency. *(General Electric Company, Fitchburg, Mass.)*

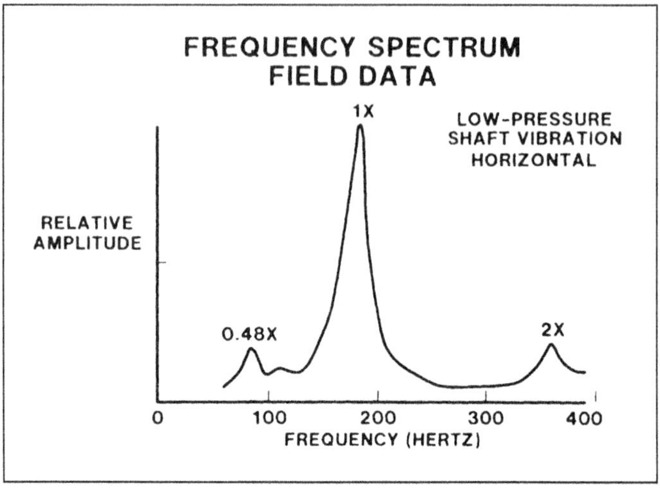

Figure 10.11 Frequency spectrum with subsynchronous component at 48 percent, probably coinciding with first critical frequency. *(General Electric Company, Fitchburg, Mass.)*

The second example is a two-stage, double-flow, noncondensing turbine, rated 20,000 hp at 11,655 r/min, which went into service in 1969. The frequency spectrum plot shown in Fig. 10.11 was made in 1979. This turbine also has tilting-pad journal bearings. The subsynchronous component frequency is about 0.48X running speed, which probably coincides with the first critical frequency.

The third example is an eight-stage extraction turbine, rated 35,640 hp at 6112 r/min, that went into service in 1969. The analysis shown in Fig. 10.12*a* was made in 1975. In this case, the subsynchronous frequency is 0.26X running speed, which is lower than the first critical frequency. This turbine has fixed-arc bearings of the axial groove type.

The fourth example is a six-stage condensing turbine, which is typical of hundreds of turbines that drive boiler feedwater pumps in central power stations. This type of turbine has been put into service in increasing numbers since 1965. The spectrum analysis shown in Fig. 10.12*b* was made in 1979. It shows a subsynchronous vibration component at a frequency corresponding to 0.35X running speed. These turbines have fixed-arc, elliptical bearings.

The frequency spectra, as shown, confirm the fact that limited amplitude, subsynchronous vibrations exist in these turbines, but there have been no incidents or evidence of classic instability during their long service times.

Extensive signature analyses have revealed that many turbines exhibit limited amplitude subsynchronous vibration behavior. However, these turbines also continue to operate reliably. To date, there has

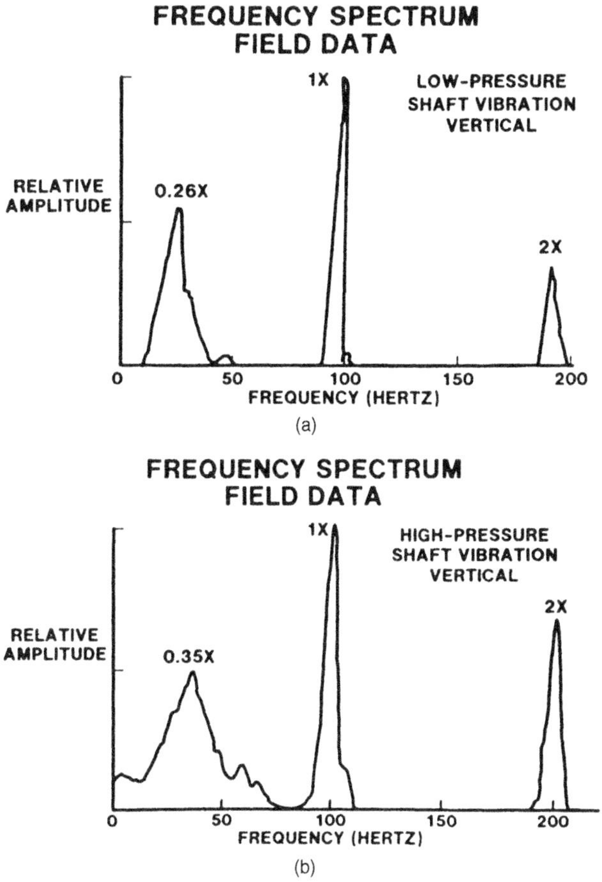

Figure 10.12 Subsynchronous frequencies of relatively high amplitude on highly successful steam turbines: (*a*) 35,640 hp at 6112 r/min; (*b*) six-stage boiler feedwater pump drive turbines. *(General Electric Company, Fitchburg, Mass.)*

been no direct correlation between the occurrence of subsynchronous vibration in the turbines and such parameters as types of bearings, operating speeds, horsepowers, steam conditions, or precision of rotor balance. Frequency spectrum data thus indicate that the presence of subsynchronous vibration components is not uncommon in turbomachines that have operated without distress for many years.

10.15 Labyrinth and Cover Seal Forces

The two elements that contribute directly to decreasing the lateral stability of a rotor are the labyrinth seals and the sealing strips on the bucket cover. Of the two, the labyrinth seals in the high-pressure end of the machine have the potential of generating the greatest destabilizing force. While rotor instabilities from bearing oil whirl are well known, instabilities from labyrinth and spill strip seals are understood to a lesser degree.

The destabilizing force on a rotor from seals is a direct consequence of the frictional drag from the circumferential swirl of the steam. If a rotor is centered in a seal, the local circumferential swirl velocity is the same all around the rotor, and there is no net force on the rotor. If the rotor is moved off center, as shown in Fig. 10.13, the chamber area is reduced at the top and increased on the bottom. The reduction in area at the top causes the swirl velocity to increase in that part of the seal. The area that has been enlarged at the bottom will locally have a lower velocity. As the steam travels from location A to C in the figure, the steam experiences a higher drag force than from going from C to A in a clockwise direction. This induces a higher pressure at A than at C. This pressure difference pushes the rotor to the right in a clockwise direction around the center of the seal. The rotor now begins to whirl around the center of the seal. The pressure difference between A and C feeds energy into the whirl orbit, which causes the orbit to increase. As it increases, the pressure difference gets even larger because the reduced area region is further decreased, causing the shaft to whirl with an increasing amplitude. This is what is meant by self-exciting vibration. The vibration itself produces forces that push the rotor to higher vibration levels.

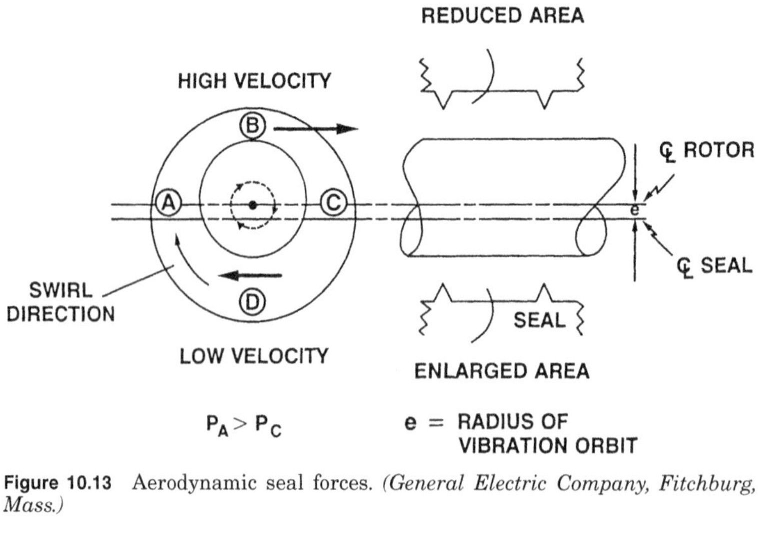

Figure 10.13 Aerodynamic seal forces. *(General Electric Company, Fitchburg, Mass.)*

10.16 Rotor Stability Criteria

After the dynamic characteristics of each component have been calculated, the components are combined into a comprehensive model where either response to mass imbalance or rotor stability can be calculated. When rotor stability is of concern, the logarithmic decrements of the critical speeds (damped natural frequencies) are calculated. Generally, the approach taken is to perform a sensitivity study of log decrement for different levels of destabilizing force. Log decrements for the first critical are calculated at different multiples of aerodynamic destabilizing force to provide a curve as shown in Fig. 10.14.

The curve represents a stable design where the log decrement is positive at the expected level of destabilizing force. In Fig. 10.14, at approximately twice the level of destabilizing force gradient, the rotor is on the verge of instability. When each of the dynamic characteristics of components in the system are accurately known, the actual level of rotor stability can be precisely predicted. To account for changes over time in turbine operation, seal and bearing wear, and other important parameters, turbines are designed to withstand at least twice the expected level of destabilizing force before the calculated log decrement goes negative. With this margin of safety, turbines will operate free from rotor instability.

10.17 Experimental Verification

Any mathematical analysis is only as good as the results it produces. The validity of the calculations must be verified by controlled tests.

The bearing oil film analysis is a key item in the overall rotor dynamics analysis. State-of-the-art manufacturers use special labora-

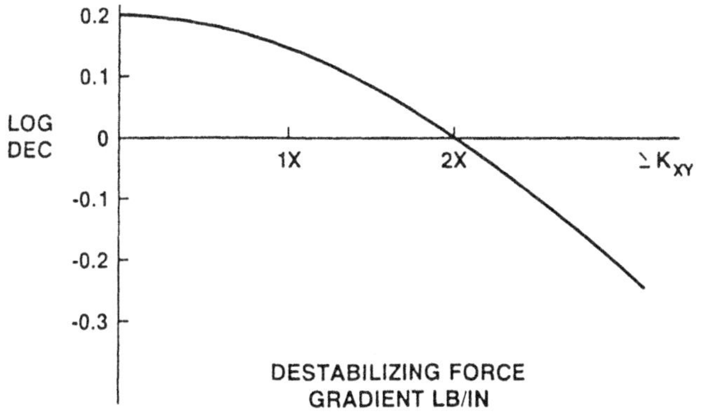

Figure 10.14 Destabilizing force vs. logarithmic decrement. *(General Electric Company, Fitchburg, Mass.)*

tory bearing test facilities to make running tests with various types of full-size bearings to confirm the analytical results. As a consequence of the testing, the accuracy of bearing analysis is very high. But experimentation and analysis have gone further. Impedance tests have been made on the bearing support structures to confirm the design stiffness calculations and to provide realistic dynamic stiffness values for the rotor dynamics model.

Finally, special rotor vibration response tests have been made on a large number of turbines of various sizes and speeds to correlate the rotor dynamics results with actual performance. Rotor response tests on new turbine designs ensure that subsequent field operation will be successful.

Chapter

11

Campbell, Goodman, and SAFE Diagrams for Steam Turbine Blades

Rotating blades of axial steam turbines are subjected to a variety of steady-state and transient mechanical loads and stresses. Steady-state stresses are due to the centrifugal force of the blade weight and the force of steam striking the blade. Transient stresses are attributable to several sources but are primarily due to blade excitation at a frequency equal to the natural frequency of the blade. Transient stresses are more rigorously defined as alternating, cyclic, or vibratory stresses.

Steady-state stresses are routinely calculated at critical points on the blade by using principles from the mechanics of deformable bodies and mechanical systems design. Both normal and shear stresses are calculated and are compared to material and historically determined acceptable limits.

11.1 Goodman Diagram

The purpose of a Goodman diagram is to serve as a stress analysis criterion of static and fatigue failure. A Goodman diagram is a graph that has the ultimate strength of a material plotted on the horizontal axis and the endurance limit plotted on the vertical axis. These two points are then connected by a straight line. This line is a *failure line* (factor of safety = 1.00), which means stress levels below it are safe and stress levels above it are likely to fail. Manufacturers such as Dresser-Rand use the yield point of the material as the steady-stress failure limit instead of the ultimate strength. By connecting the yield point with the endurance limit by a straight line, one obtains a Soderberg diagram, which is more conservative than the Goodman diagram (see Fig. 11.1).

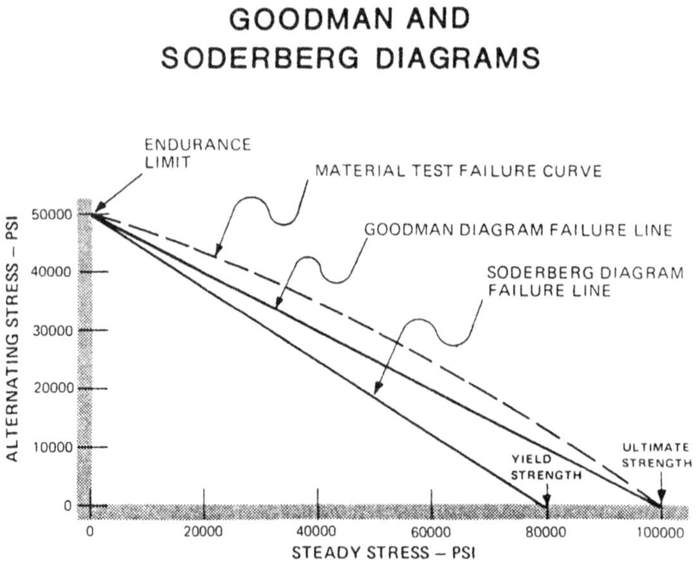

Figure 11.1 Goodman and Soderberg diagrams. *(Dresser-Rand Company, Wellsville, N.Y.)*

11.2 Goodman-Soderberg Diagram

To use the Soderberg diagram as a design criterion for blading, it will be necessary to first calculate the blade stresses. There are two types of stresses turbine blades are subjected to: steady stresses and alternating stresses. The steady stresses in a blade are tension due to centrifugal force and bending due to the steam force. The alternating, or vibratory stresses, are bending stresses caused by the disturbances in the steam flow. The steady stresses are the more accurately calculated of the two. The value that is used for vibratory stress is equal to (stress concentration factor) × (magnification factor) × (steady steam bending stress). The number used for stress concentration will vary depending on location. The magnification factor is based on experience curves, which depend on the blade frequency and the exciting frequencies. These will be explained in detail in the Campbell diagram section.

With the blade stresses now calculated, it is possible to calculate the factor of safety of the blade (see Fig. 11.2). This is done by plotting the steady stress in the horizontal direction and the vibratory stress in the vertical direction. The factor of safety is calculated by the following formula:

$$\frac{1}{\text{FS}} = \frac{\text{steady}}{\text{YP}} + \frac{\text{vibratory}}{\text{EL}} \tag{11.1}$$

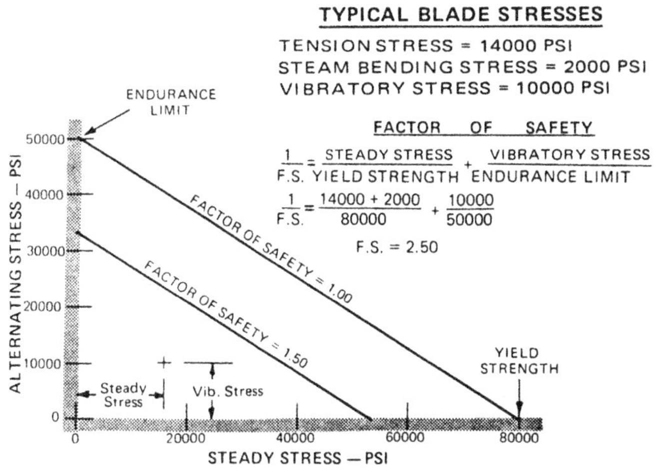

Figure 11.2 Calculating a factor of safety using the Soderberg diagram. *(Dresser-Rand Company, Wellsville, N.Y.)*

Most manufacturers limit the minimum acceptable factor of safety to 1.50 to cover variations due to machining tolerances, nonhomogenous material and other unpredictable factors. There is usually a limit to the steam bending stress which in turn helps to minimize the vibratory stresses in a stage.

If the calculated factor of safety of a turbine blade is less than 1.50, there are several alternatives: (1) Use a better blade material to improve the endurance limit and yield strength; (2) use a less dense material (like titanium) to reduce the centrifugal stress; (3) change to a different blade section to reduce stresses.

The Soderberg diagram is used to determine the factor of safety not only in blades but also in the shroud, rivet, root, and wheel rim of every turbine produced by experienced turbine manufacturers.

11.3 Campbell Diagram

The Campbell diagram, or interference diagram, is used to indicate what the vibratory stress level may be in a given stage. Since almost all blade failures are caused by vibratory stress, many reliability-conscious purchasers are requesting Campbell diagrams with turbine quotes or orders. A Campbell diagram is a graph with turbine speed (r/min) plotted on the horizontal axis and frequency, in cycles per second, plotted on the vertical axis. Also drawn in are the blade frequencies and the stage-exciting frequencies. When a blade frequency and an exciting frequency are equal, or intersect, it is called a *resonance* (see Fig. 11.3).

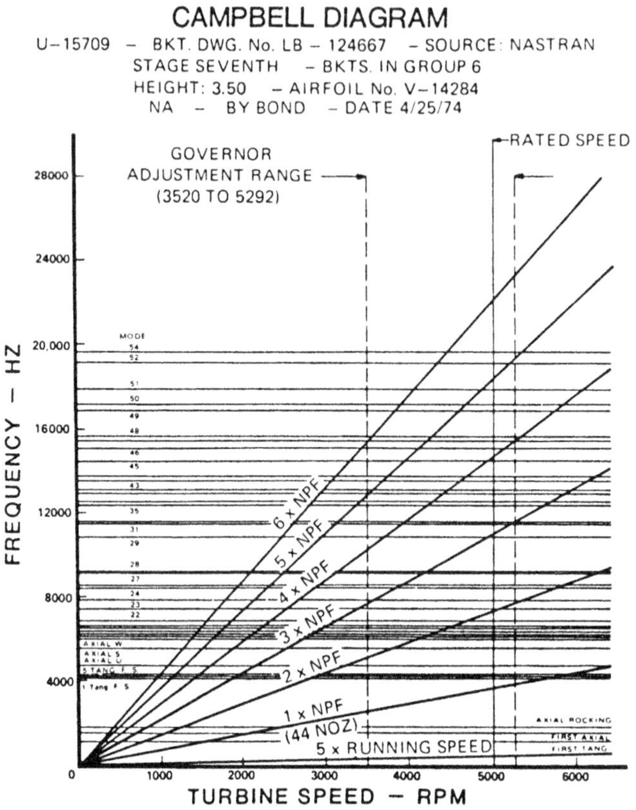

Figure 11.3 Campbell, or interference diagram. *(Dresser-Rand Company, Wellsville, N.Y.)*

The frequency that a turbine blade will have is a function of its mass and stiffness. There are three basic types of frequencies that exist in a shrouded blade group: tangential, axial, and combined axial-torsional. See Figs. 11.4 and 11.5 for illustrations of these frequency mode shapes. Blade frequency can be determined either analytically or by testing. A block diagram of the *transfer function analysis* (TFA) test system used by one major manufacturer, Dresser-Rand, is shown in Fig. 11.6. What basically is done here is a mechanical shaker (exciter) will shake the blades (structural system) at a frequency controlled by the oscillator. The input force of the shaker (F) is then compared visually to the acceleration response ($\ddot{x}$) of the blades on a plotter and oscilloscope. When the blade response signal becomes large compared to the input force, a natural blade frequency is indicated. The compliance plot of Fig. 11.7 shows the natural blade frequencies found in sweeping from 300 to 10,000 Hz. The main advantage of testing is that the exact

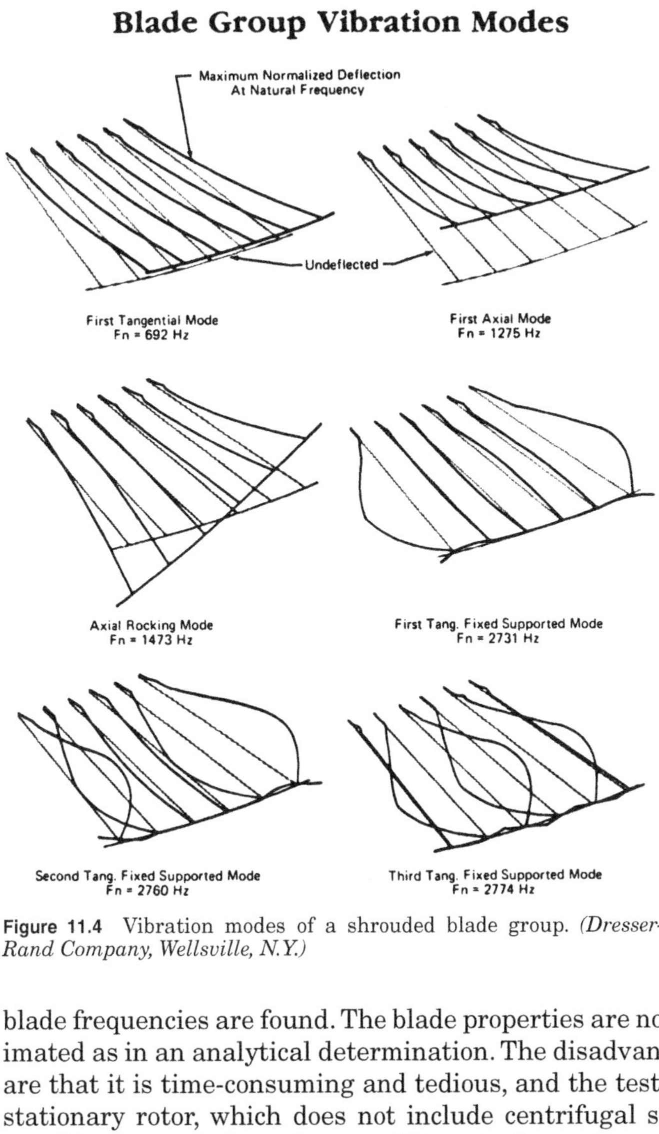

Figure 11.4 Vibration modes of a shrouded blade group. *(Dresser-Rand Company, Wellsville, N.Y.)*

blade frequencies are found. The blade properties are not being approximated as in an analytical determination. The disadvantages of testing are that it is time-consuming and tedious, and the tests are done on a stationary rotor, which does not include centrifugal stiffening and a possible change in boundary conditions in the root.

Turbine manufacturers use a number of different finite-element analysis computer programs to calculate blade natural frequencies. The blading is generally modeled by a series of beam elements to simulate the actual configuration. By supplying properties such as areas and moments of inertia to the beam elements, the natural frequency of the model can be calculated. Along with the natural frequencies, the displacements, forces, and moments within the model are given to help

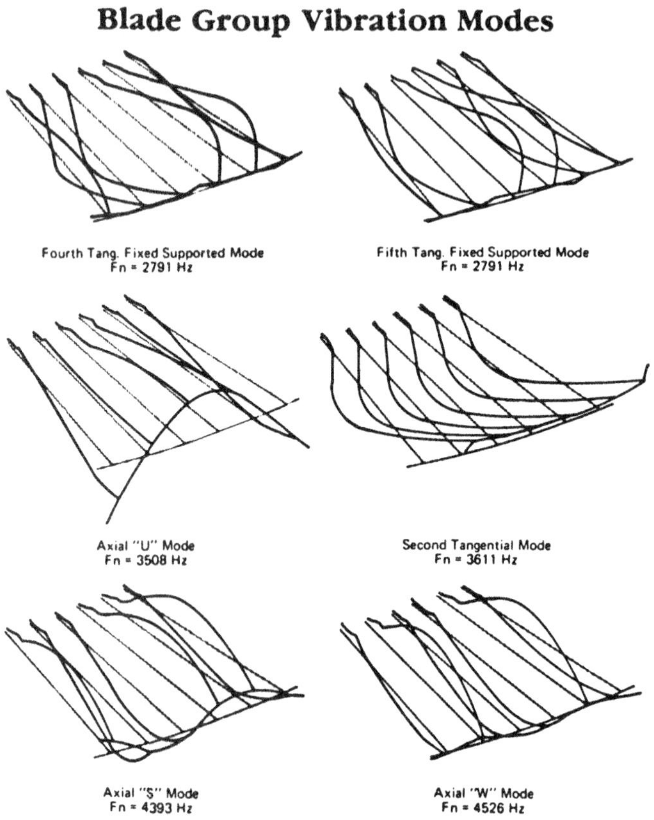

Figure 11.5 Additional vibration modes of a shrouded blade group. *(Dresser-Rand Company, Wellsville, N.Y.)*

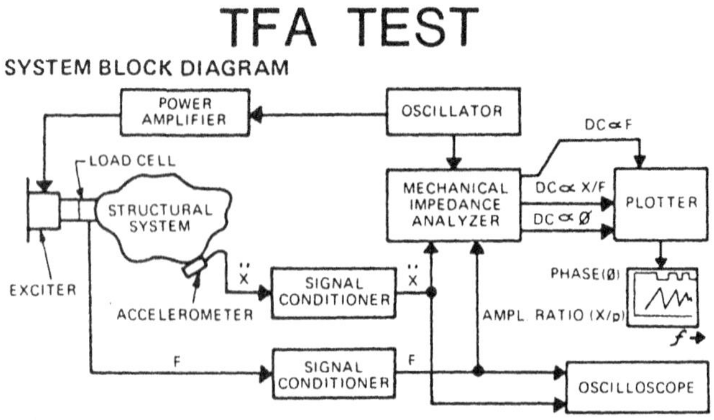

Figure 11.6 Block diagram of transfer function analysis applied to determine blade frequencies. *(Dresser-Rand Company, Wellsville, N.Y.)*

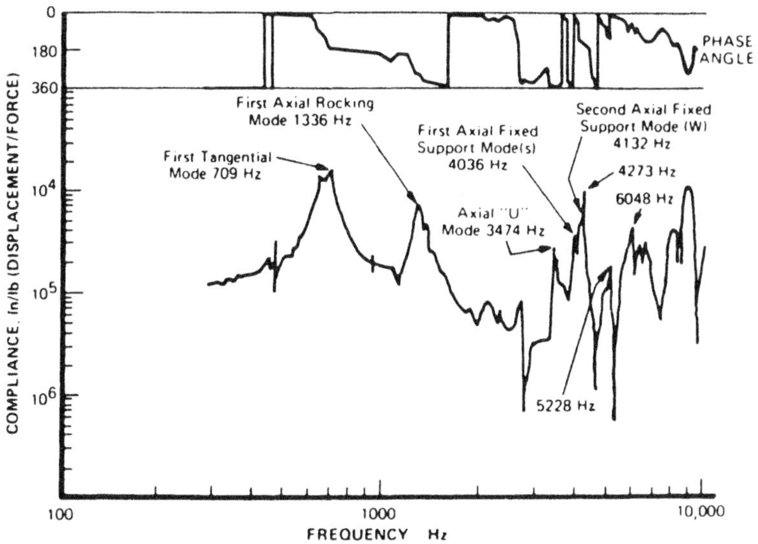

Figure 11.7 Compliance plot showing natural blade frequencies over entire frequency range. *(Dresser-Rand Company, Wellsville, N.Y.)*

determine the mode shape of each natural frequency. The finite-element analytical results are then compared to the modal test results. By making adjustments to the boundary conditions of the model, the results can be correlated. A comparison of analytical and test blade frequencies is shown in Table 11.1.

11.3.1 Exciting frequencies

The exciting frequencies shown on the Campbell diagram are variable steam forces that act on the blading and are caused by interruptions in the steam flow. The three most important exciting frequencies are:

1. Running speed or r/min excitation
2. Nozzle passing frequency
3. Partial admission excitation

Running speed excitation is caused by an unequal steam pressure acting on the blading as it rotates through one revolution. This type of condition is most prevalent in last stages where the exhaust flange is located to one side, which causes a lower steam pressure in that section. This causes one pulse of excitation per revolution as the blades pass from a high pressure to the low pressure and back to the high pressure. The first harmonic of running speed excitation is equal to (r/min/60) in Hz units. A typical design policy keeps the blade fre-

TABLE 11.1 Comparison between Vibration Testing and NASTRAN Analysis of a Particular Design

Mode	1st rotor test fn, Hz	2nd rotor test fn Hz	NASTRAN analysis fn, Hz	NASTRAN analysis Error,* %	Maximum estimated error, %
1 1st TAN	835.	790.	799.	2	5
2 1st AXIAL	1420	1423.	1274	10	10
2 AXIAL ROCKING	—	1675.	1447	14	15
4 TAN FIXED SUPPORTED	3020	2860	2879	2	5
5 ↓	↓	↓	2931	↓	↓
6			3007		
7			3047		
8	3090	3153	3058	2	5
9 Axial "U"	3580	3546	3531	2	5
10 2nd TAN			4047	—	—
11 AXIAL, FIXED SUPPORTED	— —	3880.	4483	16	20
12			4688		
13	—	4125.	4765	16	20

* Error is that compared to be average test value.
SOURCE: Dresser-Rand Company, Wellsville, N.Y.

quency above the fourth harmonic of running speed excitation for variable speed turbines. For constant speed turbines it is conceivable to have a blade frequency that is less than four times the running speed, but it will be tuned between exciting harmonics at the operating r/min.

Nozzle-passing frequency excitation is caused by the steam wakes at the trailing edges of nozzle vanes as the blade passes from one nozzle to the next. The first harmonic of nozzle-passing frequency is equal to (r/min/60 × number of nozzles in 360°) in units of Hertz.

Partial admission excitation is similar to nozzle-passing frequency excitation, except instead of each nozzle causing one pulse of excitation, a group of active nozzles will be causing the exciting force. As a blade passes from inactive nozzling to an active group of nozzles back to inactive nozzling, it will receive one pulse of excitation. Partial admission nozzling will provide a low harmonic of running speed excitation that is dependent on the number and spacing of the nozzle groups.

There are other exciting frequencies that occur in turbines besides these three most common discussed earlier. Struts or support ribs in casings can cause disturbances in the steam flow. Diaphragm manufacturing variations can cause both nozzle passing and low harmonics of running speed excitations; sharp turns in the steam path can cause random excitations.

The vibratory stress at a resonant condition depends on the strength of the exciting force, the resonant mode shape of the blades, the phase relationship between the exciting force and blade motion, and the damping of the blade-shroud structure. If there were no damping in the blades, the vibration amplitude would continuously build up until failure occurred. This sometimes happens, but the usual case is that the damping is sufficient to prevent dangerous levels of vibration to build up.

If the Campbell diagram shows that a resonance exists in the operating speed range, several design changes can be made to avoid it. The blade frequency can be changed by going to a stronger, stiffer blade section. If the resonance is with nozzle-passing frequency, the number of nozzles can be changed to avoid it. If it is impossible to avoid a resonance, then the steady stresses must be reduced to a level at which it is acceptable to operate during resonance. Methods for reducing steady stresses were discussed in the aforementioned Goodman diagram section.

A few words of caution are in order when using the Campbell diagram to evaluate blading. It should be noted that it is still difficult to accurately determine blade natural frequencies, even with improved analytical and test methods. This is especially true with short blading. Manufacturing tolerance and assembly variations can account for as much as a $\pm 10\%$ variation in blade natural frequency. Large avoidance margins must therefore be utilized to ensure that no interference exists.

In addition, all interferences shown on the Campbell diagram do not mean that the blade vibratory stresses will be high and that blade failure is imminent. Not all interferences are true resonances. The Campbell diagram only compares if the exciting force frequency and blade natural frequency match, but it does not compare if the shape of the exciting force and blade natural frequency match. For true resonance to exist, both the frequency and shape must match. Also, even if true resonance does occur, this does not mean that blade failure is certain. The level of vibratory stress occurring at resonance depends on the strength of the exciting force, the amount of damping in the structure, and the overall stress levels in the blading.

Figure 11.3 illustrates the near-impossibility of avoiding all resonances in the operating speed range.

11.4 SAFE Diagram—Evaluation Tool for Packeted Bladed Disk Assembly

The Campbell diagram is a two-dimensional projection of a three-dimensional surface; hence, it does not retain all the information for evaluation. Another two-dimensional projection of the same surface contains more information for easier evaluation. This diagram, refined for packeted bladed disk assembly by Dresser-Rand, is called a SAFE diagram.

The use of the SAFE diagram for a bladed disk assembly (not a single blade or a packet of blades as usually considered) facilitates the design evaluation process.

11.4.1 Definition of resonance

Each blade on a rotating turbine disk experiences a dynamic force when it rotates through the nonuniform flow emanating from stationary vanes. The dynamic response (e.g., stress and displacements) levels experienced by the blades depend on:

1. The natural frequencies of the bladed disk and their associated mode shapes
2. The frequency, the shape, and the magnitude of the dynamic force which is a function of the turbine speed, number of stationary vanes, and their location around the annulus and/or the number of interruptions in the flow passage, e.g., struts and their location around the annulus
3. The energy-dissipating properties called *damping* provided by blade material, frictional slip between joints, aerodynamic damping from steam, etc.

A turbine bladed disk may get into a state of vibration where the energy buildup is a maximum. This is exemplified by maxima in its response (stress, displacement, etc.) and minima in its resistance to the exciting force. This condition is called a state of resonance. There are two simultaneous conditions for the energy built up per cycle of vibration to be a maximum. These conditions are:

1. The frequency of the exciting force equals the natural frequency of vibration.
2. The exciting force profile has the same shape as the associated mode shape of vibration.

These have been demonstrated by theoretical calculations and also by measured responses of turbine bladed disks. Thus for a resonance to occur, both of the above conditions must be met.

11.4.2 Mode shape

The deflected shape attained by a vibrating bladed disk at its natural frequency is called its *mode shape*. This is shown by plotting the relative displacements of different points of the structure. For a circular symmetric system (e.g., bladed disk), there are points that remain stationary in the vibration cycle. These points fall either on a diametral line(s) or a circle(s). The characterization of mode shapes of a packeted bladed disk is done by specifying the number of these nodal diameter (D) and nodal circle (C) e.g., 0C1D, 0C2D, etc. (Fig. 11.8). The maximum number of nodal diameters in the bladed disk assembly is half the number of blades (for an even number of blades). For a disk having an odd number of blades, the maximum nodal diameter is (number of blades − 1)/2.

There are three variables describing a mode of bladed disk assembly, namely, the natural frequency of vibration, its shape (e.g., 0C1D, 0C3D, etc.), and the speed of the turbine. This information can be displayed as shown in Fig. 11.9.

The two vertical projections of this surface are shown in Figs. 11.10 and 11.11. The first plane is the Campbell plane, and the other is the SAFE plane.

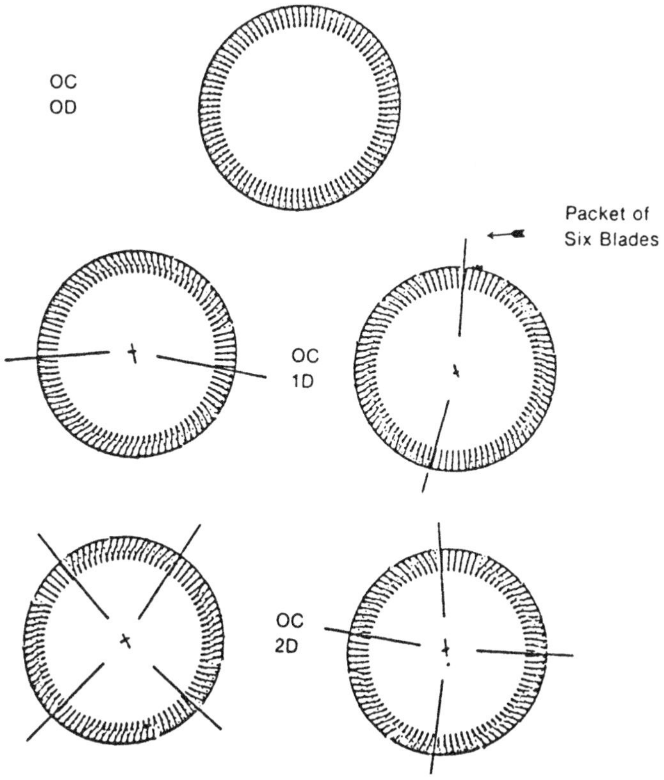

Figure 11.8 Typical tangential vibration (in-phase). *(Dresser-Rand Company, Wellsville, N.Y.)*

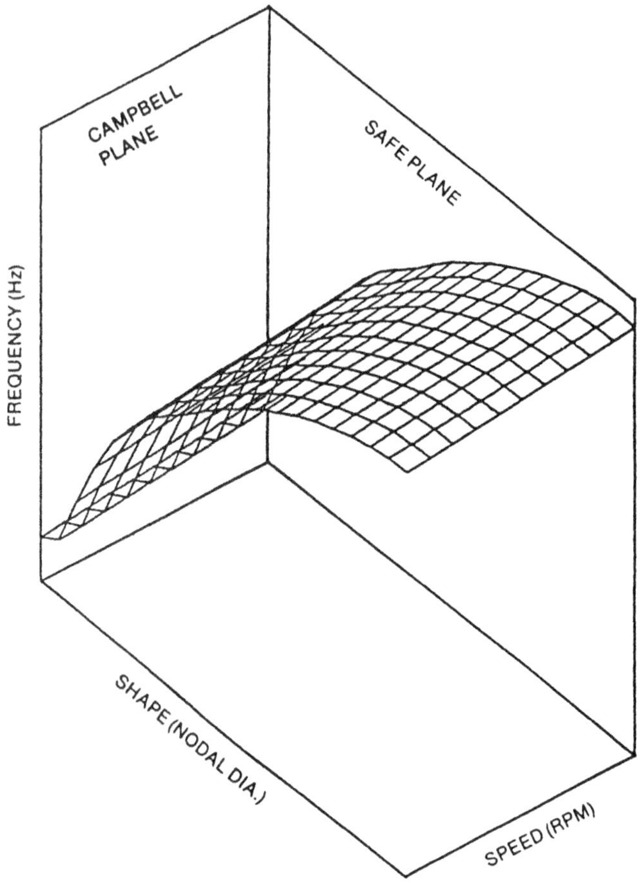

Figure 11.9 Mode shape characterization as a surface. *(Dresser-Rand Company, Wellsville, N.Y.)*

11.4.3 Fluctuating forces

Circumferentially oriented flow distortion due to stationary vanes (nozzles) is shown in Fig. 11.12.

The forces experienced by the blade depend on its relative position with respect to nozzles in one complete revolution. Under the steady-state condition, the same pattern of the force is repeated in successive revolutions making it a periodic force. Such a force is shown in Fig. 11.13.

These time-periodic forces may be composed of many harmonics. The frequencies of these harmonics are integer multiples of the angular speed of the turbine and also depend on the number of *interruptions* around the annulus.

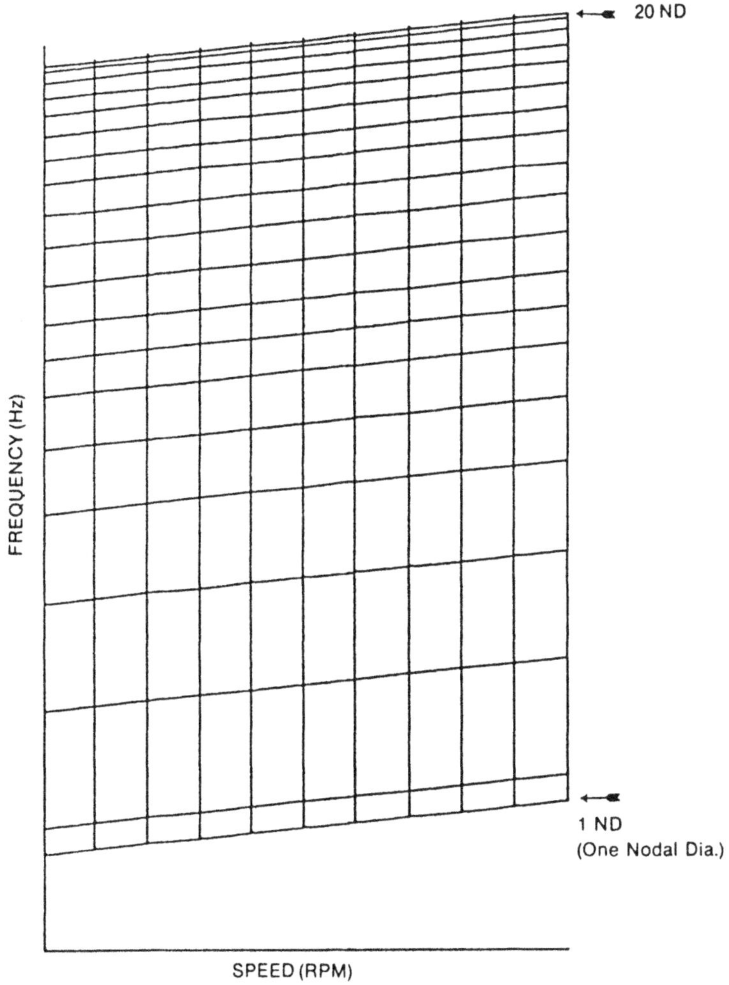

Figure 11.10 Campbell plane. *(Dresser-Rand Company, Wellsville, N.Y.)*

$$F = F_0 + F_1 \sin(\omega t + \theta_1) + F_2 \sin(2\omega t + \theta_2) + \ldots \tag{11.2}$$

The frequency (ω) of dynamic forces experienced by the blading is expressed mathematically as follows:

$$\omega = \frac{(M)(N)}{60 \text{ Hz}} \tag{11.3}$$

where ω = frequency, Hz
M = number of nozzles, number of interruptions
N = turbine speed, r/min

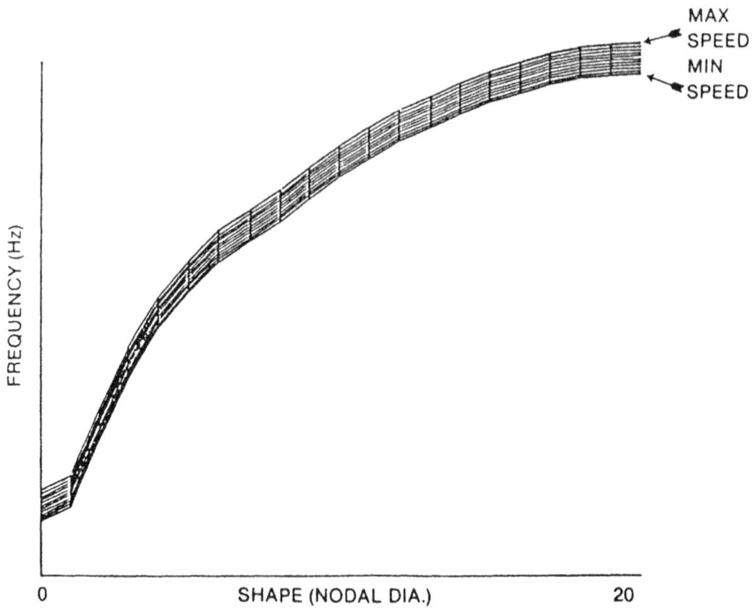

Figure 11.11 SAFE plane. *(Dresser-Rand Company, Wellsville, N.Y.)*

A three-dimensional view and two vertical projections of the surface represented by the above equations are shown in Fig. 11.14 through Fig. 11.16.

As discussed earlier, a Campbell diagram is a graph depicting turbine speed (r/min) on the horizontal axis and frequency (Hertz) on the vertical axis. The natural frequencies of the blades and the frequencies of exciting forces are plotted on the Campbell diagram. This diagram predicts where the natural blade frequencies coincide with the exciting frequencies. This condition of frequency coincidence will be termed a *possible resonance*. These points are shown by small circles in Fig. 11.17.

A further look at these coincident points with the knowledge of mode shape(s) and the shape(s) of exciting forces allows the *true resonance* points to be identified. Figure 11.18 is a representation of such an analysis where the solid circles indicate the few *true resonances* among the many *possible resonances* of the blade frequency and force frequency intersections.

The other projection (Fig. 11.19) in the vertical plane (SAFE plane) shows the same information more clearly. This is more significant in the low-speed range with high order(s) of excitation.

The explicit knowledge of the forcing function (e.g., shape and frequency) is required to find the real resonance condition. The major source of excitation is the nozzles, and the frequency of the nozzle exci-

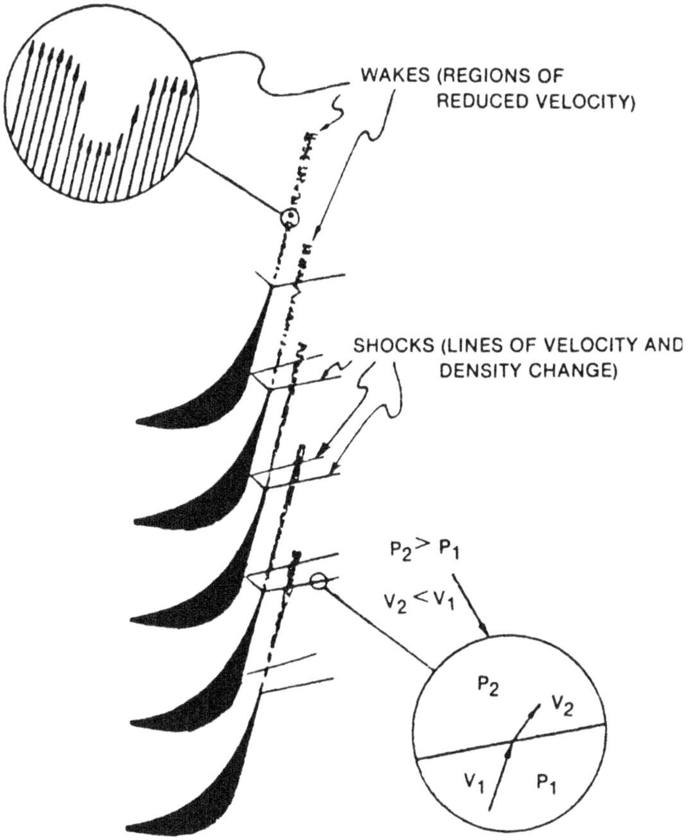

Figure 11.12 Flow distortion due to stationary vanes. *(Dresser-Rand Company, Wellsville, N.Y.)*

tation force is a function of the number of nozzles and the speed of the turbine. The shape of this force is determined by the number of nozzles. Figures 11.20 and 11.21 show such a real resonance.

11.5 SAFE Diagram for Bladed Disk Assembly

The SAFE diagram of a 40-blade disk is given in Fig. 11.22. For the same disk, the effect of shrouding is shown by three different constructions:

1. Completely shrouded (1 packet)
2. Freestanding blades with shrouding (in effect 40 packets)
3. Eight packets

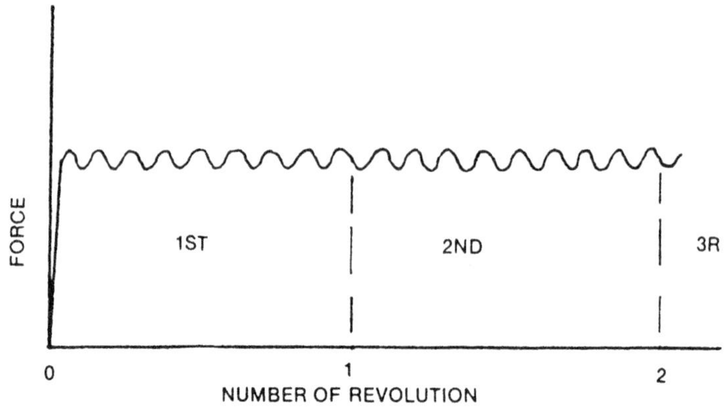

Figure 11.13 Force pattern due to nozzles. *(Dresser-Rand Company, Wellsville, N.Y.)*

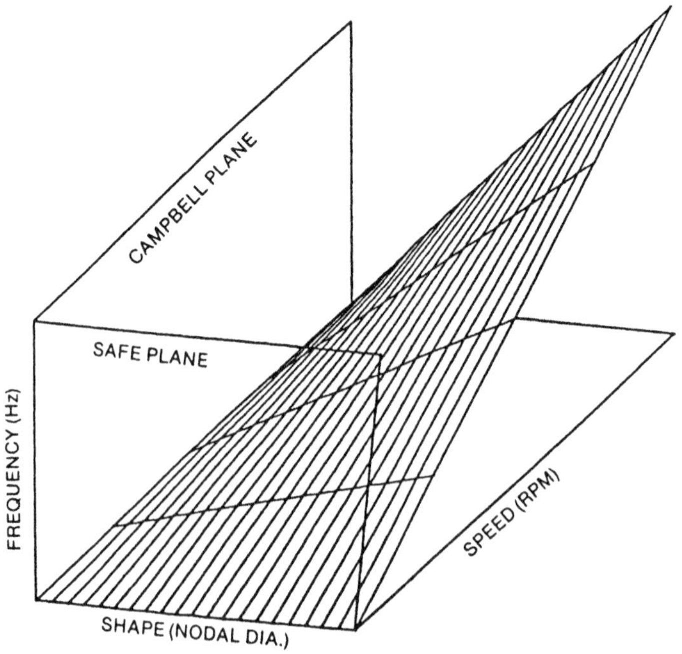

Figure 11.14 Three-dimensional view of surface represented by frequency equations. *(Dresser-Rand Company, Wellsville, N.Y.)*

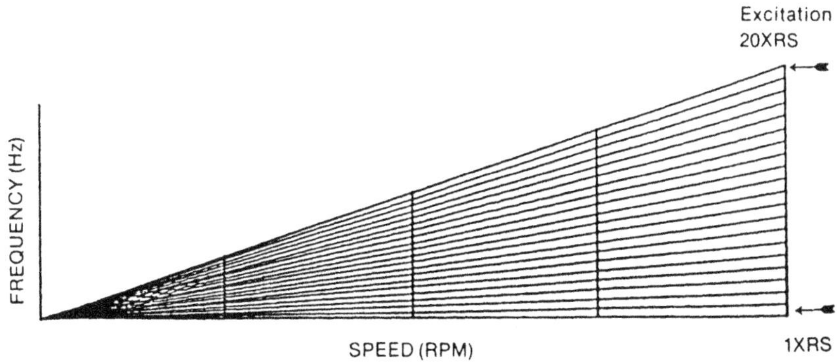

Figure 11.15 Projection on Campbell plane. *(Dresser-Rand Company, Wellsville, N.Y.)*

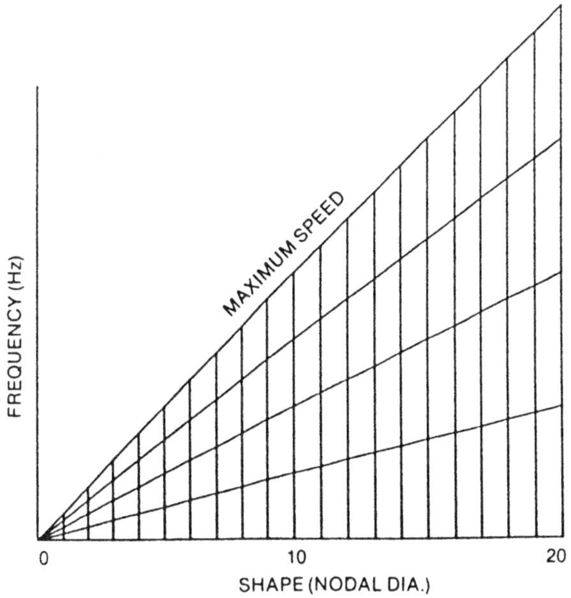

Figure 11.16 Projection on SAFE plane. *(Dresser-Rand Company, Wellsville, N.Y.)*

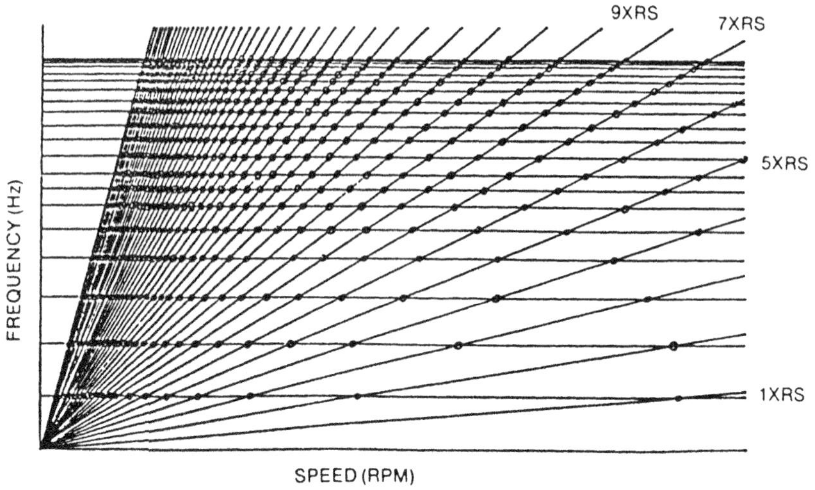

Figure 11.17 Points of possible resonance indicated on Campbell diagram. *(Dresser-Rand Company, Wellsville, N.Y.)*

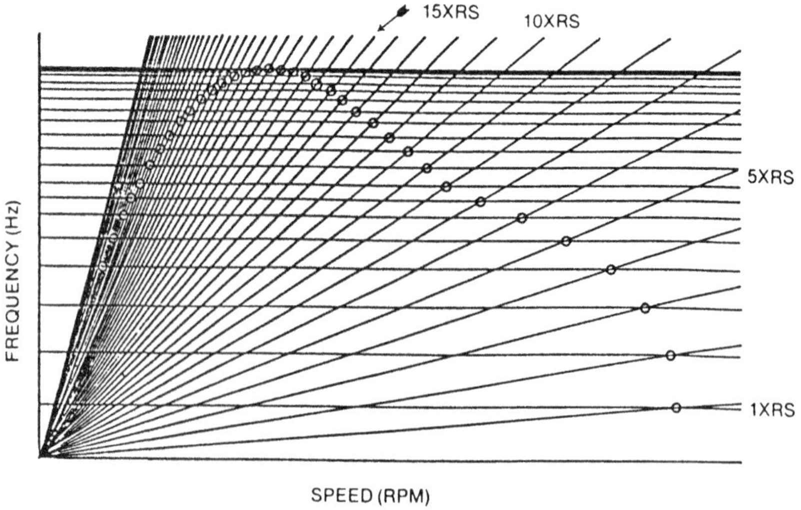

Figure 11.18 Campbell diagram with probable, actual, or true resonance points. *(Dresser-Rand Company, Wellsville, N.Y.)*

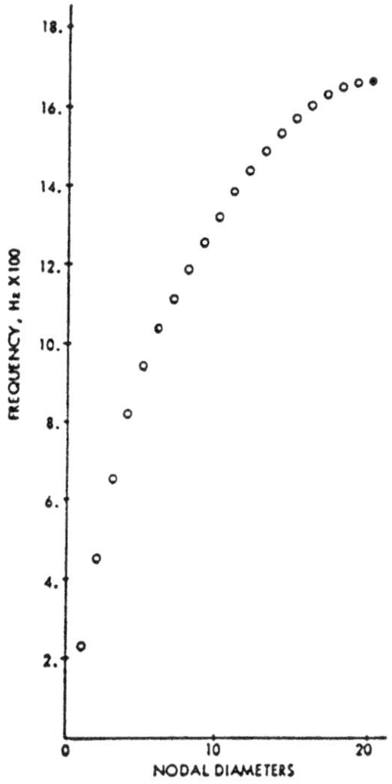

Figure 11.19 SAFE diagram—completely shrouded disk. *(Dresser-Rand Company, Wellsville, N.Y.)*

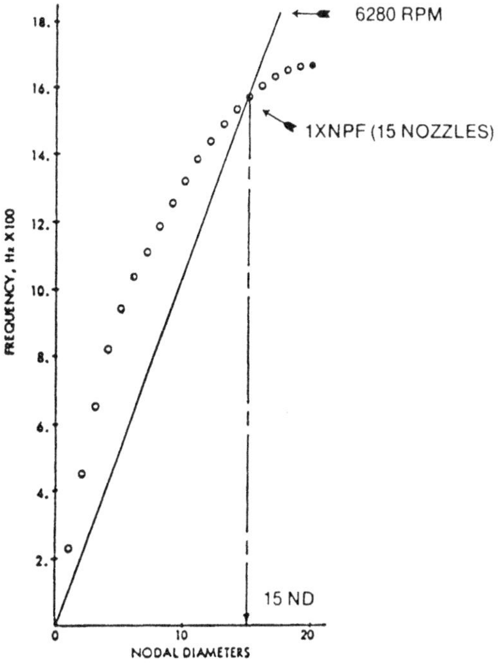

Figure 11.20 SAFE diagram—completely shrouded disk, resonant condition. *(Dresser-Rand Company, Wellsville, N.Y.)*

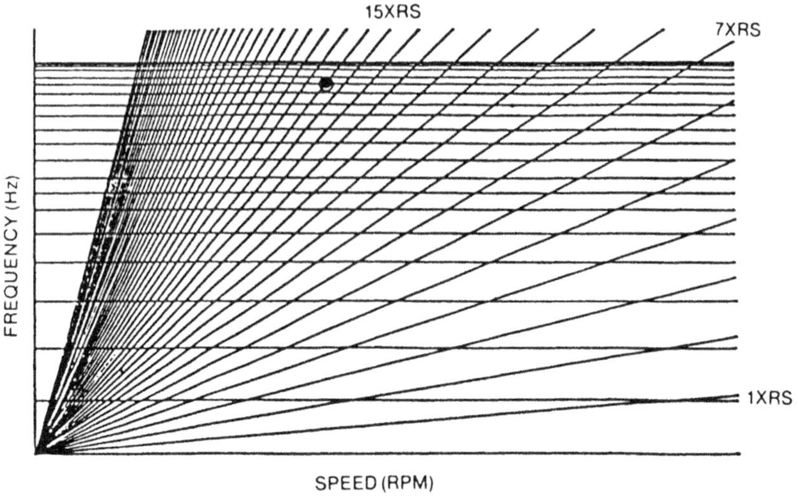

Figure 11.21 Real resonance on a Campbell diagram. *(Dresser-Rand Company, Wellsville, N.Y.)*

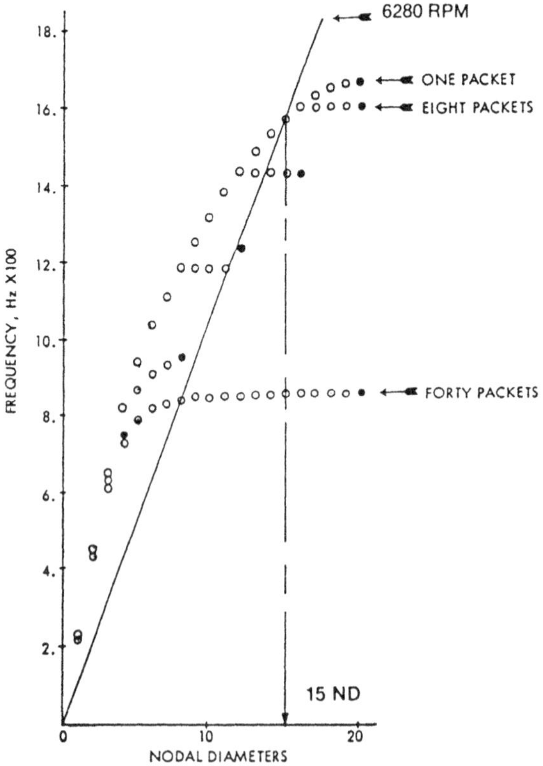

Figure 11.22 SAFE diagrams for 1 packet, 8 packets, and 40 packets system. *(Dresser-Rand Company, Wellsville, N.Y.)*

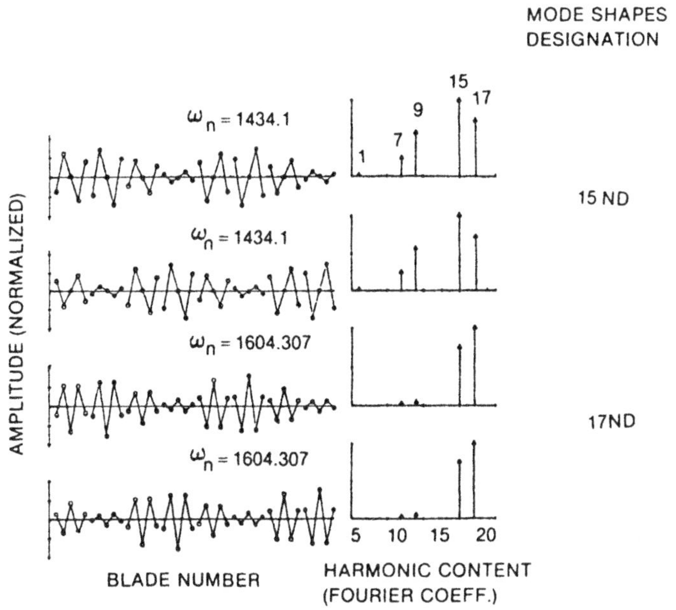

Figure 11.23 Some mode shapes (for eight packets) and their harmonic contents. *(Dresser-Rand Company, Wellsville, N.Y.)*

The 1-packet and 40-packet constructions are two limits; any other packeting, such as eight groups of five blades, will fall in between these two extremes.

11.6 Mode Shapes of a Packeted Bladed Disk

The mode shapes of a packeted bladed disk are complicated. The Fourier decomposition of the displacements of each blade does not lend to only one component (e.g., sin θ or sin 5θ) but other components as well.

For the example problem of a 40-bladed disk with eight packets, four mode shapes are plotted together with plots of different harmonic contents in Fig. 11.23.

In general, the mode shape of a packeted bladed disk can be written as:

$$Y(M) = \Sigma A_L \sin(L\theta + \theta_L) \tag{11.4}$$

for M nodal diameter mode shape

where $L = \ell \cdot n \pm M$

$\ell = 0, 1, 2, 3, \ldots$

n = number of packets

and

$0 \leq L \leq N/2$ [or $(N - 1)/2$ for N being odd]*

and

N = number of blades

A graphical depiction of this equation is given by Ewins and Imregun. The horizontal designation is harmonic content and the vertical designation is the nodal diameter. For a system having N blades and n

* Explanation: Where L is greater than or equal to zero, and where L is less than or equal to $N/2$ or N-1/2 (for N being odd).

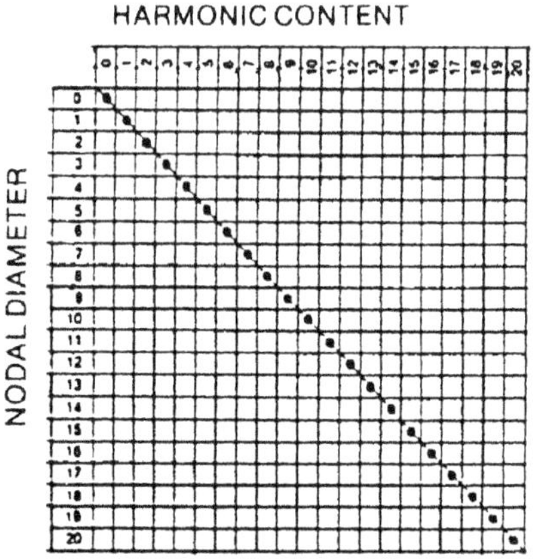

Figure 11.24 Continuously shrouded or unshrouded bladed disk (40 blades). *(Dresser-Rand Company, Wellsville, N.Y.)*

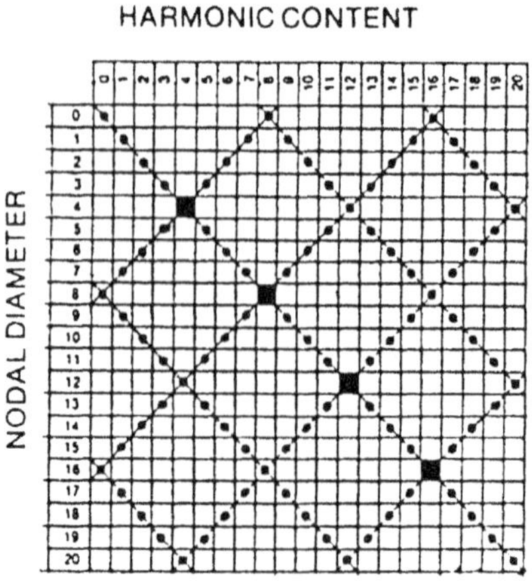

Figure 11.25 Eight packets, five blades in a packet, total forty blades. *(Dresser-Rand Company, Wellsville, N.Y.)*

packets, the maximum horizontal and vertical designations will be $N/2$ (or $(N-1)/2$ when N is odd). The diagrams are constructed by drawing $\pm 45°$ lines emanating from each integer multiple of n as shown in Figs. 11.24 and 11.25.

There are two important facts revealed by this diagram: (1) When the 45° lines cross each other and have the same horizontal and vertical designation, that mode will split, i.e., will have the same mode designation but have two different frequencies, and (2) they also indicate the harmonic contents of any mode. For a completely shrouded arrangement, each mode contains only one harmonic, but the example for 8 packets, the 15-nodal diameter mode has 17, 15, 9, 7, and 1 harmonics contents as shown in Fig. 11.23.

Implications of this are that the 15-nodal diameter mode of the above example will also respond to the 17th, 9th, 7th, and 1st harmonics of exciting force, but with different magnitude.

11.7 Interference Diagram beyond *N*/2 Limit

The maximum number of mode shapes (number of nodal diameters) plotted on an interference diagram for N blades on the disk is $N/2$ (or $(N-1)/2$, when N is odd). When the excitation is of higher harmonics than $N/2$, two questions arise:

1. Will the higher order harmonic(s) of a force excite modes of lower harmonic(s) content?
2. If the answer is Yes, then how can this information be depicted on the SAFE diagram?

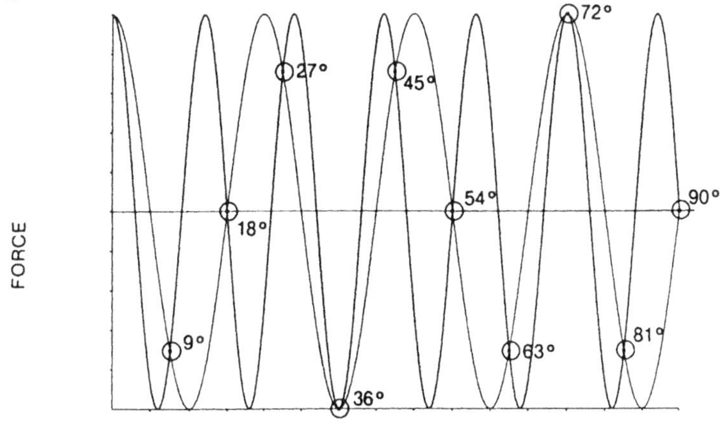

Figure 11.26 Cos 15θ and cos 25θ force shape. *(Dresser-Rand Company, Wellsville, N.Y.)*

For the 40-bladed disk, the maximum nodal diameter depicted on the SAFE Diagram is 20. A force having a 15-nodal diameter shape will excite the 15-nodal diameter mode if the frequency is equal to the natural frequency of this mode.

In Fig. 11.26, two cosine waves are plotted between 0 and 90°. They are cosine 15θ and cosine 25θ. They intersect each other at 10 points marked with asterisks. In 360° arc, they will intersect at 40 points. The spacing between consecutive intersections is equal and, in turn, it is equal to the spacing between consecutive blades.

For 40 blades on a disk, the spacing between each consecutive blade is 9°. If one measures forces on each blade, these forces will be the same for both of these waves (cos 15θ and cos 25θ). That is to say that blades cannot know whether the forces are from a cos 25θ wave or a cos 15θ wave. This information can be depicted on the SAFE Diagram.

The mode shapes of the bladed disk are plotted up to 20-nodal diameters. The force shape can be plotted to any value. It is shown to 80-nodal diameters in Fig. 11.27.

A force having cos 15θ shape will excite 15-nodal diameter mode shape at a speed of 6280 r/min (P1). A force having cos 25θ shape will excite 15-nodal diameter mode as argued earlier. The speed of rotation will be 3768 r/min for the frequency of this force to be equal to the natural frequency. This is shown as point P2 on the plot. Similarly, a force having cos 55θ and cos 65θ will also excite the 15-nodal diameter mode

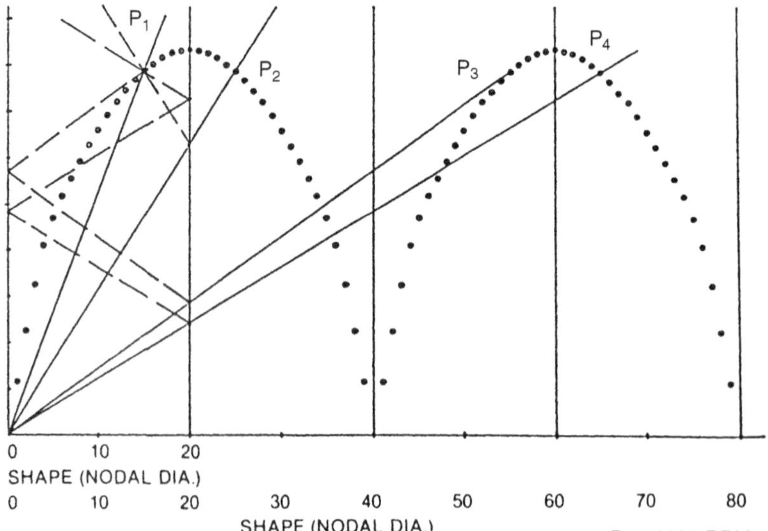

Figure 11.27 Shape (nodal diameter) P_1 = 6280 r/min; P_2 = 3768 r/min; P_3 = 1713 r/min; P_4 = 1449 r/min. *(Dresser-Rand Company, Wellsville, N.Y.)*

but at lower speeds. These are marked as P3 and P4 on the plot. By similar argument, any mode between 0- and 20-nodal diameters can be excited by force shapes higher than cos 20θ. These points are also shown up to 80-nodal diameters. It should, however, not be inferred that any force shape can excite any mode. There is a special relationship between mode shapes, shapes of the force, and total number of blades on the disk. This relationship can be described as follows: The M-nodal diameter mode will be excited by a force having a shape of cos $M\theta$, cos $(N-M)\theta$, cos $(N+M)\theta$, cos $(2N-M)\theta$, cos $(2N+M)\theta$, etc. In the above example, M is 15 and N is 40; hence, a force of cos 15θ, cos 25θ, cos 55θ, cos 65θ, cos 95θ, etc. will excite the 15-nodal diameter mode but at different speeds. Hence, the M-nodal diameter mode can be excited by a force having a cos $L\theta$ shape where $L = n \cdot N \pm M$, and where $n = 0$, 1, 2, 3–, N and N is the number of blades.

Let us identify the plot between 0- and 20-nodal diameter as Zone I, between 20- and 40-nodal diameter as Zone II and so on. Zone II is a mirror image of Zone I, Zone III is a mirror image of Zone II, etc. Hence, Zone I contains all the information about the structure. The reflections of speed lines about vertical lines passing through 0- and 20-nodal diameters are required in the first zone. It can be seen that all reflected speed lines pass through the same 15-nodal diameter point in Zone I.

These discussions can be summarized as follows:

1. Zone I as traditionally drawn contains all the information about the vibrating structure.

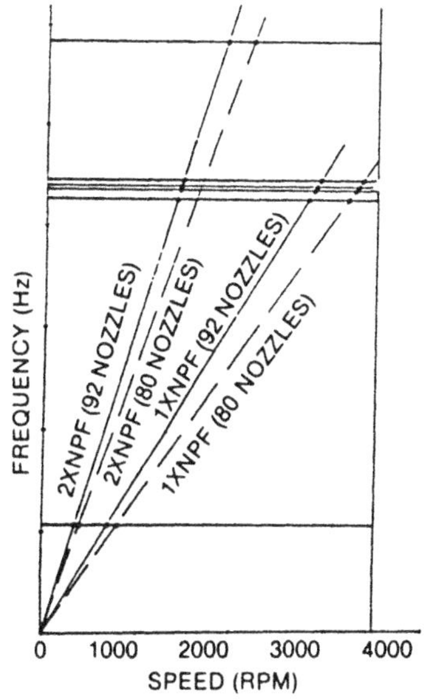

Figure 11.28 Campbell diagram. *(Dresser-Rand Company, Wellsville, N.Y.)*

2. The 15-nodal diameter mode of a 40-bladed disk can be excited by a diaphragm containing either 15, 25, 55, 65, etc. nozzles.

11.8 Explaining Published Data by the Use of Dresser-Rand's SAFE Diagram

The usefulness of the SAFE diagram is demonstrated by published results of other researchers. Data from the following two ASME publications have been used. The first one is F. L. Weaver and M. A. Prohl, "High Frequency Vibration of Steam Turbine Buckets," *ASME* Paper No. 56-A-119, 1956. The second paper is G. E. Provenzale and M. W. Skok, "A Cure for Steam Turbine Blade Failures," *ASME* Paper No. 73-Pet-17, 1973.

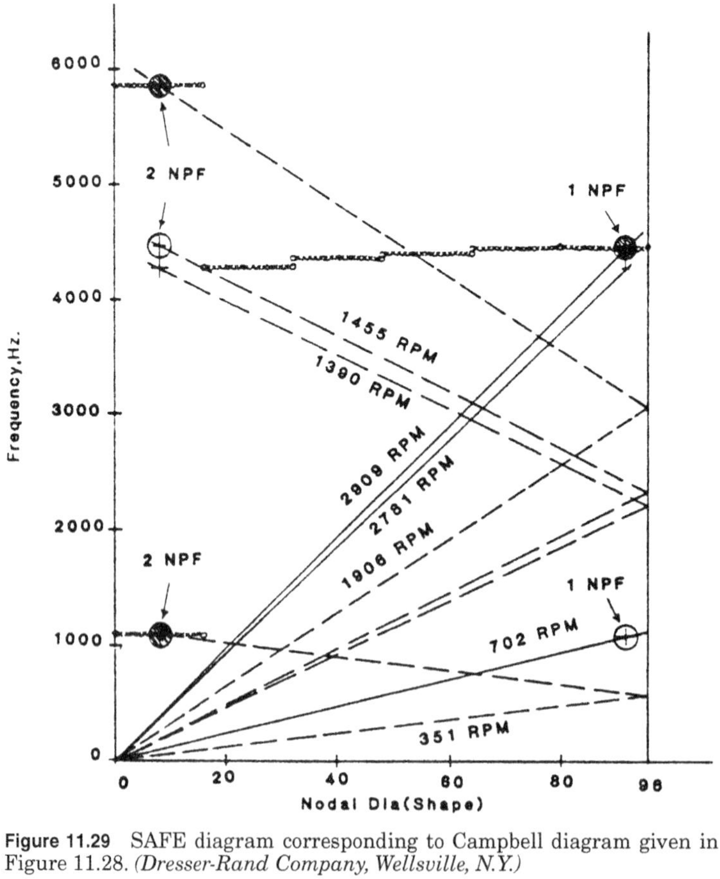

Figure 11.29 SAFE diagram corresponding to Campbell diagram given in Figure 11.28. *(Dresser-Rand Company, Wellsville, N.Y.)*

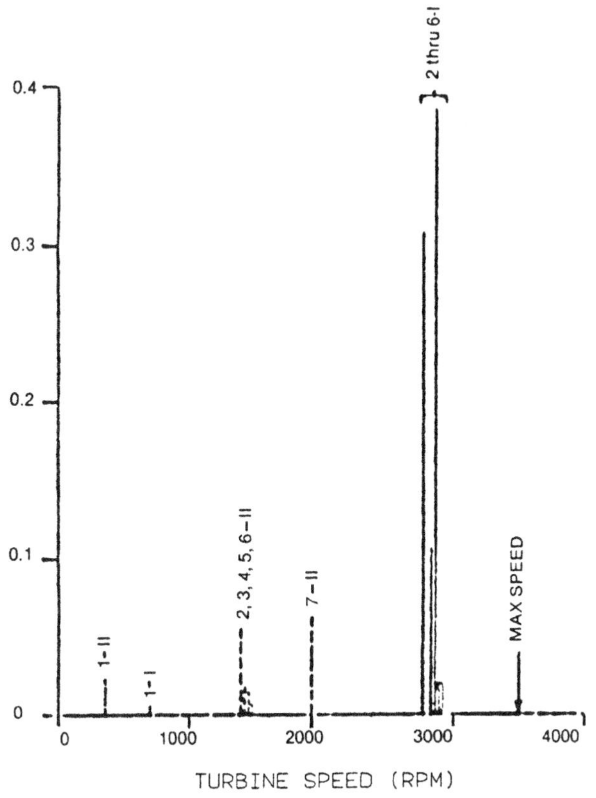

Figure 11.30 Resonant-vibration stress.

Mode 1 first tangential
2–6 tangential fixed supported mode
7 second tangential

Stimulus I—1x nozzle passing
II—2x nozzle passing

(Dresser-Rand Company, Wellsville, N.Y.)

By using an energy method, Weaver and Prohl calculated the response of a packet of blades due to nozzle-passing excitation. The turbine disk in their paper contained 6 blades per packet with a total of 192 blades. The number of nozzles was 92. Their response calculations showed that the response of the blades was highest when the nozzle passing frequency coincided with the tangential fixed supported modes. The first tangential mode response was very low compared to the tangential fixed supported mode response. A Campbell diagram is drawn using their data (Fig. 11.28). A SAFE diagram using their frequency numbers (assuming that the disk is very stiff in tangential direction) has been constructed. Assuming that the excitation comes

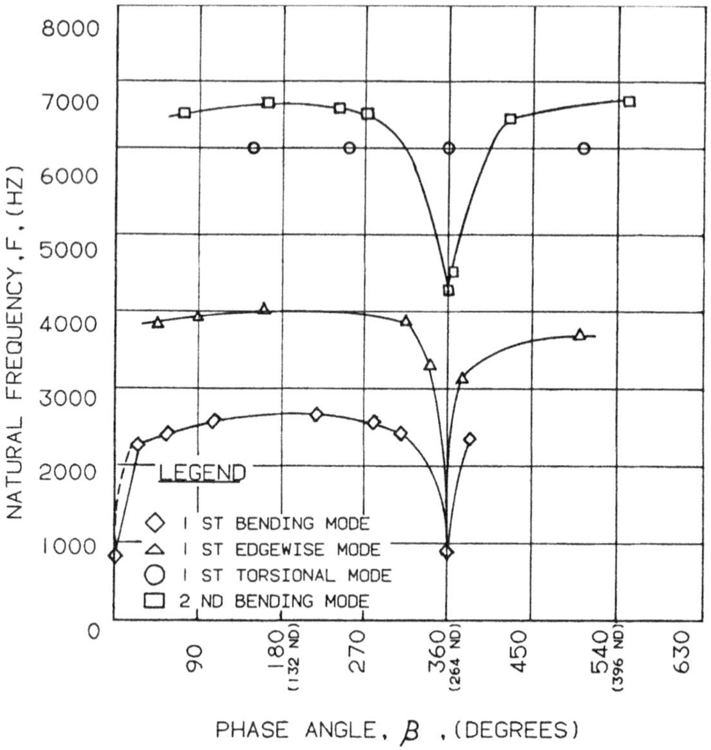

Figure 11.31 Blade natural frequency vs. phase angle β over the speed range of 0–7500 r/min (blade with integral tip shroud). *(Provenzale and Skok, ASME Paper 73-Pet-17)*

from flow distortion due to nozzles, the modes and frequency of concern have been marked by large shaded circles (Fig. 11.29).

Results from Weaver and Prohl's paper are reproduced alongside the SAFE diagram (Fig. 11.30). The shaded circles on the SAFE diagram point to the mode(s) of concern as predicted by Weaver and Prohl (after numerous response calculations).

The data provided in the paper by Provenzale and Skok (ASME Paper 73-Pet-17) were obtained by a response calculation of the blades under the action of aerodynamic forces. Forces were defined by a frequency and a shape defined by a parameter β. The relationship between β and nodal diameters is $\beta = 360° = N$ nodal diameter (N is the number of blades).

The total number of blades based on the data provided in the paper is estimated to be 264. On the basis of the theory of bladed disk vibration, there should be 264 frequencies; each of these will be excited by a force shape matching the mode shapes if frequencies coincide. Figures 2 and 3 of the paper by Provenzale and Skok are reproduced here as

Figs. 11.31 and 11.32 with the addition of definitions followed in this write-up. Figure 11.31 is a Campbell diagram with resonance points (partial). Figure 11.32 is a SAFE diagram (partial) showing the periodicity every 180° or *N*/2.

This confirms that the SAFE diagram shows all the resonant points in a bladed disk structure.

11.9 Summary

- SAFE diagrams bring in the additional required information about resonant conditions, namely mode shape and force profile for packeted bladed disk assemblies.
- A designer does not have to make numerous calculations for response but only at selected points of concern(s).
- Interferences on a Campbell diagram are very difficult to show in the low-speed range while the SAFE diagram retains clarity for every speed.

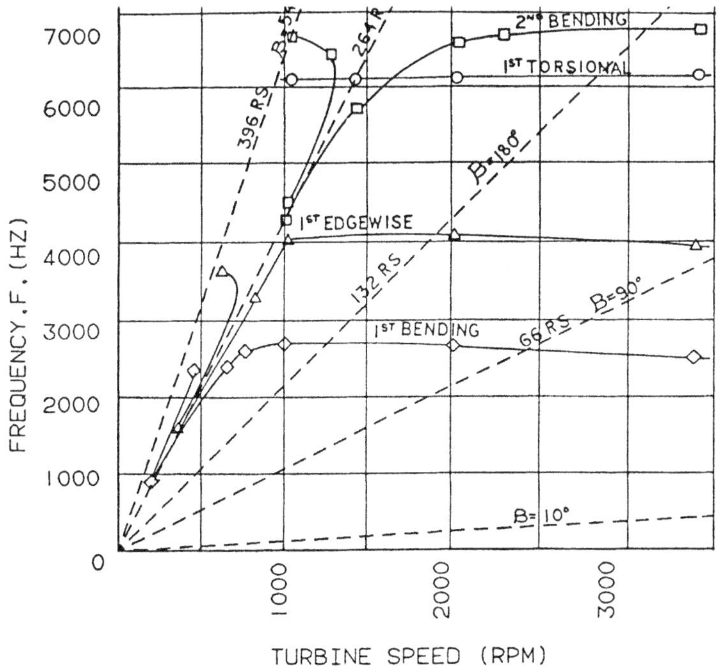

Figure 11.32 Campbell diagram showing the effect of phase angle β on blade natural frequency for 0° < β ≤ 540° (blade with integral tip shroud). *(Provenzale and Skok, ASME Paper 73-Pet-17)*

- A Campbell diagram can be constructed (if desired) by the information available from the SAFE diagram.
- Influence of turbine speed on the frequency of vibration is not easily depicted on the SAFE diagram. It is easily shown on a regular Campbell diagram.
- The reliability of steam turbines can be improved by utilizing this diagram.

Chapter

12

Reaction vs. Impulse Type Steam Turbines

12.1 Introduction

Historically, steam turbines have been divided into impulse and reaction turbines, from the thermodynamic aspect (Fig. 12.1), and into axial and radial flow turbines to denote the principal direction of the steam flow. Axial machines, and only this kind will be considered here, are also identified according to their rotor construction as disk type or drum type turbines. These are shown in Fig. 12.2. Because the terms are normally understood, the impulse turbine with little or no reaction corresponds to the disk type of construction, and the reaction turbine to the drum type. For simplicity, the terms *impulse turbine* and *reaction turbine* will be used here.

The impulse turbine, first built in 1883 by the Swedish engineer De Laval, was inevitably at an advantage in the early days of steam turbines because it is well suited to small flow volumes. The subsequent gradual but ever accelerating increase in electricity consumption led to a rapid rise in the unit capacities of power station turbines. The outputs required for mechanical drives at first remained more modest, most of these turbines being of the impulse type. This situation evidently gave rise to the view that this kind of machine is more suitable for mechanical drives.

However, the way in which a steam turbine functions has only an indirect and, as will be shown, slight influence on its reliability; when properly designed, any steam turbine can be made dependable.

In the years that followed, reaction turbines (first introduced by C. A. Parsons in 1884) also came to be used for mechanical drives. The largest units in service in the 1980s, driving pumps and compressors, were reported to be reaction machines.

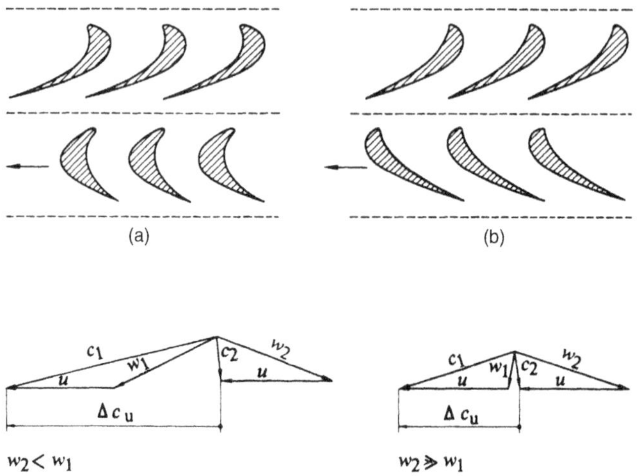

Figure 12.1 Operating principles of steam turbines: (*a*) impulse turbine; (*b*) reaction turbine. *(Asea Brown-Boveri, Baden, Switzerland)*

12.2 Impulse and Reaction Turbines Compared

The essential differences between the two types can be considered in two categories, thermodynamic and constructional. Differences of the latter kind are largely determined by the former. The thermodynamic differences have been described earlier and, except for the question of efficiency, are not repeated here. For turbine reliability, it is the constructional differences that are most important, and so these will be examined more closely. The main differences between the two designs can be seen in Table 12.1.

12.3 Efficiency

A variety of factors, of course, have to be accounted for when calculating efficiency. To compare impulse and reaction designs, therefore, it is helpful to take a dimensionless characteristic, the most suitable being the volume coefficient δ. This is defined in Fig. 12.3, which shows the efficiencies of the two types.

If we first compare impulse blading (curve *a*) to reaction blading with tip sealing, i.e., with no shroud, (curve *b*), we arrive at the relationships stated in Table 12.1. It can also be seen from Fig. 12.3 that with the reaction turbine a substantial improvement in efficiency is possible, especially at low volume coefficients, by using shrouds (curve *c*). On economic grounds this technique has become standard practice for power station turbines above a certain capacity. Whether there is

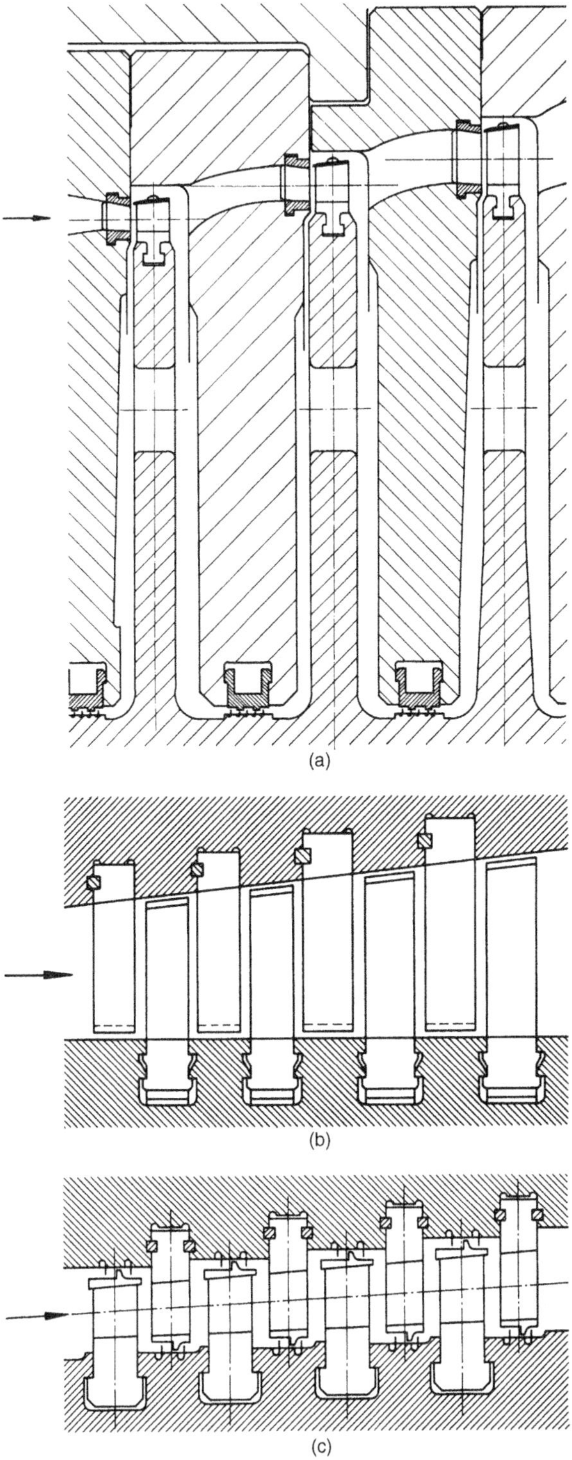

Figure 12.2 Construction of axial steam turbines: (*a*) disk type rotor; (*b*) drum type rotor with tip-sealed blades; (*c*) drum type rotor with shrouded blades. *(Asea Brown-Boveri, Baden, Switzerland)*

TABLE 12.1 Basic Comparison Between Impulse and Reaction Turbines

Feature	Impulse turbine	Reaction turbine
Efficiency	Better with small volume coefficients, poorer with medium- and high-volume coefficients	Better with medium- and high-volume coefficients
Rotor	Disk construction	Drum construction
Blading (Fig. 12.2)	Few stages, wide in axial direction Fixed blades mounted in diaphragms Moving blades on disks of rotor	More stages, narrow in axial direction Fixed blades mounted in casing or blade carrier Moving blades on drum
Maintenance	Longer time elapse between major overhauls	Somewhat shorter elapsed time between major overhauls

SOURCE: Asea Brown-Boveri, Baden, Switzerland.

any advantage in using shrouds for mechanical drives as well must be judged from the economic circumstances of each case. Cost and variable speed experience are possible issues. To find the overall efficiency of a reaction turbine, one must also allow for losses due to the thrust balance piston. Viewed as a whole, however, the fact remains that the impulse machine has its advantages when capacities are low, while for medium and high ratings the merits of the reaction type of construction are of interest to specifying engineer and purchaser.

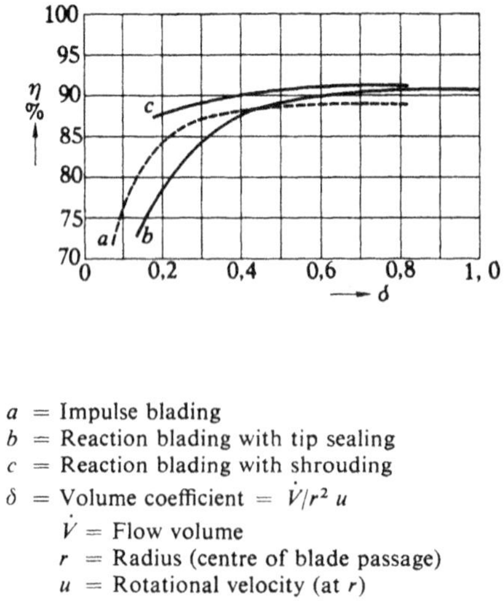

Figure 12.3 Comparing the efficiency η of impulse and reaction blading; a = impulse blading; b = reaction blading with tip sealing; c = reaction blading with shrouding; δ = volume coefficient = $\dot{V}/r^2\,u$; $\dot{V}$ = flow volume; r = radius (center of blade passage); u = rotational velocity (at r). *(Asea Brown-Boveri, Baden, Switzerland)*

12.4 Design

12.4.1 Rotor

The construction and behavior of the rotor are vital in determining whether a steam turbine performs properly. The differences in the two kinds of machines can be summarized as follows:

Impulse type

- Rotors are composed of disks, generally made from the solid.
- From the vibration standpoint the disks are independent structures, and their behavior must be accounted for in the design. In service these disk vibrations are superimposed on the rotor vibration. This can result in a complex vibration pattern.
- Material defects may require scrapping the whole rotor.
- With disk rotors of relatively small diameter, rapid temperature changes, on starting for example, are accompanied by a greater tendency to distortion.
- The principal mass of the rotor is at its center. Since the blades are fitted on the outside of the thin disks, a rotor of this kind, when balanced, is somewhat more sensitive to uniform distribution of the blade mass than a reaction type rotor.

Reaction type

- Rotors are executed in the form of a drum machined from the solid or composed of separate forgings. Since 1930, the rotors of all medium and large turbines produced by at least one major manufacturer have consisted of sections welded together. The advantages of this construction have been reported and are largely fabrication related.
- The rotor can be uniformly heat treated, hence excellent strength properties are achievable throughout.
- The rotor is very easy to test.
- Sections can be replaced easily and quickly in the event of rejects.
- The rotor has evenly distributed stresses.
- If the rotor of a reaction turbine is made from the solid, the same situation arises in the case of material defects as with an impulse type rotor.

Critical speed. Just as basic as the question of which is the most suitable kind of turbine is the question whether the rotor speed should be above or below the critical value. Both types have been built in subcritical and supercritical form, and no clear preference can be logically spelled out for one or the other. Which to choose depends, among other factors, on the design tradition of the manufacturer and the optimum configuration of the machine.

About turbines for driving pumps and compressors, the issue is to some extent influenced by past practice, since rotors for low steam throughput capacities can be made subcritical without any serious drawbacks. However, larger subcritical rotors may result in a design that might be quite impractical. All large power station turbines in the 100 to 1000-MW range, for example, run at supercritical speed. Also, knowledge in this field has now advanced so far that it is no longer sufficient to design for rigid bearings, without taking into consideration oil film, bearing and foundation stiffness.

In comparing the two kinds of machines, it can be assumed that the efficiency penalty with subcritical operation is greater in the case of the impulse turbine. The reason for this is that to raise the critical speed of the rotor, the hub diameters have to be enlarged, whereupon leakage losses, otherwise a point in favor of the impulse construction, are greatly increased.

In addition, the following requirements are essential to the reliable operation of any rotor:

- Proper design of rotor and bearings
- Correct choice of bearing type
- Excellent balance

12.4.2 Blading

Apart from the rotor, the blading of impulse and reaction turbines show the most marked differences. By its nature, the reaction machine requires 75 to 85 percent more stages than an impulse turbine, for the same heat drop. Despite this, the casings of both types are roughly the same length because a single-reaction stage is much shorter in the axial direction than an impulse stage, where the moving blades and the space between the blades take up more length.

The fact that there are fewer stages often leads to the mistaken view that on strength grounds or for operational reasons the reaction stage cannot withstand such severe stresses, and therefore more stages are needed. This is not the case at all; the number of stages is governed by thermodynamic differences.

Vibration. If one wishes to examine the behavior of the blading in normal operation, it is sufficient to consider the dynamic stresses on the blades. The static forces (centrifugal force and the force exerted by the flow) can be predicted with sufficient accuracy; blade failures are almost entirely due to dynamic stresses.

Although in theory no set of blades can be made free of resonance, it is possible to design and manufacture blading so that dangerous resonances are avoided and no unacceptably high stresses occur. The necessary conditions can be considered separately for stages with partial and full admission.

Control stage. Conditions in the control stage, with partial admission, are essentially the same for both types of machines. Extensive studies show that attention must be paid to the following points to ensure that this stage functions reliably:

- Low stress level
- Careful design of the blade fixing
- Blades joined to form segments, or packets

Blade fixings of the mechanical kind are usually adequate, but welding (as shown in Fig. 12.4), which has performed well in large turbines, is also possible if required. In either case the blades are joined to form

Figure 12.4 Welded moving blades of a control stage. *(Asea Brown-Boveri, Baden, Switzerland)*

segments by welding the shroud portions together. The effect of joining the blades is allowed for mathematically by means of the bonding factor described later. Figure 12.5 shows how the dynamic loading on the blades, i.e., the cyclic stress, can be drastically reduced by this segment arrangement.

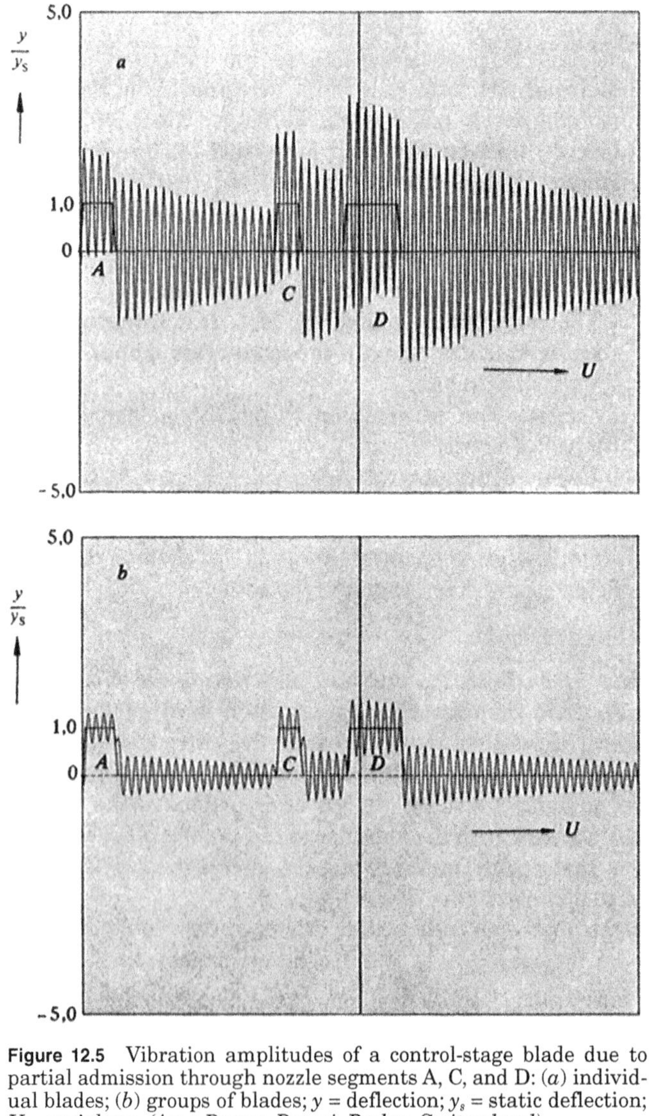

Figure 12.5 Vibration amplitudes of a control-stage blade due to partial admission through nozzle segments A, C, and D: (*a*) individual blades; (*b*) groups of blades; y = deflection; y_s = static deflection; U = periphery. *(Asea Brown-Boveri, Baden, Switzerland)*

Full-admission stages. The dynamic stress on the remaining stages, with full admission (Fig. 12.6), is estimated with the aid of the following equation:

$$\sigma dyn = \pm S \times V \times H \times \alpha \times \sigma_b \tag{12.1}$$

where S = stimulus
V = resonance rise
H = vibration form factor
α = bonding factor
σ_b = static bending stress due to flow forces

Here it is assumed that the dynamic loading on a blade, i.e., the cyclic stress, is proportional to the static loading due to the flow forces exerted by the steam. This is a perfectly logical assumption when one considers merely the wake behind the trailing edges of the blades, e.g., or the pressure differences in exhaust casings, which can lead to fluctuations in flow-induced forces.

Figure 12.6 Reaction stages with full admission. *(Asea Brown-Boveri, Baden, Switzerland)*

It is an established fact that a sound design must avoid resonance due to the preceding blade row causing excitation at the fundamental bending frequency (nozzle excitation). With that in mind, the relative vibration behavior of impulse and reaction stages has been summarized as follows:

1. Excitation of a vibration is often described by stimulus *S*. Substantive investigations confirm the orders of magnitude for *S* as stated by texts such as Traupel (see Bibliography). For small ratios of moving blade pitch/fixed blade pitch (Fig. 12.7), *S* is typically in field *G* and somewhere near line 1. For values of this ratio of roughly 1, *S* is typically in field *U* and somewhere between lines 1 and 2. Field *G* applies primarily to impulse turbines, and field *U* to reaction machines. If account is also taken of the fact that the stimulus diminishes with increasing axial spacing, then it is higher with impulse stages than with reaction stages (Fig. 12.8). Thus, for the same dynamic stress, i.e., the same factor of safety, the reaction stage allows higher specific loading.
2. Another advantage often cited for reaction blading is that resonance between the fundamental oscillation and the excitation due to the preceding row does not occur in practice (Fig. 12.9), whereas in an impulse turbine close attention must be paid to this phenomenon. Resonances between the first harmonic and the preceding row are either avoided or accompanied by much smaller dynamic stresses.
3. If resonances causing possible dynamic stresses are unavoidable, steps must be taken to prevent blade failure. Damping wires and lashing wires are shown in Fig. 12.6. They have for a long time been used to

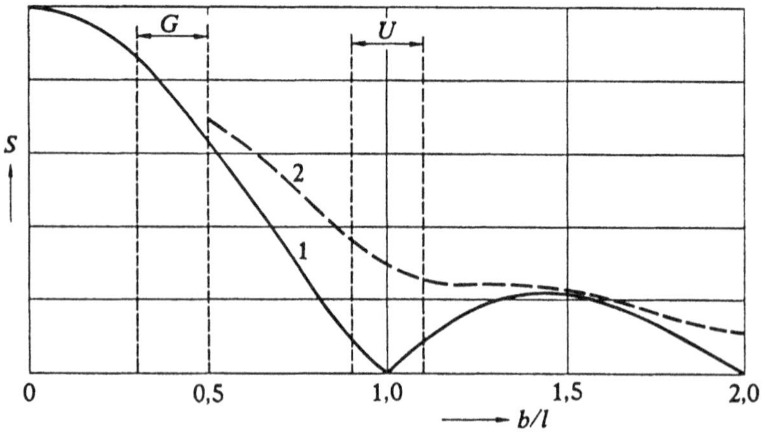

Figure 12.7 Stimulus *S* in relation to moving blade pitch/fixed blade pitch: *b* = moving blade pitch; *l* = fixed blade pitch; *G* = impulse blading; *U* = reaction blading. *(Asea Brown-Boveri, Baden, Switzerland)*

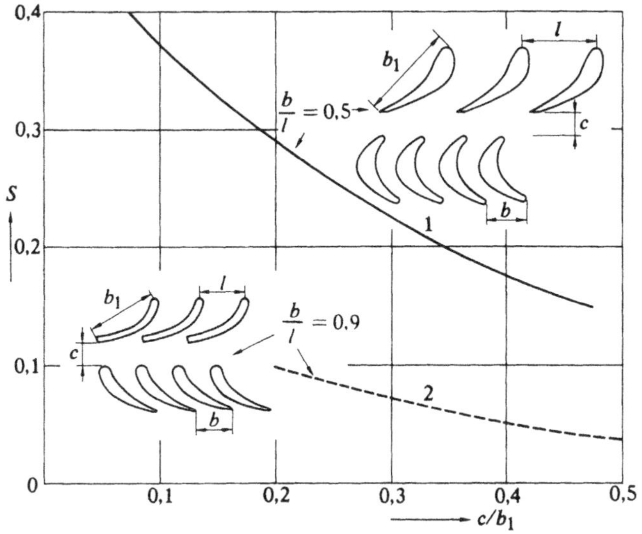

Figure 12.8 Stimulus S in relation to reference axial gap c: 1 = impulse turbine; 2 = reaction turbine. *(Asea Brown-Boveri, Baden, Switzerland)*

prevent this resonance effect. The damping wire is inserted through, but not attached to, the blades, while the lashing wire is firmly fixed to them. The moving blades of impulse stages are joined by shroud bands, as illustrated earlier. All in all, the reaction type of construction claims to have certain advantages of dynamic stressing of the blades.

Blade damage. It has been suggested that the moving blades of an impulse turbine are subject to a greater inducement to vibrate than the blades of a reaction machine. This contention has occasionally surfaced in statistics on blade damage. One of these lists one case of damage to impulse blades for every million component running hours, while the rate for reaction turbines was only once per 3.8 million component hours. If the number of components (number of stages) is also taken into consideration, the average for an impulse turbine would appear to be one case of blade damage in 25,000 h of operation, as against one in 56,000 h for the reaction turbine. Whether these statistics compare only reputable, experienced manufacturers of the two turbine types or are perhaps biased in a number of ways is a matter of conjecture.

Blade clearances. Axial clearance is an important factor in the efficiency of an impulse stage; however, in reaction machines the radial clearance must be as small as possible. To avoid blade damage if the

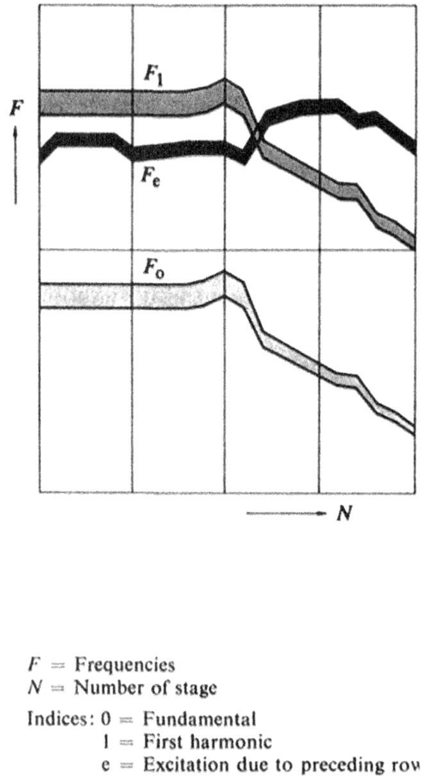

Figure 12.9 Natural and excitation frequencies of reaction blading: F = frequencies; N = number of stage; Indexes: 0 = fundamental; 1 = first harmonic; e = excitation due to preceding row. *(Asea Brown-Boveri, Baden, Switzerland)*

rotor becomes distorted, the tips of unshrouded blades are sharpened to a thin edge. Shrouded blades are sealed by means of thin labyrinth strips so that rubbing is essentially harmless. Strip-to-tip clearances are sufficient to avoid rubbing contact under moderate conditions of rotor bow and vibration.

12.5 Erosion

With condensing turbines attention must be paid to the behavior of the blades in the region of increasing wetness. Under unfavorable conditions of water content, peripheral velocity and geometry, the water droplets that form can erode the moving blades, i.e., metal is worn away from the leading edges (Fig. 12.10). Water deposited on the stator blades is dragged by the steam flow to the trailing edge where it collects to form large drops. When torn away from the trailing edge, these drops, which have a diameter of about 1 mm, are broken up into smaller droplets.

Figure 12.10 Blades of the final stage of a medium-sized condensing turbine after completing more than 100,000 h of operation. *(Asea Brown-Boveri, Baden, Switzerland)*

The causes of erosion are shown in Fig. 12.11. The moving blades collide with the small droplets, and the resulting impact removes blade material. The water droplets are accelerated to only about 10 to 20 percent of the steam velocity and thus strike the blades at an unfavorable angle. The most serious erosion occurs in the region b_e. The means of preventing erosion damage include provision of casing internal channels or troughs (Fig. 12.12) and special diffusion coatings (titanium nitride) for turbine rotor blades.

The opinion is often expressed that reaction turbines are more susceptible to erosion. Recent technological advances have made the final stages of both impulse and reaction condensing turbines almost equivalent in this respect. In stages where there is likelihood of erosion, the degree of reaction is increased over the length of the blade. Consequently, modern final stages operate more or less as impulse stages on the inside, with slight or moderate reaction, but very definitely as reaction stages toward the tip.

There is, therefore, no reason to expect a meaningful difference in the erosion behavior of the two turbine types, impulse vs. reaction. To this extent, earlier ideas may have to be corrected.

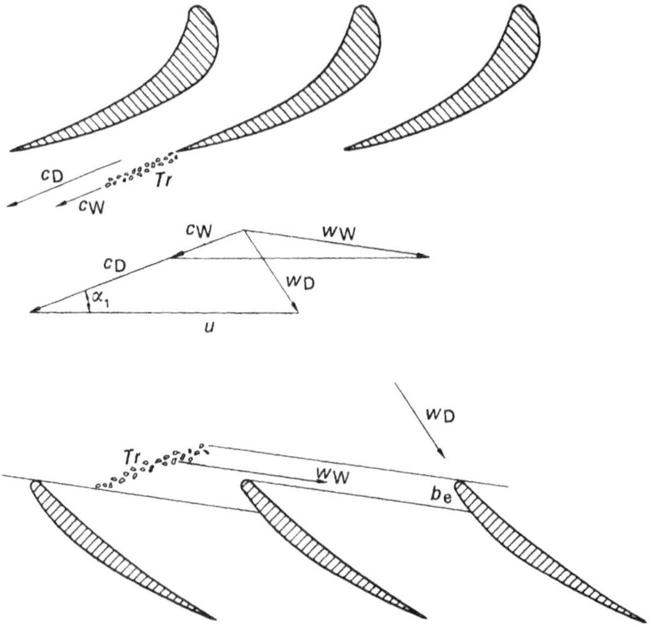

Figure 12.11 Velocities in low-pressure blading: (*top*) fixed blades; (*center*) velocity triangles; (*bottom*) moving blades; C_D = absolute steam velocity; C_W = absolute water velocity; W_D = relative steam velocity; W_W = relative water velocity; u = peripheral velocity; b_e = wetted width = erosion width; Tr = path of water droplets; α_1 = outlet angle of fixed blades. *(Asea Brown-Boveri, Baden, Switzerland)*

12.6 Axial Thrust

Under normal circumstances, axial thrust in an impulse turbine is small because none, or only a small part, of the pressure drop occurs in the moving blades. However, the pressure distribution in the turbine can change, e.g., because of solid deposits on the blades. In that case, considerable thrust forces can arise, and these may present a greater danger for the impulse turbine than for the reaction machine because the higher pressure in front of the moving row acts on the whole disk. Holes machined in the disks tend to balance out these pressure differences.

Thrust variations present no problems in reaction turbines if each turbine casing is provided with its own balance piston, which is always the case with single-casing turbines. The balance piston is usually made with a number of stepped diameters so that abnormal thrust conditions, caused either by mineral deposits or by changes in extraction flow rates, have little or no effect on the reliability of the turbine.

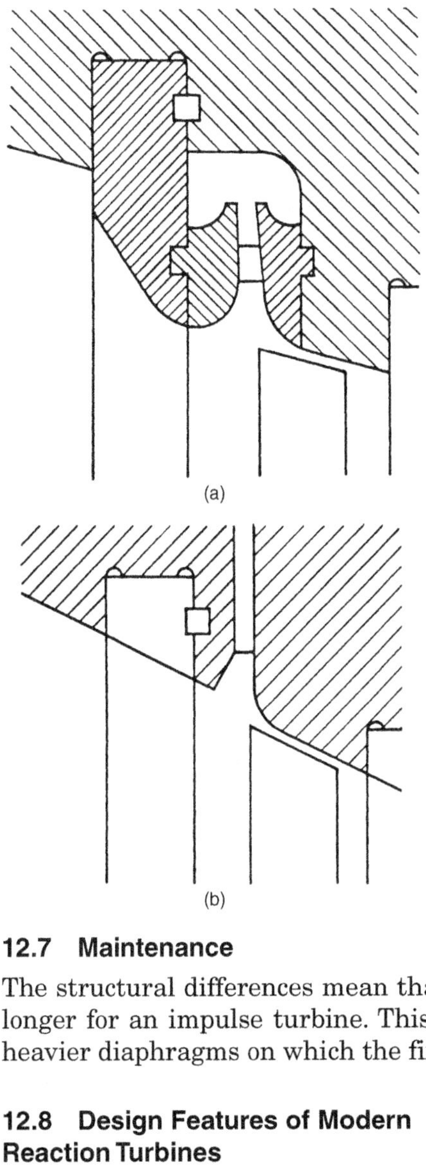

Figure 12.12 Drainage channels inside the reaction turbine: (*a*) in large turbines; (*b*) in small turbines. *(Asea Brown-Boveri, Baden, Switzerland)*

12.7 Maintenance

The structural differences mean that times between overhauls can be longer for an impulse turbine. This is primarily due to the generally heavier diaphragms on which the fixed blades are mounted.

12.8 Design Features of Modern Reaction Turbines

It has already been mentioned that the reliability of a steam turbine today depends only to a small extent on its basic constructional philosophy. Machines that satisfy the design, operational, and reliability expectations of modern user facilities are available from several manufacturers. The main features of these machines can be listed briefly as follows (Fig. 12.13):

- The nozzle chest at the steam inlet provides flexibility and, in particular, short starting times (Fig. 12.14).
- If necessary, separate fixed blade carriers are fitted to facilitate thermal expansion (Fig. 12.15).
- Welded rotor construction and its possible advantages have been discussed earlier. One-piece rotors can also be built for small machines.
- Movable segments in the shaft seals at balance piston and casing penetrations for the higher ratings.
- Securely fixed control-stage blades, welded if necessary.
- Reaction blading of rugged construction.

Reaction type mechanical drive turbines have been performing satisfactorily for decades, proving the suitability of the design. This is further supported by the fact that some of the largest mechanical drive units in the world are the reaction type. A good example is the 50-MW feedpump turbine for a 1350-MW Asea Brown-Boveri steam turbine at a power generating plant in the United States that has been running successfully since 1973. In the late 1970s several 100-MW compressor drive turbines were put in service at Skikda, Algeria. They, and others since then, have run without problems.

It should be noted that since the earliest days of the steam turbine the impulse and reaction types of construction have existed side by side. That neither has been able to supersede the other in any field of application—whether for mechanical drives or power-generating plants—implies that neither type is fundamentally better than the

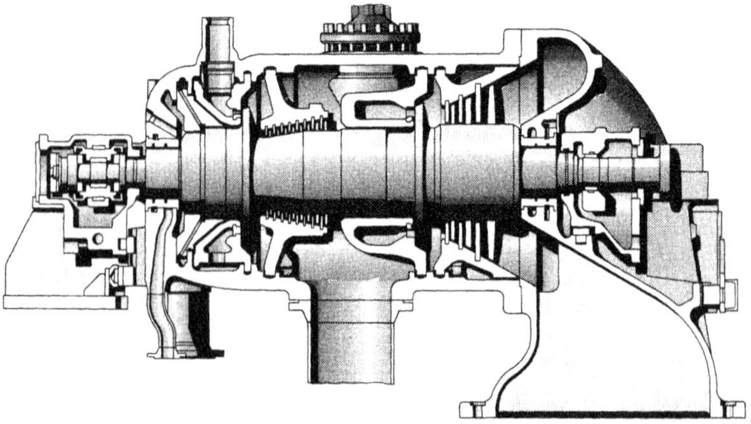

Figure 12.13 Automatic extraction turbine, reaction type. *(Asea Brown-Boveri, Baden, Switzerland)*

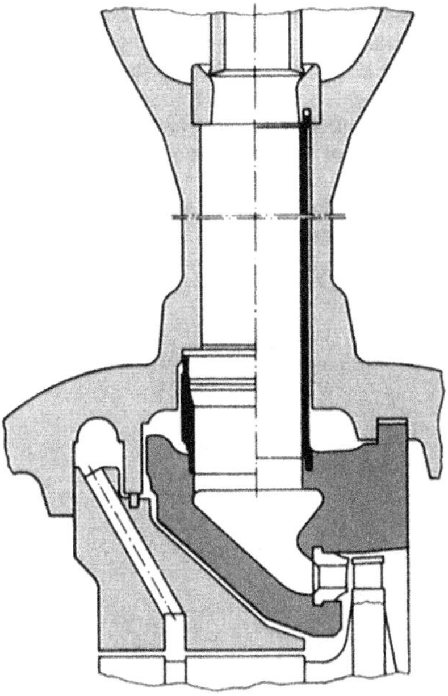

Figure 12.14 Inlet section into nozzle box of a reaction turbine. *(Asea Brown-Boveri, Baden, Switzerland)*

other. This fact was reaffirmed by one of the world's largest multinational petrochemical companies whose worldwide affiliates have liberally applied both impulse as well as reaction steam turbines. Through decades of operating experience and collection of applicable failure statistics, even the blade failure frequencies assembled by *others,* and seemingly favoring reaction turbines, were put in question. The bottom line is simply that thoroughly well-designed steam turbines of *either* type will give highly satisfactory long-term service, while construction shortcuts can weaken *either* type to the point of yielding unacceptable performance.

12.9 Deposit Formation and Turbine Water Washing

Occasions may arise when deposits form on the internal parts of steam turbines. The accumulation of these deposits may be indicated by a gradual increase in stage pressures over a period of time with no evidence of vibration, rubbing, or other distress. Such deposits have a marked detrimental effect on turbine efficiency and capacity. When deposits cause extensive plugging, thrust bearing failure, wheel rubbing, and other serious problems can result.

Deposits are classified as water insoluble and water soluble. The characteristics of the deposits should be determined by analysis of

Figure 12.15 Guide blade carrier accommodates thermal expansion. *(Siemens Power Corporation, Milwaukee, Wis. and Erlangen, Germany)*

samples, and corrective measures should be taken to eliminate these deposits during future operation.

When it has been determined that deposits have formed on the internal parts of the turbine, three methods may be employed to remove the deposits:

1. Turbine shut down, casing opened, and deposits removed manually
2. Turbine shut down and allowed to cool, the deposits cracking off because of temperature changes

3. *Water washing,* the on-line (running) removal of deposits, when deposits are water soluble

Since deposits tend to accumulate to a greater extent during steady high-load conditions (e.g., base load generator drive, ethylene process drive), the application and plant operating conditions will dictate which method will best serve to restore the turbine to optimum performance. In a generator-drive service, shutting down the unit or water washing at low speed and reduced load may create minimum plant upset. The size of process drivers and plant-operating conditions present different circumstances.

Turbine washing at full speed (on-stream cleaning) can be and has been successfully accomplished on many mechanical drive steam turbines. Considerable hazard attends any method of water washing and full-speed washing is *more hazardous* than washing at reduced speed. But this can be accomplished provided great care and judgment is exercised. While we know of no steam turbine manufacturer who would guarantee the safety of turbines for any washing cycle, capable manufacturers recognize that deposits do occur. They will, therefore, help operators as much as possible in dealing with the problems until effective prevention is established.

Saturated steam washing by water injection is the conventional and well-tried method of removing water soluble deposits from turbines. The amount and rate of superheat to be removed and the amount of steam flow required for operation determine the water injection rate. *It is the injection of large quantities of liquid (such as may be required on process drivers) that create potential problems.*

The nature of a typical impulse turbine lends itself to full-speed water washing. Axial clearances between first-stage buckets and nozzles and between moving buckets and diaphragms will range from 0.050 to 0.090 in. The NiResist labyrinth packing radial clearance when the unit is cold will be approximately 0.007 in. The labyrinth will seal on the shaft only; the moving blades will not require seals. With impulse turbines, these liberal clearances help minimize the hazards associated with water washing. Nevertheless, numerous reaction turbines have also been successfully water washed.

Water injection is accomplished by a piping arrangement for the atomization and injection of water into the steam supply to ensure a *gradual* and uniform reduction in the temperature of the turbine inlet steam until it reaches 10 to 15°F superheat. It is probably a safe rule that the temperature should not be reduced faster than 25°F in 15 min or 100°F/h. Figures 12.16 through 12.18 show suggested piping arrangements for the admission of water and steam as well as a simple assembly of fabricated pipes to form a desuperheater. Failure of water injection pumps presents a great hazard, especially at maximum injection rates. To guard against pump failure, untreated boiler feedwater is used since these pumps are usually the most reliable in a plant.

If plant-operating conditions allow, the vacuum on a condensing turbine should be reduced to 5 to 10 mmHg; for noncondensing turbines, the exhaust pressure should be reduced to atmospheric pressure. Note that on any noncondensing unit requiring full-speed washing, the manufacturer should be consulted about minimum allowable exhaust pressure. Extraction turbines should be run with the extraction line shut off.

A steam gauge and thermometer should be installed between the trip-throttle valve and the governor-controlled valves. The thermometer should preferably be a recording type and should be very responsive to small changes in temperature.

Low speed wash, as illustrated in Fig. 12.16, represents a well-understood method of deposit removal.

To start the washing procedure, it is normally recommended to operate the turbine on the trip-throttle valve at one-fifth to one-fourth normal speed with no load. The live-steam valve to the mixer would now be opened and the boiler stop valve closed, after which the trip-throttle valve and the governor valves may be opened wide and the speed controlled by the small live-steam valve to the mixer. Water is then supplied to the mixing chamber in quantities sufficient to reduce the steam temperature at the recommended rate until 10 to 15°F (6 to 9°C) superheat at turbine inlet is reached. During the washing cycle the exhaust steam should be discharged to the sewer.

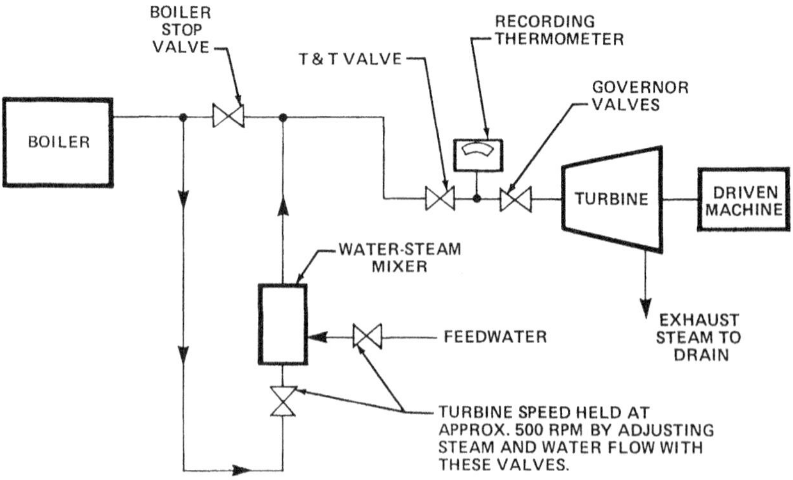

Figure 12.16 Schematic of low-speed water wash system. *(Dresser-Rand Company, Wellsville, N.Y.)*

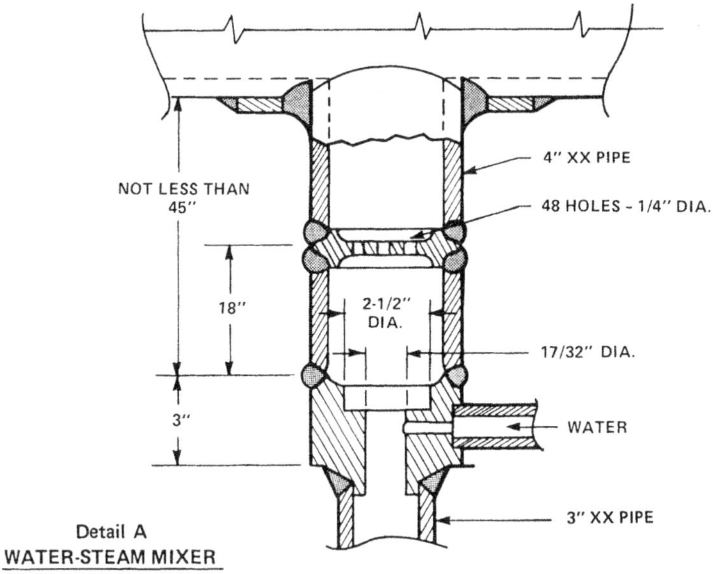

Figure 12.17 Water-steam mixing arrangement used in low-speed water wash systems. *(Dresser-Rand Company, Wellsville, N.Y.)*

The washing may be continued until conductivity measurements of the condensate or exhaust steam indicate that no further deposits are being dissolved. Additionally, it is appropriate to note that the trip-throttle valve should be checked to be certain it is operating satisfactorily since washing may not thoroughly clean deposits from this valve and that if carryover deposits are not also cleaned from the piping between the boiler and the turbine, this carryover may be deposited in the turbine in a very short period after washing.

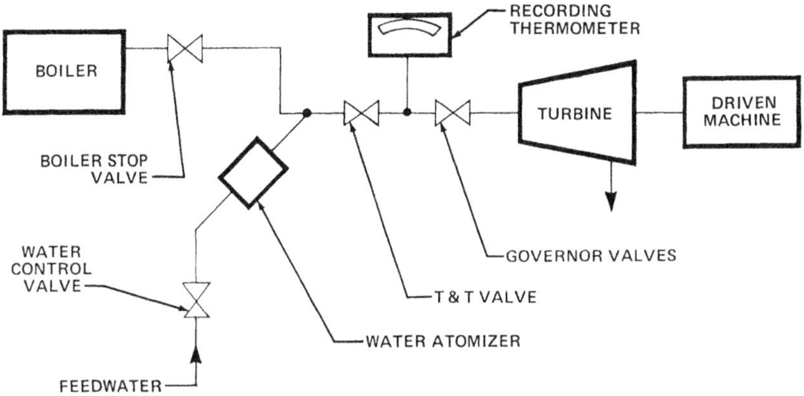

Figure 12.18 Full-speed water wash arrangement. *(Dresser-Rand Company, Wellsville, N.Y.)*

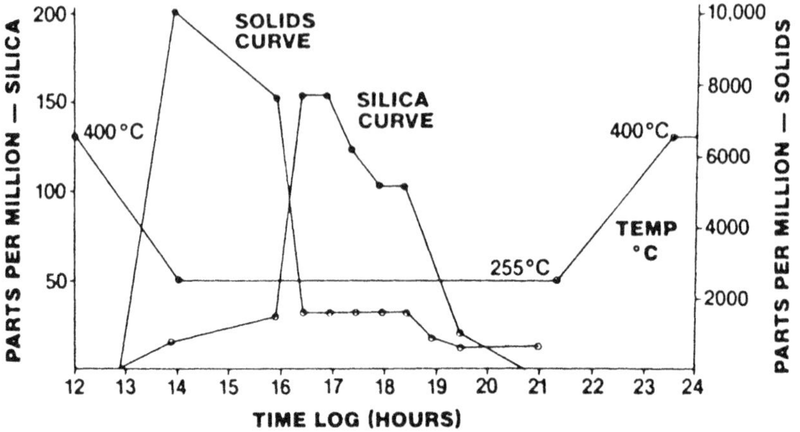

Figure 12.19 Actual water wash results and analysis of condensate effluent from a 4860 kW, 6230 r/min back-pressure turbine (550/55 psig, 750°F; 38/4 Bar, 400°C). *(Dresser-Rand Company, Wellsville, N.Y.)*

The turbine may be put on the line by increasing the steam temperature at the previously recommended rate (25°F or 14°C increase per 15 min) by slowly closing the feedwater valve to the mixer. When normal steam temperature is reached, the trip-throttle valve may be closed to a point where the pressure ahead of it is brought up to line pressure, still holding speed at one-fifth to one-fourth normal. The

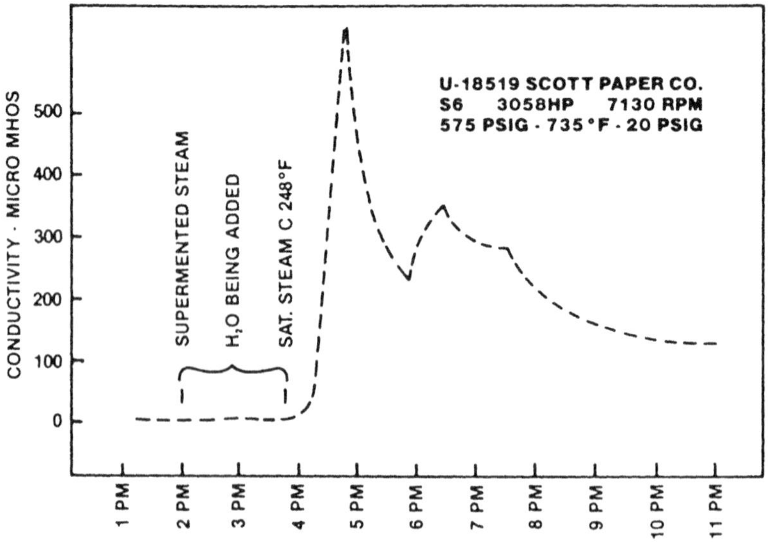

Figure 12.20 Conductivity vs. time plot obtained while water washing a turbine. *(Dresser-Rand Company, Wellsville, N.Y.)*

boiler stop valve may then be opened and the turbine brought up to speed. The turbine may also be shut down entirely and started up by normal procedure.

During the entire procedure, the unit should be carefully watched for signs of rubbing or other distress.

To accomplish *full-speed wash* (Fig. 12.18), the operator would slowly open the feedwater valve to the atomizer chambers. This valve would then be adjusted to supply water to the atomizer chamber in quantities to reduce the steam temperature at the recommended rate until 10 to 15°F (6 to 9°C) superheat is reached at the turbine inlet. During the washing cycle the exhaust steam would be discharged to the sewer.

The washing may be continued until tests of either the condensate or exhaust steam indicate that no further deposits are being dissolved. This is neatly illustrated in Fig. 12.19 and 12.20.

The same precautions concerning the trip-throttle valve and piping between boiler and turbine apply.

The turbine may be restored to normal steam conditions by increasing the steam temperature at the previously recommended rate (25°F or 14°C increase per 15 min) by slowly closing the feedwater valve to the atomizer chamber.

During the entire procedure, the unit should be carefully watched for signs of rubbing or other distress. Instrumentation such as vibration probes and thrust temperature detectors are invaluable during this period.

Chapter

13

Transmission Elements for High-Speed Turbomachinery

Efficiency is of decisive importance in the selection of a prime mover. The efficiency of a steam turbine is determined by the area of the flow cross section and the expansion clearance between rotor and casing.

To keep the tip losses low, the blades are made as long as possible, consistent with a small rotor diameter. With small and medium flow rates this results in a high rotational speed. A high rotational speed has two advantages, assuming the same stage pressure drop. The rotor diameter can be kept small and hence a compact turbine is feasible. Also, relatively long blades can be incorporated, which results in efficient blade channels with low tip losses and other leakage losses.

Turbine efficiencies are a function of turbine blade geometry, flow, and speed, as shown in Figs. 13.1 through 13.3. Operating the turbine at optimum efficiency and using a geared speed-modifying unit to accommodate the required driven equipment speed will often make economic sense.

13.1 Spur Gear Units

For small and medium outputs the advantages mentioned above give the high-speed turbine a clear superiority. But when they are used to drive generators and other slow-running machines such as reciprocating compressors and pumps, a gearbox must be incorporated to reduce the high speed of the turbine. In power generation, e.g., the lower speed enables a moderately priced four-pole generator to be used. The higher efficiency and resulting gain in output of the high-speed turbine by far outweigh the losses incurred in the speed-reducing or step-down gearbox. And the additional cost of the gearbox is made up for by the lower

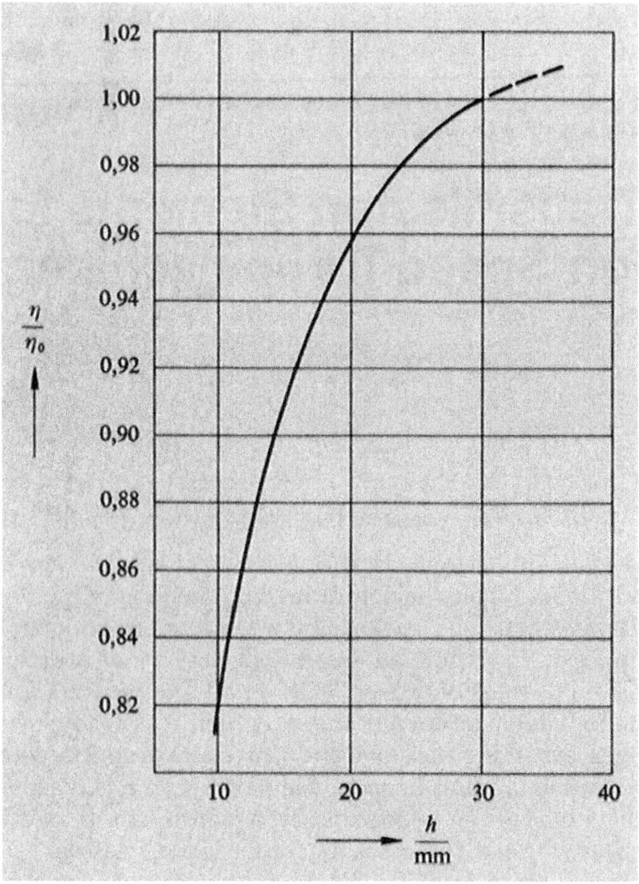

Figure 13.1 Dependence of the efficiency η of a control stage on the blade height h, at constant U/c_0. (η_0 = reference efficiency; U = peripheral speed; c_0 = steam velocity.) *(Asea Brown-Boveri, Baden, Switzerland)*

cost of the turbine and generator. This will be intuitively evident from Tables 13.1 and 13.2.

The high-speed steam turbine is making inroads into both the low-power range, where it is competing with the reciprocating steam engine that previously dominated this field, and into the high-power range where a new series of machines are providing an alternative to the traditional low-speed sets.

Geared turbosets require less space so that costs for foundations and buildings are reduced.

However, gear units are not limited to steam turbine-generator sets. Thousands are in service as speed increasers, or step-up gears between electric motor drivers and turbocompressors. Most of these incorporate

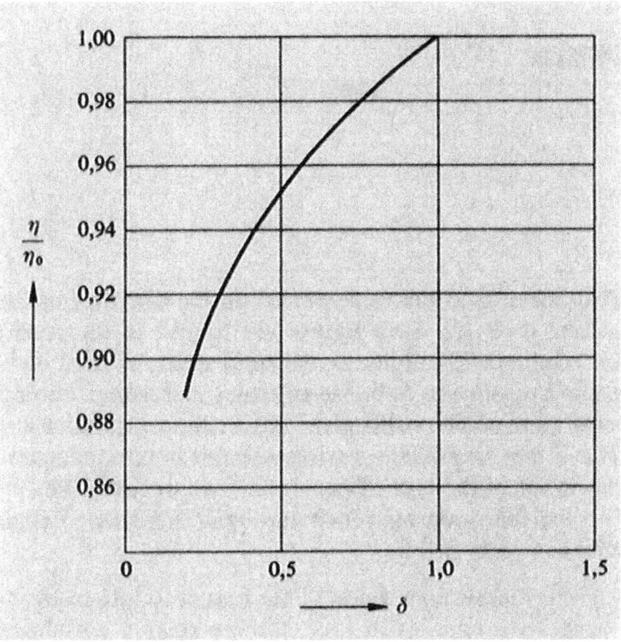

Figure 13.2 Variation of reaction stage efficiency as a function of the volume coefficient δ. (η_0 = reference efficiency; δ = volume coefficient = $\dot{V}/r^2u$; r = radius, at mean diameter of the blade path; u = peripheral speed at radius r; $\dot{V}$ = volume flow.) *(Asea Brown-Boveri, Baden, Switzerland)*

either single- or double-helical gears as shown in Figs. 13.4 and 13.5, but some are designed as epicyclic or planetary gear sets.

13.2 Epicyclic Gears

Epicyclic gearing represents a special case of the family of gear drives known as split train gears. Power is transmitted through multiple paths (Fig. 13.6) to reduce the size of the split train drive compared to that of conventional single-path drives. Figure 13.7 illustrates how in epicyclic gears the transmitted power is also split through multiple paths.

The four basic elements of epicyclic gears are an internally toothed flexible annulus ring, a plant or star carrier, a central sun wheel, and planet or star wheels. Holding stationary any one of the first three elements (internally toothed annulus, star carrier, or sun wheel) results in the three types of epicyclic gear configurations shown in Fig. 13.8*a* through *c*. With the fixed annulus system we obtain what is generally called a *planetary gear unit*. The input vs. output shaft speed ratios can

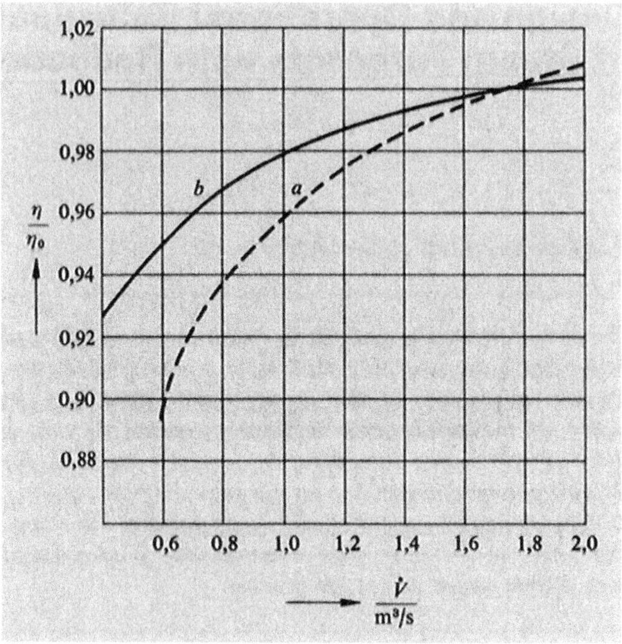

Figure 13.3 Variation of the efficiency at the coupling, η, as a function of the live-steam volume flow $\dot{V}$. (a = direct drive: efficiency at the turbine coupling; b = drive through reduction gear: efficiency at the coupling between gearing and generator; η_0 = reference efficiency.) *(Asea Brown-Boveri, Baden, Switzerland)*

range from 3:1 to about 12:1. This configuration (Fig. 13.8*a*) is generally the most compact and cost-effective for a given torque capacity and is understandably the most common. The carrier speed range of the planetary is limited by the resultant centrifugal loading of the planet bearings. Both shafts rotate in the same direction.

Figure 13.8*b* depicts the so-called *star gear setup.* Input and output shafts rotate in opposite directions and ratios range from 2:1 to 11:1. This configuration is often preferred for higher speed applications where centrifugal loading of the planet bearings in the planetary configuration exceeds design limits.

In the *solar gear configuration* of Fig. 13.8*c* input and output shafts rotate in the same direction. Because of the rather limited ratio range from 1.1:1 to 1.7:1, the solar gear is not used very often.

13.3 Clutches

Certain energy recovery, combined cycle technology, and cogeneration services are among the many applications for synchronous clutches.

TABLE 13.1 Technical Data of a Back-Pressure Turboset with and without Reduction Gear

Variant*	Without reduction gear	With reduction gear
Turbine speed, r/min	3000	9600
Generator speed, r/min	3000	1500
Control stage:		
nominal diameter, mm	1020	450
Reaction stages:		
number	58	19
Turbine casing:		
mass, t	28.9	8.2

* Both variants are designed for the same efficiency at the generator coupling. Live steam = 80 bar/500°C; back pressure = 5 bar; Output = 14 MW.
SOURCE: Asea Brown-Boveri, Baden, Switzerland

Figure 13.9 shows a typical energy recovery application of a Maag synchronous clutch coupling in a compressor installation where excessive process gases are available. These gases, instead of being relieved by a valve, are now used to drive a gas expander that is coupled via synchronous clutch with a centrifugal compressor. In the case when no excessive process gases are available, the gas expander will be disconnected automatically, thus avoiding windage losses in the gas expander.

Many turbomachinery clutches are known as *SSS clutches*. The manufacturer and licensor, SSS Gears Limited of Sunbury-On-Thames, England, chose SSS to denote the "Synchro-Self-Shifting" action of the clutch whereby the clutch teeth are phased and then automatically shifted axially into engagement when rotating at precisely

TABLE 13.2 Technical Data of a 20-MW Extraction/Condensing Turboset with and without Reduction Gear

Variant*	Without reduction gear	With reduction gear
Turbine speed, r/min	3000	7500
Generator speed, r/min	3000	1500
Control stages:		
number	2	2
Type	2-row	Single row
Nominal diameter, mm	1060 and 800	each 700
Reaction stages:		
number	61	23
Turbine casing:		
mass, t	56	35

* Both variants are designed for the same efficiency at the generator coupling. Live steam = 108 bar/520°C; extraction pressure = 10 bar; exhaust pressure = 0·1 bar.
SOURCE: Asea Brown-Boveri, Baden, Switzerland

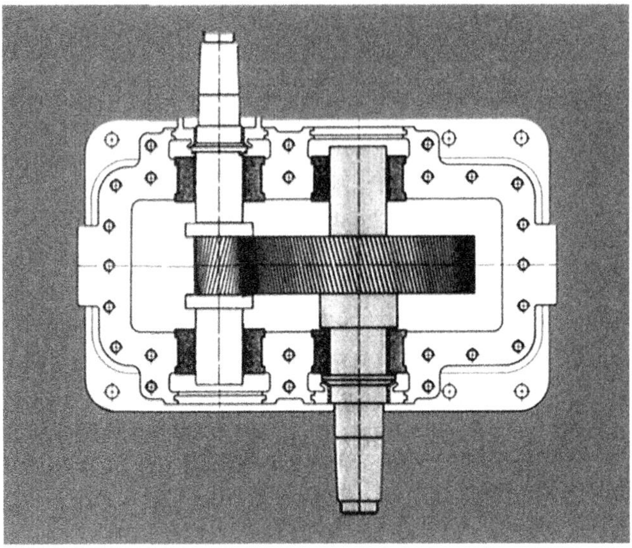

Figure 13.4 Parallel shaft single-helical high-speed gear with thrust collar. *(Cincinnati Gear Company, Cincinnati, Ohio)*

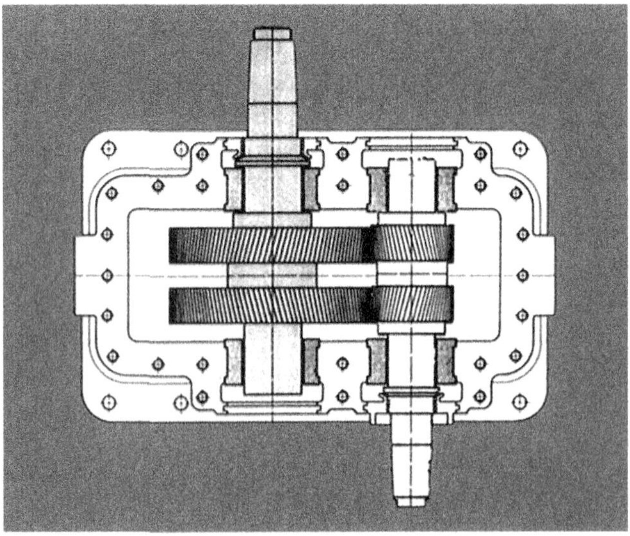

Figure 13.5 Parallel shaft double-helical high-speed gear unit. *(Cincinnati Gear Company, Cincinnati, Ohio)*

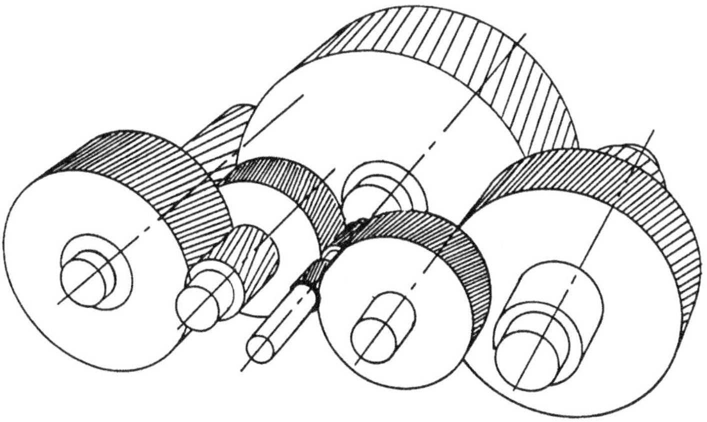

Figure 13.6 Split train principle embodied in epicyclic gears. *(Philadelphia Gear Corporation, King of Prussia, Pa.)*

the same speed. The clutch disengages as soon as the input speed slows down relative to the output speed.

The basic operating principle of the SSS clutch can be compared to the action of a nut screwed on to a bolt. If the bolt rotates with the nut free, the nut will rotate with the bolt. If the nut is prevented from rotating while the bolt continues to turn, the nut will move in a straight line along the bolt.

In an SSS clutch (Fig. 13.10), the input shaft (E) has helical splines (D) that correspond to the thread of a bolt. Mounted on the helical splines is a sliding component (C) that simulates the nut. In the diagram, the sliding component has external clutch teeth (B) at one end and external ratchet teeth (G) at the other.

When the input shaft rotates, the sliding component rotates with it until a ratchet tooth contacts the tip of a pawl (A) on the output clutch ring (F) to prevent rotation of the sliding component relative to the output clutch ring. This position is shown in Fig. 13.10 (1).

As the input shaft continues to rotate, the sliding component will move axially along the helical splines of the input shaft. When a ratchet tooth is in contact with a pawl tip, the clutch engaging teeth are perfectly aligned for interengagement and will thus pass smoothly into mesh in a straight line path.

As the sliding component moves along the input shaft, the pawl passes out of contact with the ratchet tooth, allowing the clutch teeth to come into flank contact and continue the engaging travel as shown in Fig. 13.10 (2). Note that the only load on the pawl is that required to shift the lightweight sliding component along the helical splines.

Driving torque from the input shaft will only be transmitted when the sliding component completes its travel by contacting an end stop on

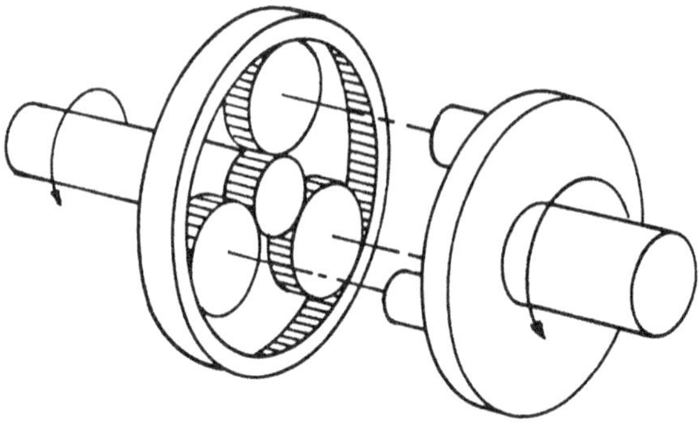

(a)

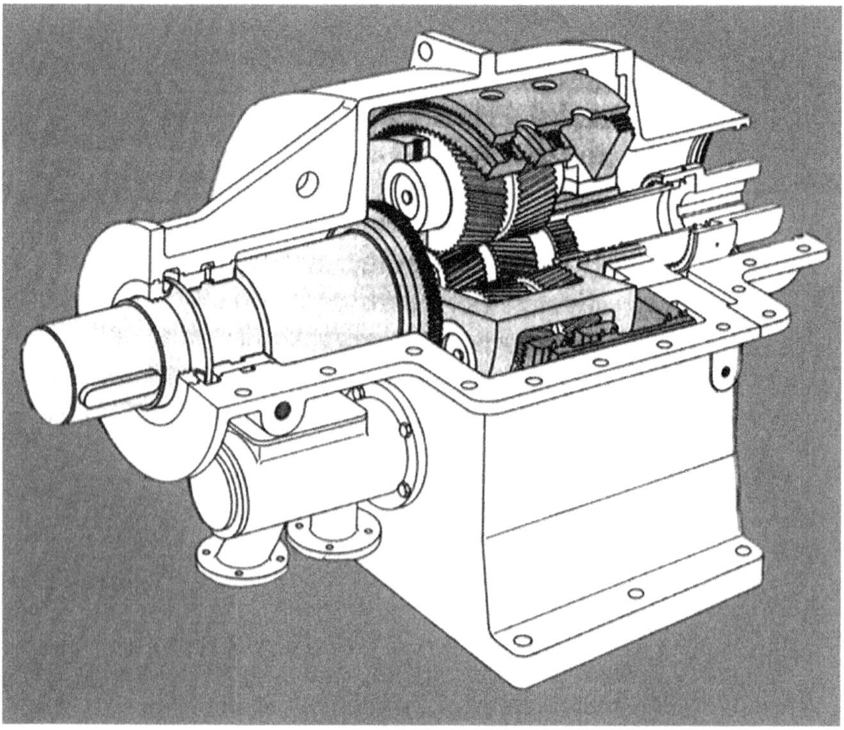

(b)

Figure 13.7 Epicyclic gears distributed to optimize available space ([a] *Philadelphia Gear Corporation, King of Prussia, Pa.;* [b] *Cincinnati Gear Corporation, Cincinnati, Ohio)*

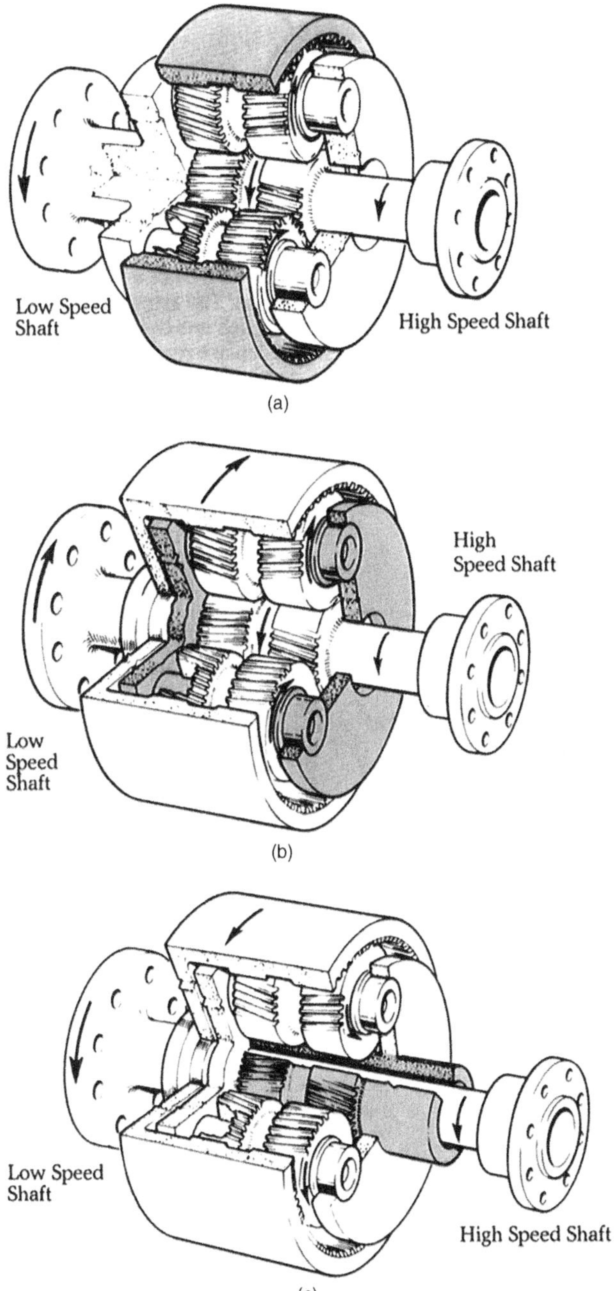

Figure 13.8 Epicyclic gear types: (*a*) planetary gear, fixed annulus system, input and output shafts rotate in same direction, ratios from 3:1 to 12:1; (*b*) star gear, fixed carrier assembly, input and output shafts rotate in opposite directions, ratios from 2:1 to 11:1; (*c*) solar gear, fixed sunwheel, input and output shafts rotate in same direction, ratios from 1.1:1 to 1.7:1. *(Philadelphia Gear Corporation, King of Prussia, Pa.)*

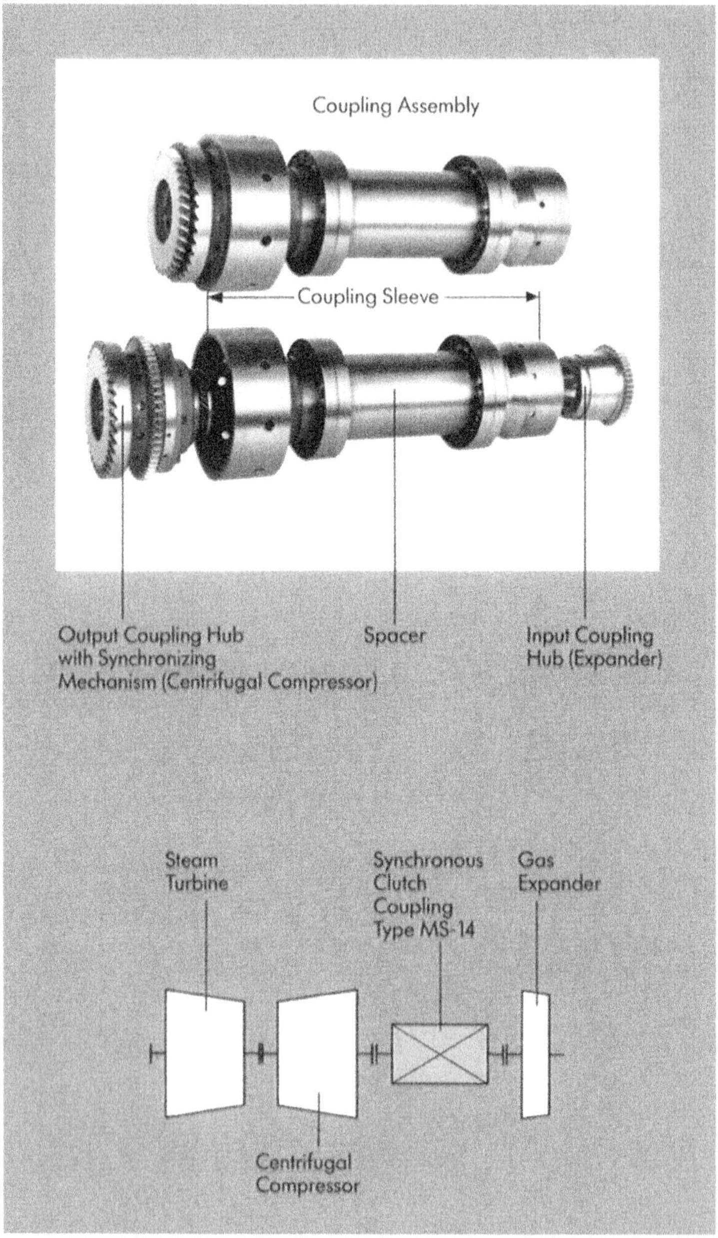

Figure 13.9 Maag synchronous clutch coupling. With no gas available at the gas expander, the clutch coupling will disengage the expander. *(Maag Gear Company, Zurich, Switzerland)*

the input shaft, with the clutch teeth fully engaged and the pawls unloaded as shown in Fig. 13.10 (3).

When a nut is screwed against the head of a bolt, no external thrust is produced. Similarly, when the sliding component of an SSS clutch contacts its end stop and the clutch is transmitting driving torque, no external thrust loads are produced by the helical splines.

Where necessary, an oil dashpot is incorporated in the end stop to cushion the clutch engagement.

If the speed of the input shaft is reduced relative to the output shaft, the torque on the helical splines will reverse. This causes the sliding component to return to the disengaged position, and the clutch will overrun. At high overrunning speeds, pawl ratcheting is prevented by a combination of centrifugal and hydrodynamic effects acting on the pawls. The basic SSS clutch can operate continuously engaged or overrunning at maximum speed without wear occurring.

13.4 Hydroviscous Drives

Together with hydrokinetic or fluid couplings, and magnetic or eddy current couplings, hydroviscous drives make up the principal variable slip

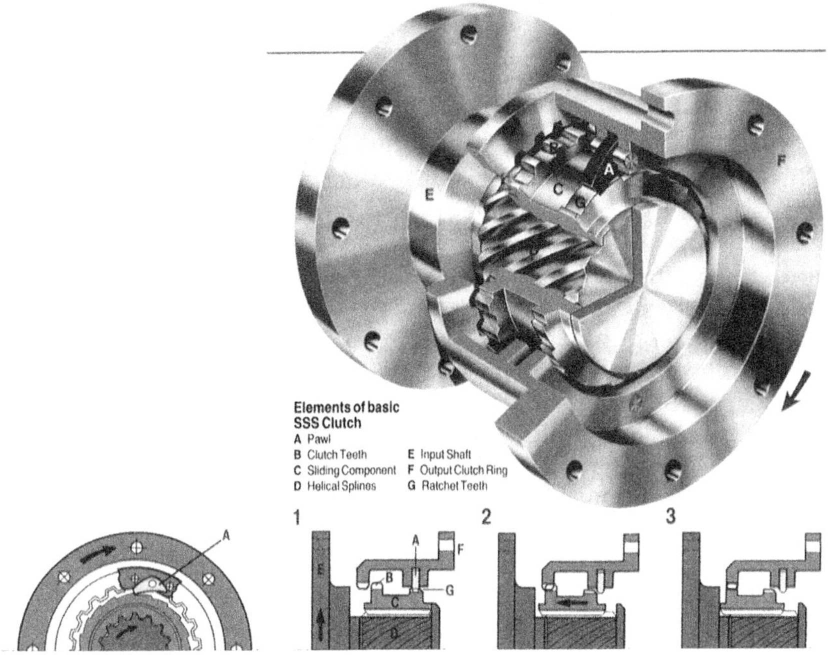

Figure 13.10 SSS clutch. *(SSS Gears, Ltd., Sunbury-On-Thames, Middlesex, England)*

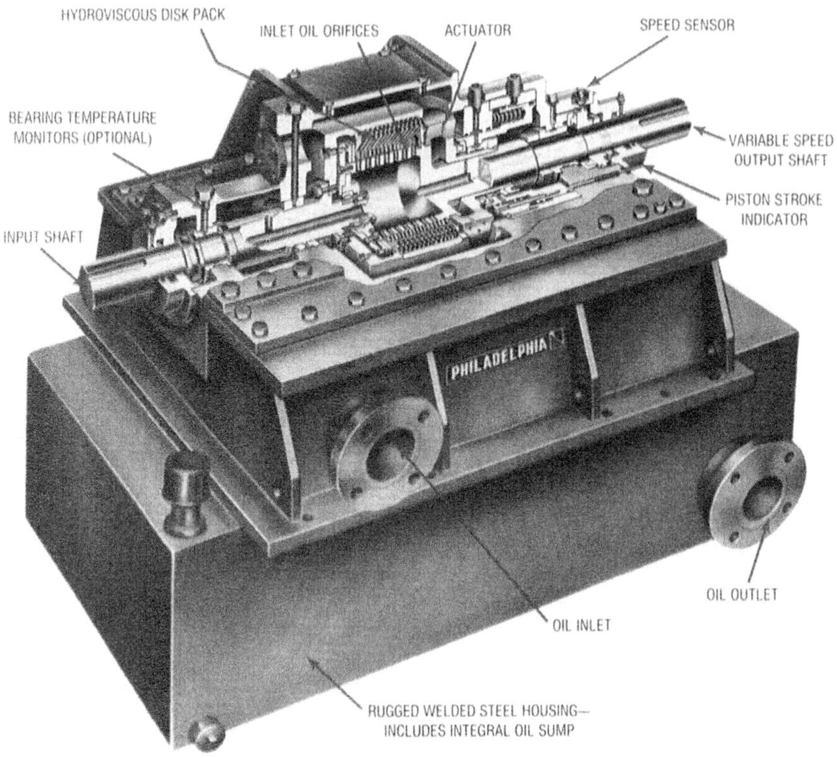

Figure 13.11 Hydroviscous drive. *(Philadelphia Gear Corporation, King of Prussia, Pa.)*

drives. Figure 13.11 shows a typical hydroviscous drive marketed by the Philadelphia Gear Corporation under the trade name *Synchrotorque.*

The Synchrotorque variable-speed drive operates on the principle of shearing an oil film to transmit torque. This hydroviscous (hydrodynamic shearing) effect transmits torque in proportion to a variable clamping force. For unrestrained loads, the higher the clamping force, the faster the output speed. The input drive plate functions as the driver. The output friction disk is faced with a suitably grooved resilient material. The material couple between input drive plates and output friction disks allows virtually infinite speed control right up to 100 percent of input speed. This simple phenomenon (based on established hydrodynamic bearing principles) is the basis of controlled torque transmission in hydroviscous drives.

More specifically, the input member accelerates the oil particles tangentially, and a hydrodynamic film is established through the action of the friction disk oil grooving. The oil particles also are acted upon by centrifugal force, which accelerates them outwardly. This natural pumping action ensures a uniform oil film across disk faces for controlled shearing. The oil film is continuously replenished by a circulating oil pump, which pumps oil from the reservoir through a cooler and back to the center of the rotating members where the cycle is repeated (refer to Fig. 13.12).

Figure 13.13 will assist in understanding how torque is controlled.

The variable clamping force is obtained by varying the pressure to a piston. This piston force is transmitted to the actuator, which rotates with the output shaft. The actuator clamps the disk pack in proportion to the pressure applied to the piston. Piston pressure is controlled by the synchrodrive control system which responds to the command signal. The synchrodrive control system is a closed-loop system using output shaft speed as feedback. Any input command signal is automatically compared to output shaft speed. The synchrodrive control system quickly and accurately regulates piston pressure (clamping force) the proper value. Response is virtually instantaneous and the speed feedback feature ensures linearity between input command and output shaft speed. The standard source of actuation pressure is hydraulic (*electrohydraulic control*). For certain special applications, the source of actuation pressure may be pneumatic (*electropneumatic control*). The input command signal may come from a customer provided source (*automatic control*) or from within the controller (*manual control*).

Hydroviscous drives have been applied in sizes ranging from 100 to 20,000 hp (approximately 75 to 15,000 kW). They are frequently chosen because of their lockup capabilities.

Eddy current and fluid couplings exhibit fixed slip loss ranging from 2 to 5 percent. This means highest attainable efficiency is only 95 to 98

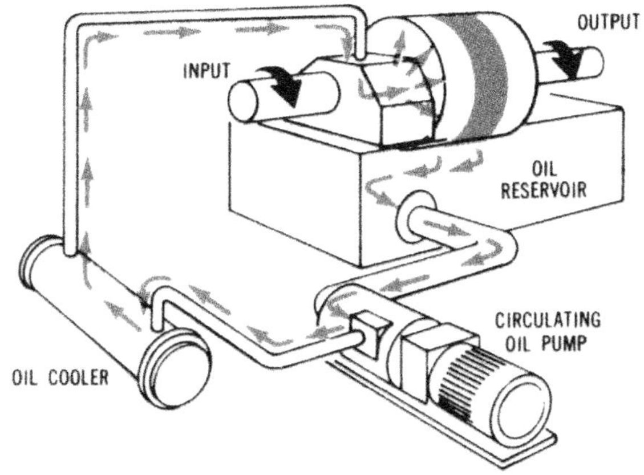

Figure 13.12 Oil circuitry of a hydroviscous drive. *(Philadelphia Gear Corporation, King of Prussia, Pa.)*

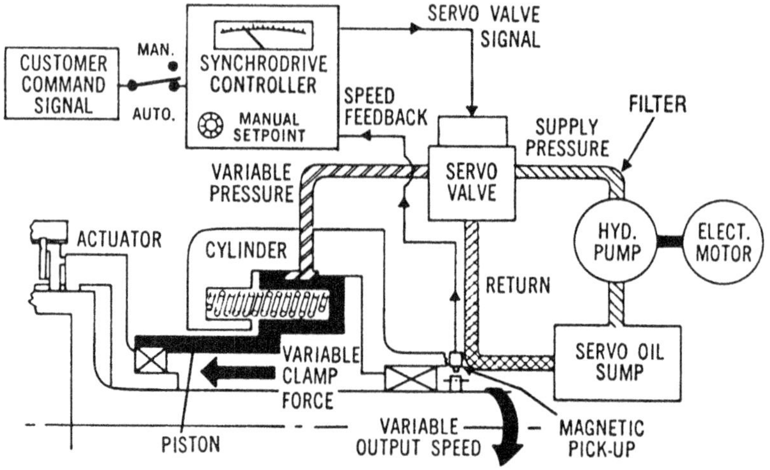

Figure 13.13 Torque control schematic of a hydroviscous drive, showing actuator pressing down on disk pack. Torque is transmitted in proportion to a variable clamping force. *(Philadelphia Gear Corporation, King of Prussia, Pa.)*

percent with corresponding maximum attainable output speed of 95 to 98 percent of input speed. Unlike eddy current and fluid couplings, hydroviscous drives transmit torque up to 100 percent of input speed. This savings of 2 to 5 percent compared to eddy current and fluid couplings means important power savings, additional capacity, and/or compensation for pump, fan, or other driven equipment wear.

The amount of torque that can be transmitted varies directly with the number of disk surfaces over which the shearing action occurs. To increase the number of working surfaces, a hydroviscous disk pack is built up by alternately stacking disks splined to the input member between disks splined to the output member. All disks are free to slide axially but must rotate with the member to which they are splined. The variable clamp force is applied to one end of the disk pack and is distributed to the remainder of the disks because of their freedom to slide axially. The working oil is introduced (orificed) to each set of surfaces by oil passages appropriately drilled in the output member. This is illustrated in Fig. 13.14.

The design principles embodied in hydroviscous and other fluid type drives lead to many or all of the following features or attributes. They

- Provide vibration free, virtually infinite variable speed.
- Represent one of the most efficient drives, with no fixed slip losses. Output speed equals input speed upon command.
- Accept any standard process control signal.
- Are easily adaptable for parallel operation.

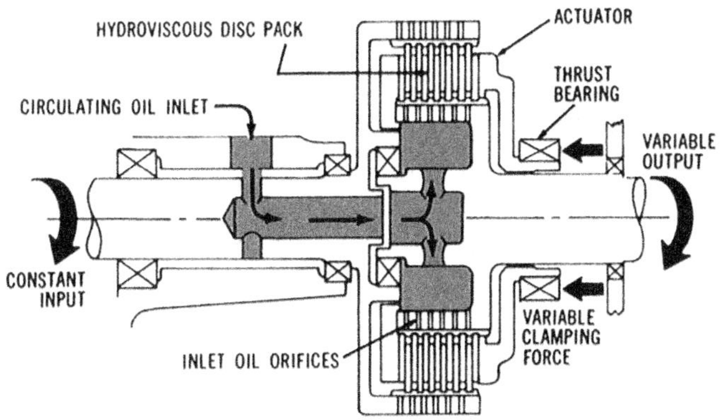

Figure 13.14 Oil passageways to each set of disk surfaces in a hydroviscous drive. *(Philadelphia Gear Corporation, King of Prussia, Pa.)*

- Provide instantaneous response upon command.
- Transmit torque in either direction.
- Allow prime mover to be started under no load. Eliminate the need for reduced voltage starting and expensive electrical accessories when using synchronous motors.
- Allow prime mover to remain running while disconnected from load.
- Provide controlled acceleration for very large inertia loads.
- Protect drive train from excessive loads with automatic torque limiting feature.
- Represent an economical solution for variable-speed operation of fans, pumps, blowers, and compressors, rather than wasteful throttling for those processes with variable-flow demands.
- Extend fan blade liner life when running in abrasive atmospheres by varying fan speed rather than throttling flow.

The estimated overall performance capabilities of hydroviscous drives are represented in Fig. 13.15.

13.5 Hydrodynamic Converters and Geared Variable-Speed Turbo Couplings

Hydrodynamic and other proven mechanical components are combined in the design of such variable-speed devices as the Voith Transmission Company's VORECON, Fig. 13.16. The power savings that can be realized from these multistage variable-speed drives and from hydrodynamic geared variable-speed couplings can be visualized from Fig. 13.17.

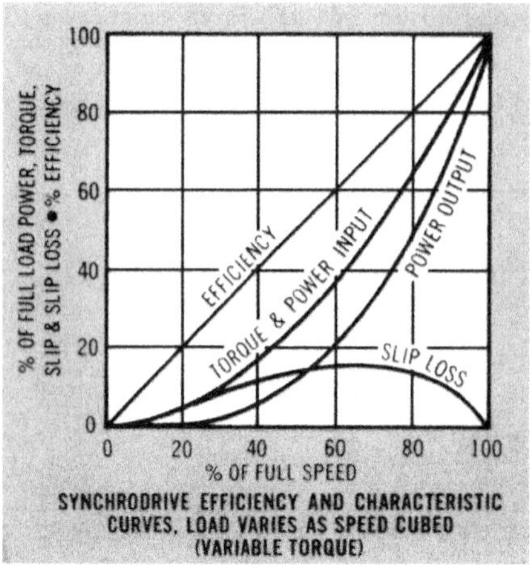

(a)

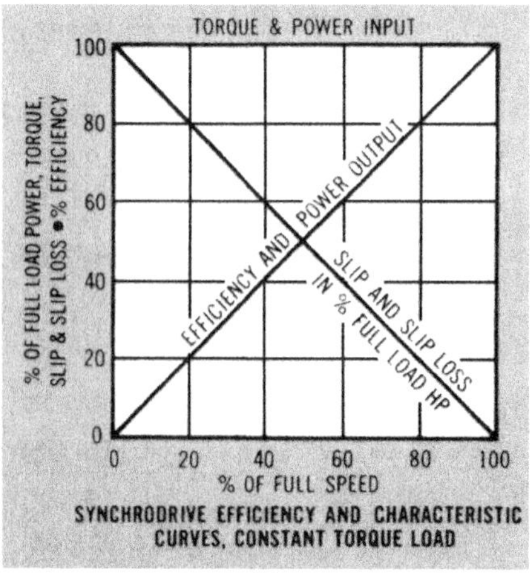

(b)

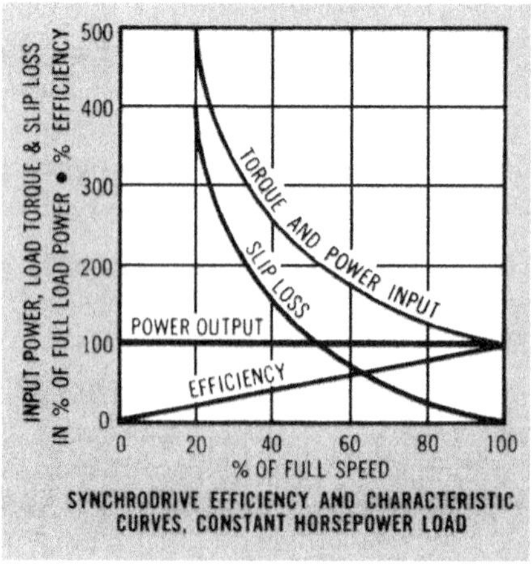

(c)

Figure 13.15 Efficiency and characteristic performance curves for hydroviscous drives: (*a*) variable torque; (*b*) constant torque load; (*c*) constant horsepower load. *(Philadelphia Gear Corporation, King of Prussia, Pa.)*

Figure 13.16 Hydrodynamic converter between electric motor and centrifugal compressor in a European oil refinery. *(Voith Transmissions, Inc., York, Pa.)*

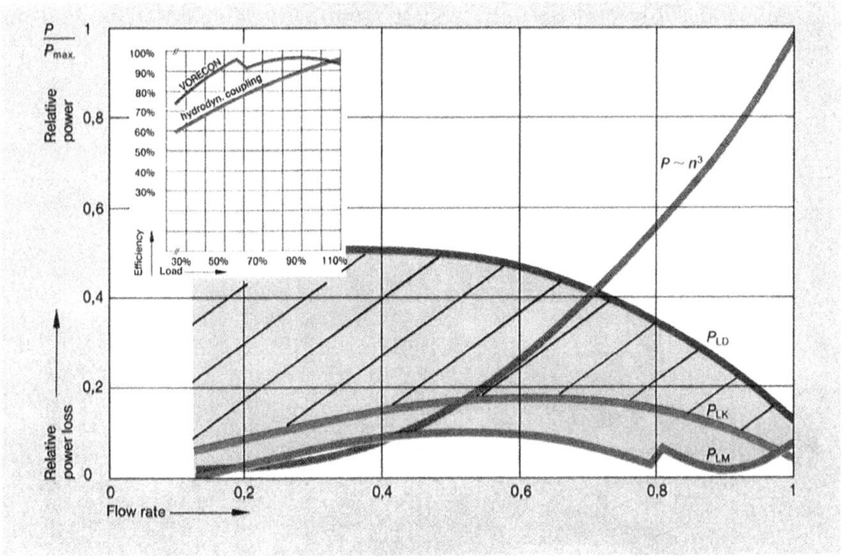

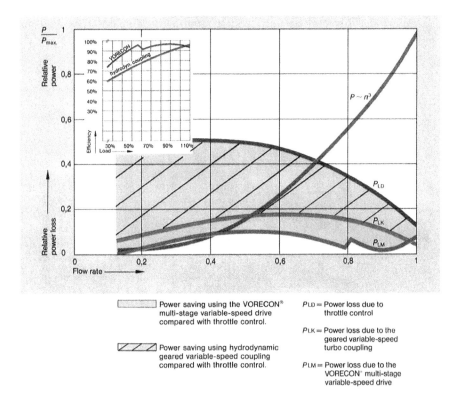

Power saving using the VORECON® multi-stage variable-speed drive compared with throttle control.

Power saving using hydrodynamic geared variable-speed coupling compared with throttle control.

P_{LD} = Power loss due to throttle control

P_{LK} = Power loss due to the geared variable-speed turbo coupling

P_{LM} = Power loss due to the VORECON® multi-stage variable-speed drive

Figure 13.17 Power savings available from multistage variable-speed drives and geared variable-speed turbocouplings. *(Voith Transmissions, Inc., York, Pa.)*

The VORECON works on the principle of power splitting using a superimposed planetary gear unit. The design with horizontally split housing is compact and space saving with the following individual elements highlighted in Fig. 13.18:

A. Hydrodynamic variable-speed coupling

B. Hydraulically controlled lockup clutch

C. Hydrodynamic torque converter

D. Hydrodynamic brake

E. Planetary gear, fixed

F. Planetary gear, revolving

The individual components lie coaxially on the shaft center line. Working and lube oil is pumped from a common oil tank by a mechanically driven pump, as is current practice for variable-speed units. The oil tank can be freestanding. Since all the functional changes within the VORECON are made solely by filling or draining the hydrodynamic components, smooth changeover from one component to another is ensured. This can be achieved by using fixed solenoid valves and standard electromechanical or pneumatically operated actuator. The

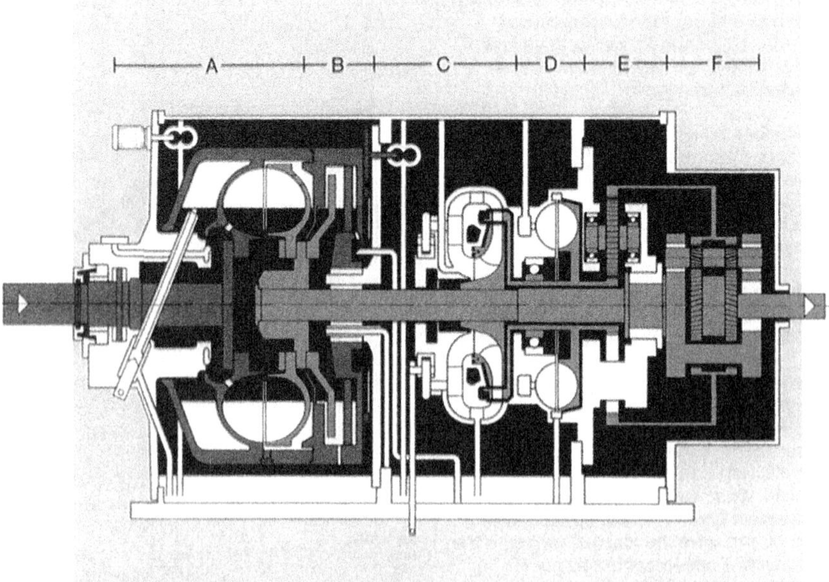

Figure 13.18 VORECON hydrodynamic variable-speed drive. *(Voith Transmission, Inc., York, Pa.)*

VORECON requires the usual external input signal (for example, 0/4 to 20 mA). An electronic logic unit handles the internal control of the VORECON. The logic unit is speed-dependent. Instrumentation for monitoring bearing and oil temperatures as well as oil pressure is provided as standard. It is designed so that it can be incorporated in a plant-monitoring system.

13.5.1 Function of the multistage variable-speed drive

The operation of the hydrodynamic drive can be divided into two ranges. In the first range the power is transmitted by the variable-speed coupling directly through the planetary gear. The speed is controlled by changing the level of oil in the coupling using the adjustable scoop tube. The regulating range is from 0 to approximately 80 percent speed. The torque converter is drained and therefore has no function in the first range. The hydrodynamic brake is filled and generates the counter torque for the planetary gear, which is designed as a superimposing device. A hydrodynamic rather than a mechanical brake was chosen because it allows the planetary gear to rotate continuously at slow speed, preventing brinelling or chatter marking of the gear teeth.

In the second range the impeller and turbine wheel of the hydraulic coupling are locked together by the clutch, bridging the input and output of the variable-speed coupling so that the drive motor is now coupled mechanically to the driven machine. The regulating range from 80 to 100 percent speed is covered using the torque converter. Speed regulation is effected by changing the guide vane position of the torque converter.

An additional drive via the planetary gear fixed is superimposing the speed of the revolving planetary gear. The hydrodynamic brake is drained.

13.5.2 Design and operating details

As depicted in Fig. 13.19, variable-speed turbo couplings can often be combined with one or more gear stages in a common housing. The bottom part of this compact unit forms an oil sump. From the basic concept consisting of a speed-increasing gear followed by a variable-speed turbocoupling, other models have been derived to provide stepless speed control for both high-power, high-speed machines such as boiler feed pumps and compressors and with a speed-reducing gear low-speed machines such as coal mills, ID fans, and crude oil pumps. The Voith turbocoupling is a hydrodynamic fluid coupling. Power developed by the prime mover is converted into kinetic energy in the impeller (primary wheel) of the turbocoupling and converted back into mechanical energy in the turbine wheel (secondary wheel) which is connected to the driven

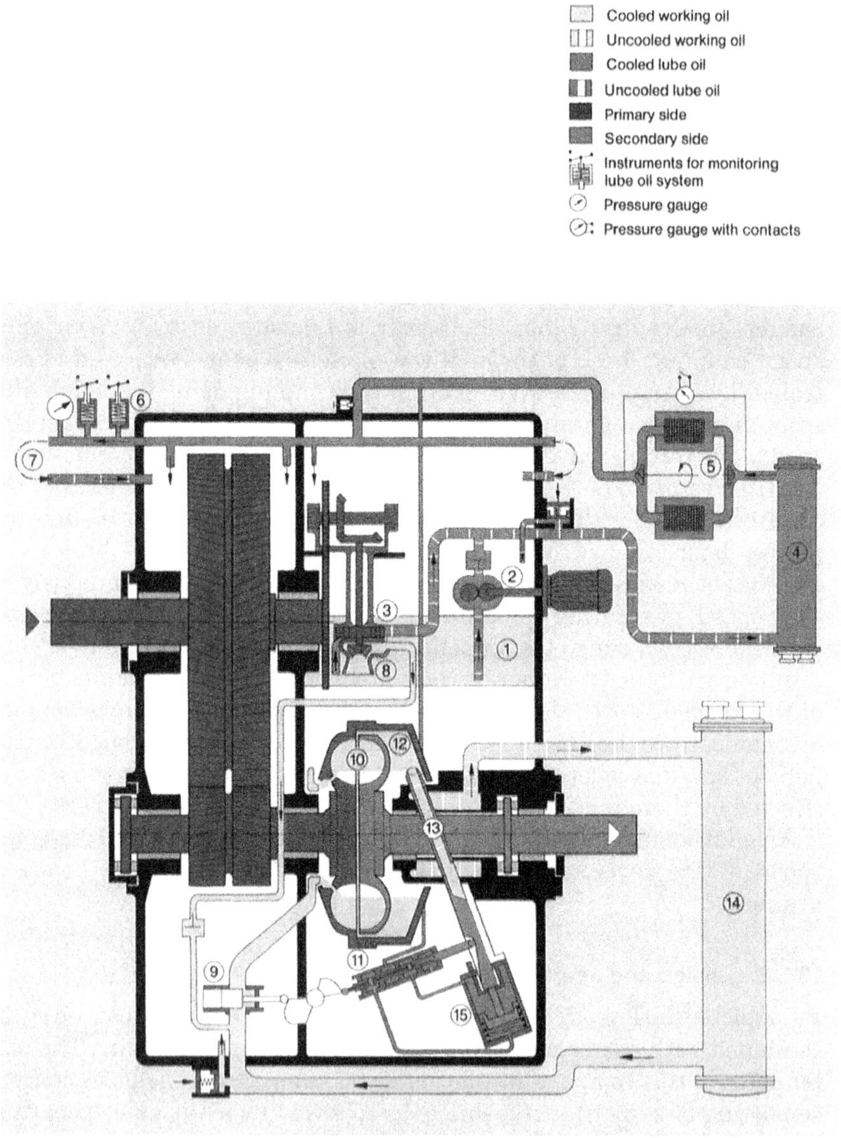

Figure 13.19 Design and oil flow arrangement of variable-speed turbocoupling. *(Voith Transmission, Inc., York, Pa.)*

machine. Because there is no metal-to-metal contact between primary and secondary wheels, there is no wear. Hydraulic oils with additives are used for power transmission. The amount of oil in the coupling can be varied during operation using the scoop tube. This in turn regulates the power-transmitting capability of the coupling and provides stepless speed control dependent on load of the driven machine.

The coupling has a regulating range of 4:1 to 5:1 for driven machines with increasing parabolic torque load characteristics such as centrifugal pumps and fans. For machines with approximately constant torque load characteristics, the regulating range is 3:1. With centrifugal machines, this method of speed regulation is much more efficient than throttling the machine output, giving considerable power savings. The motor is started under no-load conditions with the coupling drained. When running the load on, the motor can be controlled by the coupling.

Moreover by draining the fluid coupling, the prime mover can be disconnected from the driven machine while the prime mover is still running.

A mechanically driven oil pump on the primary side of the coupling pumps oil from the reservoir underneath the coupling through a control valve into the working chamber of the turbocoupling. The level of the oil in the working chamber and therefore the power that the turbocoupling can transmit depends on the radial position of the adjustable sliding scoop tube. The scoop tube can pick up more oil than the pump can deliver. The oil picked up by the scoop tube passes through an oil cooler/heat exchanger and control valve back to the working chamber and/or the oil reservoir. The heat exchanger dissipates the heat originating from the slip of the turbocoupling. The scoop tube actuator can be operated either electrically, hydraulically, or pneumatically.

Geared variable-speed turbocouplings are connected to the driving and driven machines through torsionally flexible couplings (gear couplings lubricated by circulating oil or diaphragm couplings). All rotating parts are housed in an oil-tight casing. A mechanical gear stage on the input side of the variable-speed turbocoupling matches the speed of the high-speed driven machine to that of a standard two- or four-pole squirrel cage motor. The compact space-saving design is a characteristic feature of this gear stage and turbocoupling combination.

The rotating parts of the variable-speed turbocoupling—primary wheel, secondary wheel, and casing—are usually made of cast steel or special steel. All Voith variable-speed turbocouplings have plain bearings. The highly loaded radial and thrust bearings are white metal lined. Remote thermometers monitor the temperature of the bearings. The housing is split along the shaft center-line giving ready access to the bearings, gears, and turbocoupling. Only minor dismantling and reassembly is required for maintenance.

13.5.3 Working oil and lube oil circuits

In Fig. 13.19 the main oil pump (8) which is driven mechanically from the gear input shaft draws the working fluid from the oil reservoir (1) in the lower part of the housing and delivers it through the flow-control valve (9) into the coupling's working chamber (10). This chamber formed by the primary and secondary wheels connects with the space inside the coupling casing surrounding the secondary wheel. Because of centrifugal force, the fluid inside the casing forms a ring (12) around the periphery. The inside diameter of this ring is determined by the radial position of the adjustable scoop tube (13).

The oil picked up by the scoop tube passes through a heat exchanger (14) to the flow-control valve (9) and from there back to the working chamber or the oil reservoir. The flow-control valve whose function depends on the position of the scoop tube regulates the oil flow rate relative to the heat due to slip generated in the coupling. The excess oil flows back to the oil reservoir. As a result the response time to a control signal is very fast. On the signal *increased speed* the full flow of the oil pump passes into the working chamber; on the signal *reduce speed* oil flow to the working chamber is cut off.

13.5.4 Lubricating system

The lubricating system is separate from the working circuit although the same oil is used for both. The main lubricating pump (3) is fitted to the main working oil pump drive shaft. Lube oil is pumped from the oil reservoir through an oil cooler (4) and a double filter (5) to the bearings and the gears. A motor-driven auxiliary pump (2) is switched on before start-up to ensure that all bearings are lubricated before the turbocoupling runs. The auxiliary pump operates until the turbocoupling reaches its rated speed when the mechanically driven main lube oil pump (3) takes over. The auxiliary pump is stopped automatically on a signal from the instruments monitoring the lube oil circuit (6). If the pressure in the lube oil circuit drops, e.g., if the drive unit is shut down, the auxiliary pump is switched on again.

A double filter is fitted in the lubricating system to allow for continuous operation. Only one of the two filters is in use at any time, and if this becomes clogged a differential pressure gauge with alarm contacts provides a visual or audible alarm to indicate the need to change filters. The dirty filter must then be shut down and cleaned.

The lubricating system of the geared variable-speed coupling can be used if required for lubricating the driving and/or driven machine (7).

13.5.5 Lubricant oil containment on gear and variable-speed units

Measures have to be taken to limit contaminant ingress into lubricating oil. Whenever a gap exists between the rotating shaft and the surrounding stationary housing components, these housings will "breathe," as illustrated in Fig. 13.20. Gearbox manufacturers have used lip seals for bearing protection and oil containment for many decades. However, lip seals are subject to wear and must be replaced on a time-based preventive maintenance schedule. If lip seals are not replaced in this manner, they will allow contaminants to enter because of pressure differences between the gearbox interior space and the ambient atmosphere. (See Fig. 13.20.) Thus, there is serious risk of water vapor entry and condensation, with the likely outcome shown in Fig. 13.21. A wear groove in the shaft is also clearly evident in this illustration.

Recognizing the limitations of lip seals for longer-term contamination control prompted the American Petroleum Institute (API) and many equipment users to seek and recommend superior preventive measures. Alternative sealing devices include both rotating *noncontacting* labyrinth bearing housing seals and rotating *contacting* dual-face seals. All styles are collectively called bearing isolators or bearing protector seals. Please to refer to the discussion of bearing protector seals in Chap. 7.

Various factory tests and a thorough review of field experience have established the viability and effectiveness of cartridge-type magnetic

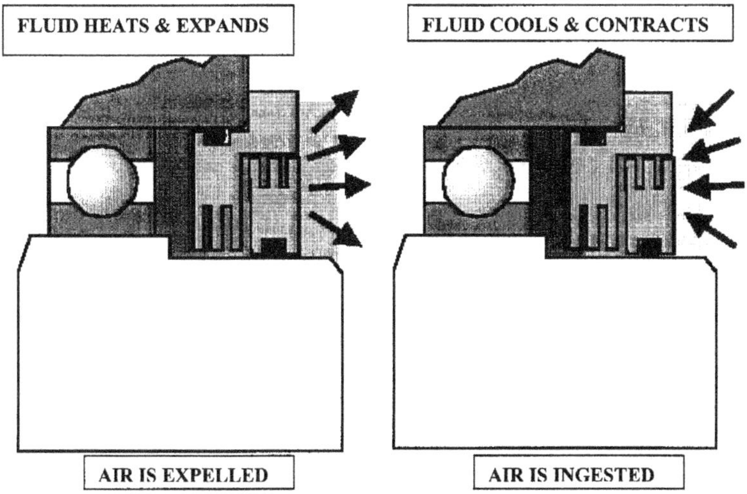

Figure 13.20 Expulsion and induction of surrounding air in an elementary labyrinth seal.

Figure 13.21 Failure of bearing due to water intrusion in lip seal–equipped shaft system.

dual-face bearing protectors on gear units. The bearings or component housings of gear speed reducers and increasers are among the hundreds of shaft-driven and bearing-supported types of equipment that can be effectively sealed with dual- (or double-) face magnetic seals. In particular, many gear speed reducers have benefited substantially from the virtual elimination of lubricant oil contamination. Cost justifications for the projected gear life extension are readily available.

In particular, replacing lip seals with dual-face magnetic seals often results in extending oil replacement intervals from one year to as many as four years. Based on feedback involving cooling towers in petrochemical plant locations where moisture intrusion has been eliminated, cooling tower gearbox failures should no longer occur. The benefits often exceed conversion and upgrading costs by an order of magnitude, and payback periods ranging from a few weeks up to three months are not unusual.

Chapter

14

Shortcut Graphical Methods of Turbine Selection

14.1 Mollier Chart Instructions

As will be seen, the water rate of a steam turbine can be obtained from a Mollier Chart, Fig. 14.1 and 14.2, if steam conditions and turbine efficiency are known. On the simplified Mollier Diagram, Fig. 14.3, we observe the following:

(A) Enthalpy A thermodynamic property that serves as a measure of heat energy in a system above some datum temperature (for water it is 32°F or 0°C). In this case it represents the energy in one pound of steam (Btu/lb or kJ/kg).

(B) Entropy (BTU per lb)/ °R) or (kJ per kg/K) An indication of the unavailable energy in a system.

(C) Constant pressure line (psia or bar, absolute) Represents the absolute steam pressure at varying enthalpy values.

(D) Constant temperature line Represents the total steam temperature at varying steam pressures.

(E) Constant moisture line Represents the percent moisture content in the steam.

(F) Vapor dome This line represents the dry and saturated state of steam. The area above the vapor dome represents superheated steam, and the area below the vapor dome represents wet steam.

To determine the power output of a steam turbine, the entrance and exhaust conditions must be determined. This process is explained in Fig. 14.4.

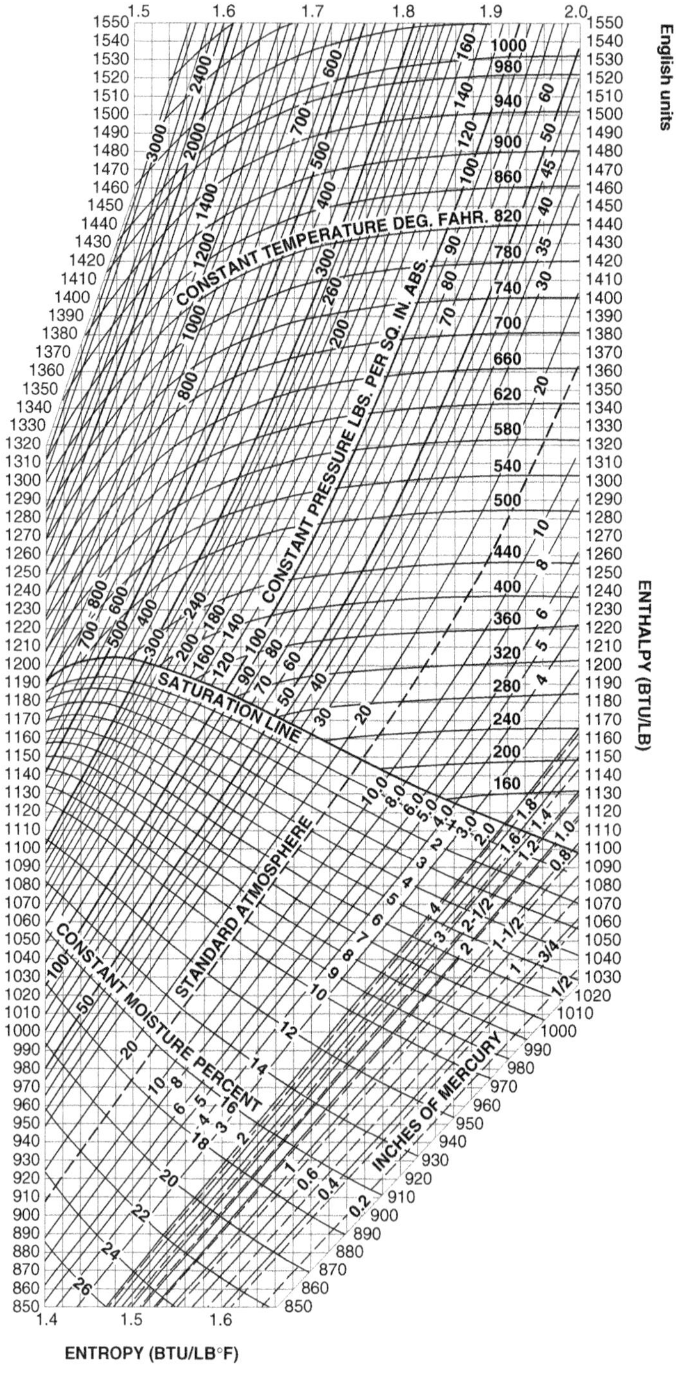

Figure 14.1 Mollier Chart for Steam, English Units *(Adapted from Fig. 22, page 311, of 1967 ASME Steam Tables. Copyright 1967 by The American Society of Mechanical Engineers, New York, N.Y.)*

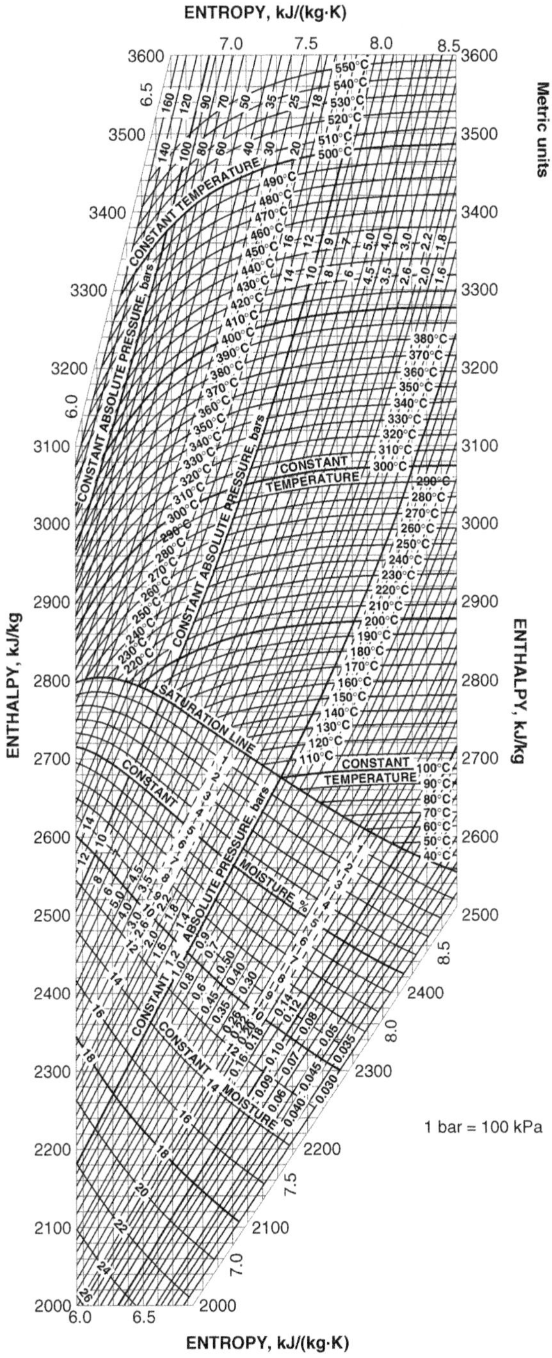

Figure 14.2 Mollier Chart for Steam, Metric Units (Steam Tables, International Edition, *by J. H. Keenan, F. G. Keyes, P. G. Hill and J. G. Moore. Copyright 1969 by John Wiley and Sons, Inc.*)

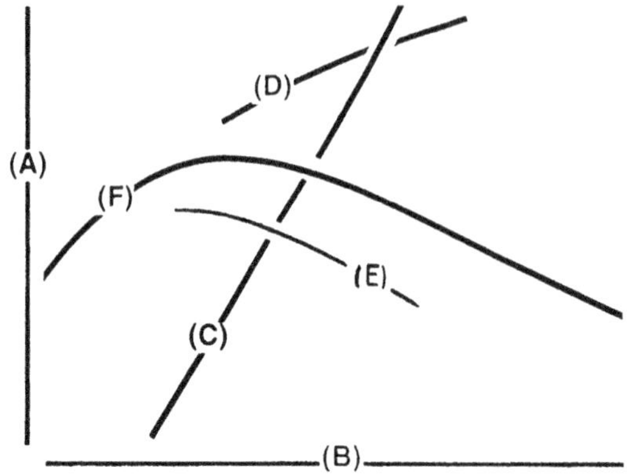

Figure 14.3 Simplified Mollier diagram indicating (A) enthalpy, (B) entropy, (C) constant pressure line, (D) constant temperature line, (E) constant moisture line, (F) vapor dome. *(Trane Murray Company, LaCrosse, Wis.)*

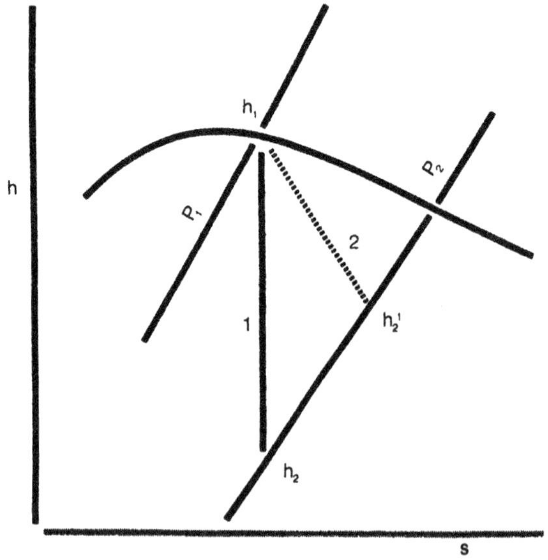

Figure 14.4 Mollier diagram showing determination of theoretical steam rate (TSR).

$$\text{TSR} = \frac{2547}{h_1 - h_2} \text{ lb/hp·h}$$

(Trane Murray Company, LaCrosse, Wis.)

Using an actual Mollier diagram (Fig. 14.1 or 14.2), find (h_1) at the intersection of inlet pressure and temperature lines. Extend a line vertically down until it crosses the pressure line associated with the exhaust pressure. This is indicated by line 1. Line $h_1 - h_2$ corresponds to the theoretical enthalpy drop. The actual exhaust enthalpy will be at a higher value because of the turbine efficiency. Therefore, the actual process will follow line 2. From the equations on this chart, a theoretical steam rate (TSR) can be determined. The actual steam rate (ASR) can be determined by knowing the turbine efficiency. Efficiencies for single-stage turbines will range from 30 to 60 percent, and multistage turbines will range from 50 to 80 percent depending on operating conditions.

14.2 Estimating Steam Rates

Estimating steam rates is best illustrated by using the example of a plant that has a surplus of 50,000 lb/h of 300 psig steam and requires 15 psig steam elsewhere in its process operations. The pressure reduction could be accomplished across a turbine and it is assumed that this turbine would have an efficiency of 65 percent. It would drive an electric generator through a speed-reducing gear; their respective efficiencies are assumed to be 94 and 98 percent. Here would be the result:

$$h_1 = 1203.5 \text{ Btu/lb}$$

$$h_2 = 1026.4 \text{ Btu/lb}$$

$$\text{TSR} = \frac{2547^*}{1203.3 - 1026.4} = 14.38 \text{ lb/hp·h}$$

$$\text{ASR} = \frac{14.38}{0.65} = 22.12 \text{ lb/hp·h}$$

$$\text{kW produced} =$$

$$\frac{50{,}000 \text{ lb/h} \times 0.98 \text{ gear eff.} \times 0.94 \text{ gen. eff.}}{22.12 \text{ lb/hp·h} \times 1.341 \text{ hp/kW}}$$

$$= 1550 \text{ kW}$$

Steam turbine selection is also possible by making extensive use of vendor-supplied data.

For *simple single-stage machines,* Tables 14.1 and 14.2 give performance and physical data (dimensions, etc.) that closely approximate

* 1 hp·h = 2547 Btu.

TABLE 14.1 General Specifications for Simple Single-Stage Steam Turbines

Frame	AYR	BYR (& BYRIH)	CYR	DYR (& DYRM)	BYRH (& BYRHH)
Maximum initial gauge pressure (psi/bar)	700/48	700/48	700/48	700/48	700/48
Maximum initial temperature (°F/°C)	750/399	750/399	750/399	750/399	750/399
Maximum exhaust pressures (gauge, psi/bar)	vac-100/6.9	vac-100/6.9*	vac-90/6.2	vac-100/6.9	250/17†
Speed range (r/min)	1000–7064	1000–6675	1000–6950	1000–5770	1000–7090
Wheel pitch diameter (in/mm)	14/360	18/460	22/560	28/710	18/460
Number of stages (impulse type)	1	1	1	1	1
Number of rows of rotating blades	2	2	2	2	2
Inlet sizes (ANSI, in)	3	2, 3, 4	2, 3, 4, 6	2, 3, 4, 6	2, 3, 4, 6
Inlet location (facing governor)	right	right	right	right	right
Exhaust size (ANSI, in)	6	8	10	12‡	8
Exhaust location (L.H. Standard)	R.H. optional	R.H. optional	R.H. optional	R.H. optional§	R.H. optional
Approximate range of capacities (hp/kW)	750/560	to 1400/1050	to 2500/1850	to 3500/2600	to 3000/2250
Approximate shipping weight (lb/kg)	870/400	1275/580	2050/930	2600/1180	2300/1050

* BYRIH: 160 psi/11 bar.
† BYRHH: 375 psi/25 bar.
‡ DYRM: 14 in, max. exhaust pressure 75 psig.
§ DYRM: Optional R.H. not available.
SOURCE: Elliott Company, Jeannette, Pa.

the offers of several major equipment manufacturers. Used in conjunction with Figures 14.5 through 14.19, steam rates can be easily calculated within ±5 percent accuracy. As shown, a simple nine-step procedure will give the desired answers.

Step 1: Determine *theoretical steam rate* (TSR) using a Mollier diagram (see Figs. 14.1 through 14.4), Tables 14.3 and 14.4, or using the Keenan, Keyes, Hill & Moore Steam Tables (International Edition-Metric Units) or other steam tables or Mollier charts in accord with the International Skeleton Tables of 1963 of the International Conference on the Properties of Steam. Note TSR = $2547/h_2 - h_1$ lb/hp·hr, where h_1 is the enthalpy in Btu/lb of steam at the turbine inlet, and h_2 is the enthalpy in Btu/lb of steam at the exhaust pressure and inlet entropy.

TABLE 14.2 **Approximate Dimensions (in/mm) of Simple Single-Stage Steam Turbines**

	A	AA	FA	DA	EA	B*	AB	BB	CB	FB*	C	AC	BC	CC	D
AYR	40.56	12.00	15.74	12.82	15.18	31.12	13.74	13.74	12.62	18.50	24.00	12.00	5.00	6.50	1.9325
	1030	305	400	325	386	791	349	349	321	470	610	305	127	165	49.09
BYR, BYRIH	45.88	12.26	21.18	12.44	17.30	36.12	16.00	16.00	13.62	22.50	28.00	14.00	7.74	7.00	1.9325
	1165	311	538	316	440	918	406	406	346	572	711	356	197	178	49.09
CYR	51.24	12.26	21.18	17.82	16.06	40.12	19.24	19.24	16.62	23.50	34.74	17.00	8.50	8.50	2.4325
	1301	311	538	452	408	1019	489	489		597	883	432	216	216	61.78
DYR	50.88	12.26	21.68	16.94	16.56	43.12	22.24	22.24	19.62	23.50	42.50	22.00	13.50	12.00	2.9325
	1292	311	551	430	421	1096	565	565	499	597	1080	559	343	305	74.48
DYRM	50.88	12.26	22.68	15.94	17.56	43.12	23.00	†	19.62	23.50	42.50	22.00	13.50	11.24	2.9325
	1292	311	576	405	446	1096	584	†	499	597	1080	559	343	286	74.48
BYRH, BYRHH	55.38	12.26	27.00	16.12	21.88	39.88	18.74	18.74	16.38	23.50	31.12	17.00	8.50	9.00	2.4325
	1407	311	686	410	556	1013	476	476	416	597	791	432	216	229	61.78

* AYR, BYR, BYRIH 3-in inlet. 4-in inlet for all other frames.

† Optional right hand exhaust is not available. Left hand exhaust (AB dimension) is standard.

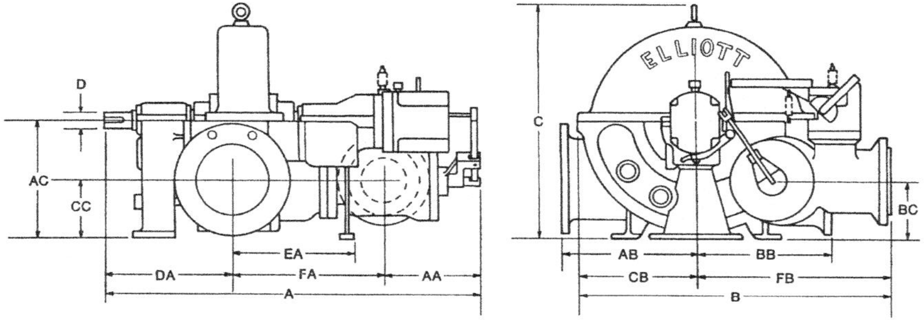

SOURCE: Elliott Company, Jeannette, Pa.

TABLE 14.3 Condensed Table of Theoretical Steam Rates (lb/kWh)

Exhaust pressure	125 #G 500* FTT 147* FS	150 #G 500* FTT 134* FS	200 #G 388* FTT 0* FS	200 #G 500* FTT 112* FS	250 #G 406* FTT 0* FS	250 #G 500* FTT 94* FS	400 #G 448* FTT 0* FS	400 #G 750* FTT 302* FS	600 #G 575* FTT 85* FS	600 #G 750* FTT 261* FS	600 #G 825* FTT 336* FS	850 #G 750* FTT 223* FS	850 #G 825* FTT 298* FS	850 #G 900* FTT 373* FS	1250 #G 825* FTT 251* FS	1250 #G 900* FTT 326* FS	1500 #G 825* FTT 228* FS	1500 #G 900* FTT 302* FS	1800 #G 1000* FTT 378* FS
3 in HgA	10.57	10.28	10.62	9.87	10.22	9.59	9.48	7.70	8.34	7.40	7.05	7.20	6.85	6.53	6.69	6.36	6.64	6.30	5.86
4 in HgA	11.10	10.78	11.12	10.31	10.66	10.00	9.85	7.99	8.64	7.65	7.28	7.42	7.06	6.73	6.88	6.54	6.83	6.48	6.01
5 in HgA	11.57	11.21	11.54	10.70	11.05	10.35	10.17	8.22	8.89	7.86	7.48	7.61	7.24	6.89	7.04	6.69	6.98	6.62	6.14
0 #G	18.55	17.50	17.51	16.08	16.30	15.16	14.26	11.19	11.96	10.40	9.82	9.87	9.32	8.81	8.90	8.40	8.76	8.25	7.54
10 #G	23.18	21.48	21.10	19.28	19.31	17.90	16.43	12.71	13.49	11.64	10.96	10.94	10.30	9.71	9.75	9.18	9.56	8.99	8.16
20 #G	27.96	25.47	24.56	22.34	22.12	20.44	18.34	14.02	14.79	12.68	11.90	11.82	11.10	10.43	10.43	9.80	10.20	9.56	8.64
30 #G	33.3	29.73	28.11	25.46	24.91	22.95	20.15	15.23	15.98	13.62	12.76	12.60	11.80	11.07	11.03	10.34	10.75	10.07	9.06
50 #G	47.2	39.9	35.99	32.3	30.8	28.21	23.67	17.56	18.19	15.36	14.31	14.01	13.07	12.21	12.07	11.28	11.72	10.94	9.77
60 #G	57.2	46.4	40.55	36.2	34.0	31.07	25.45	18.74	19.26	16.18	15.06	14.66	13.66	12.74	12.55	11.71	12.16	11.34	10.08
75 #G	79.1	59.1	48.5	42.9	39.4	35.8	28.21	20.57	20.85	17.40	16.17	15.61	14.51	13.51	13.22	12.32	12.77	11.85	10.53
100 #G	—	96.6	66.6	58.1	50.1	45.21	33.14	23.86	23.53	19.43	18.07	17.14	15.88	14.77	14.28	13.26	13.74	12.75	11.21
150 #G	—	—	—	—	—	—	45.21	31.93	29.24	23.83	22.15	20.15	18.62	17.33	16.28	15.05	15.52	14.33	12.44
250 #G	—	—	—	—	—	—	—	—	43.97	35.40	32.91	26.84	24.80	23.08	20.22	18.67	18.89	17.37	14.78
400 #G	—	—	—	—	—	—	—	—	—	—	—	—	—	—	27.05	24.99	24.30	22.32	18.39

NOTE: Interpolate where necessary for intermediate values.
SOURCE: "Theoretical Steam Rates Tables" by J. H. Keenan and F. G. Keyes, published in 1938 by ASME.

TABLE 14.4 Condensed Table of Theoretical Steam Rates (kg/kWh)

Inlet pressure, gauge		5 bar		10 bar		20 bar		30 bar			40 bar			62 bar		100 bar		135 bar	
Inlet temp., °C		200	250	200	250	250	350	250	350	400	350	400	450	400	480	400	480	400	510
Exhaust pressure																			
Abs. mbar	Gauge bar																		
50		4.96	4.69	4.51	4.27	3.95	3.52	3.82	3.39	3.20	3.31	3.12	2.96	3.03	2.77	2.99	2.70	3.02	2.59
75		5.31	5.01	4.78	4.52	4.15	3.70	4.00	3.54	3.34	3.45	3.26	3.08	3.15	2.87	3.10	2.80	3.12	2.67
100		5.60	5.28	5.01	4.73	4.31	3.83	4.15	3.66	3.45	3.56	3.36	3.17	3.24	2.95	3.19	2.87	3.21	2.74
150		6.08	5.72	5.37	5.07	4.57	4.05	4.38	3.85	3.63	3.74	3.52	3.32	3.39	3.08	3.32	2.98	3.33	2.84
200		6.49	6.10	5.67	5.34	4.78	4.23	4.56	4.01	3.77	3.88	3.65	3.44	3.50	3.18	3.42	3.07	3.43	2.92
300		7.21	6.75	6.17	5.81	5.13	4.52	4.86	4.26	4.00	4.11	3.86	3.63	3.69	3.33	3.58	3.21	3.59	3.04
1013.25	0	11.27	10.40	8.72	8.14	6.73	5.83	6.19	5.36	4.99	5.09	4.74	4.42	4.44	3.98	4.25	3.77	4.22	3.53
	2	26.72	23.88	15.05	13.83	9.85	8.30	8.56	7.26	6.68	6.71	6.19	5.71	5.62	4.96	5.23	4.59	5.13	4.23
	3	44.22	39.49	18.97	17.26	11.34	9.45	9.60	8.09	7.40	7.38	6.78	6.23	6.08	5.33	5.60	4.89	5.46	4.48
	4	—	—	24.00	21.61	12.92	10.66	10.64	8.89	8.11	8.02	7.34	6.73	6.50	5.68	5.93	5.16	5.76	4.70
	6	—	—	—	—	16.48	13.17	12.79	10.53	9.58	9.28	8.43	7.72	7.29	6.32	6.53	5.65	6.30	5.09
	15	—	—	—	—	—	—	27.39	21.63	19.69	16.02	14.48	13.27	10.78	9.28	8.93	7.54	8.33	6.54
	20	—	—	—	—	—	—	45.68	35.63	32.45	21.86	19.78	18.12	13.06	11.25	10.25	8.60	9.39	7.29
	25	—	—	—	—	—	—	—	—	—	31.38	28.38	26.00	15.85	13.65	11.63	9.73	10.45	8.05
	40	—	—	—	—	—	—	—	—	—	—	—	—	31.09	26.76	16.70	13.95	13.94	10.58

NOTE: Interpolate, where necessary, for approximate values.
SOURCE: Computed from "Steam Tables, International Edition" by Keenan, Keyes, Hill and Moore, copyright 1969 by John Wiley & Sons, Inc.

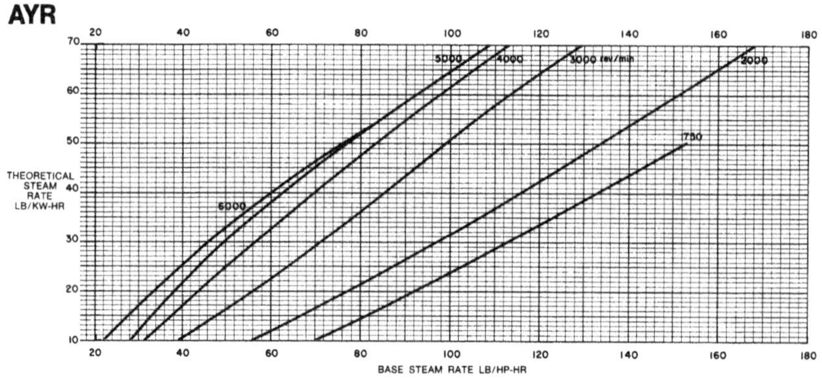

(a)

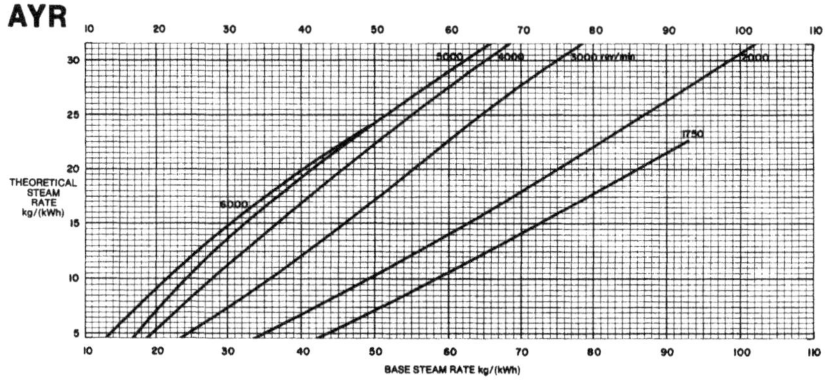

(b)

Figure 14.5 Base steam rates of single-valve, single-stage steam turbines, 6 in/150 mm exhaust, 750 hp/560 kW. *(Elliott Company, Jeannette, Pa.)*

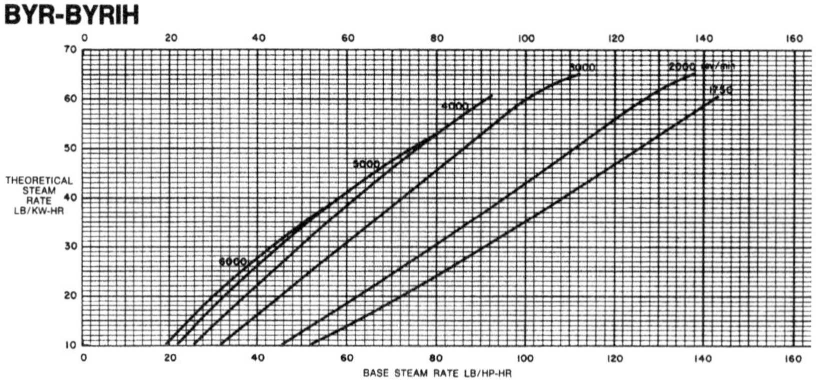

(a)

Figure 14.6 Base steam rates of single-valve, single-stage steam turbines, 8 in/200 mm exhaust, 1400 hp/1050 kW. *(Elliott Company, Jeannette, Pa.)*

BYR-BYRIH

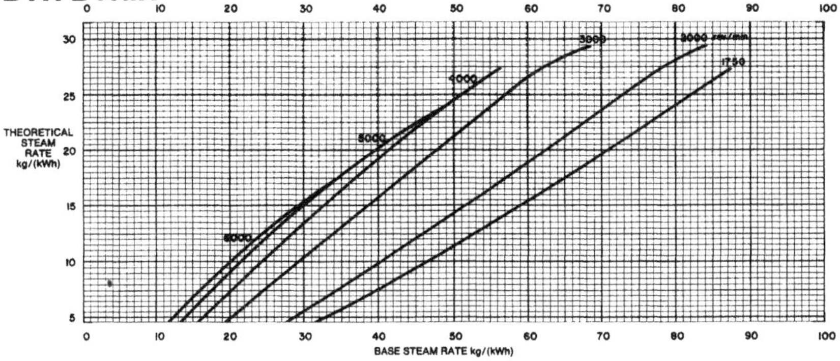

(b)

Figure 14.6 *(Continued)*

CYR

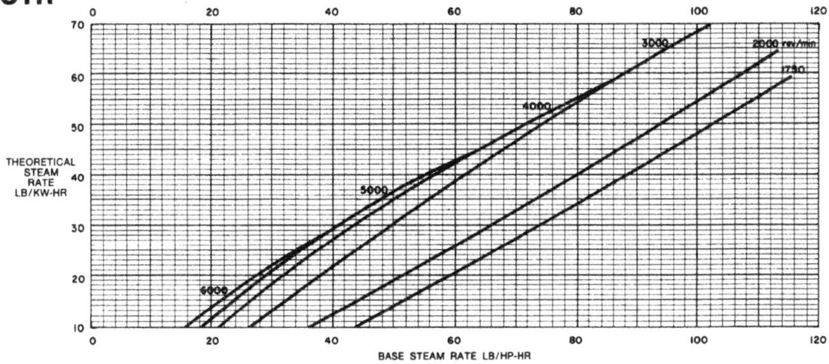

(a)

CYR

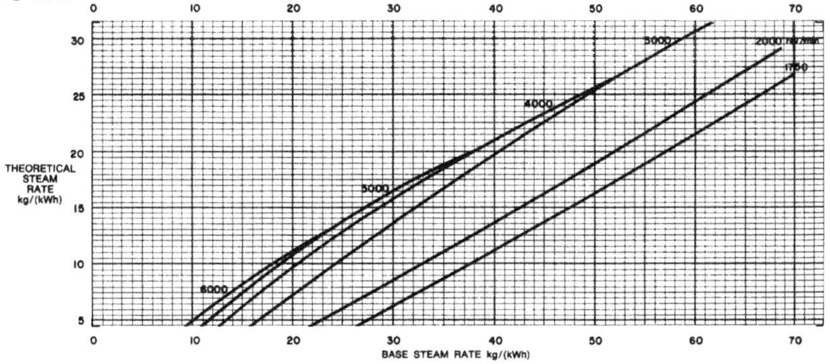

(b)

Figure 14.7 Base steam rates of single-valve, single-stage steam turbines, 10 in/250 mm exhaust, 2500 hp/1850 kW. *(Elliott Company, Jeannette, Pa.)*

DYR-DYRM

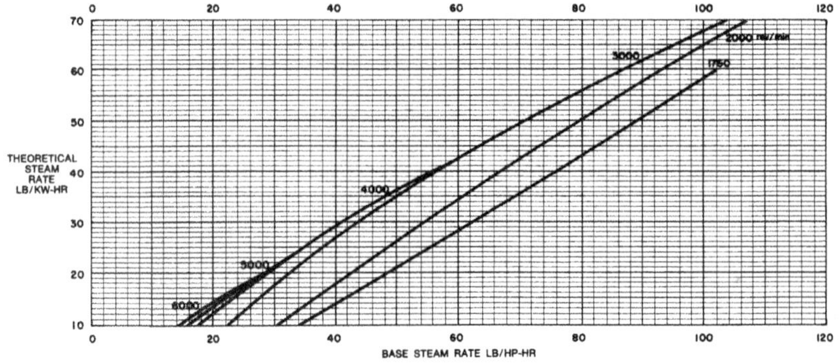

(a)

DYR-DYRM

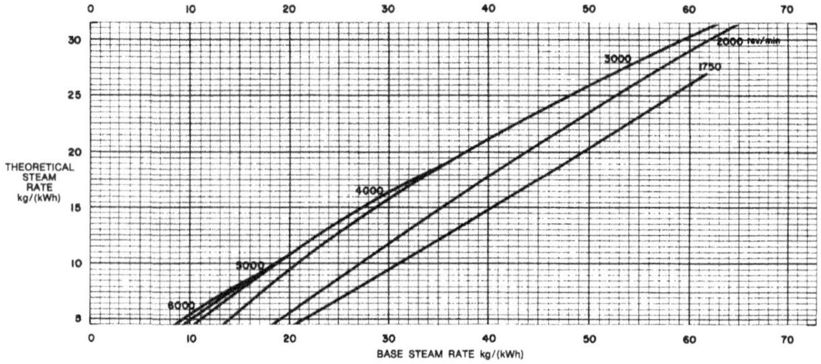

(b)

Figure 14.8 Base steam rates of single-valve, single-stage steam turbines, 12 in/300 mm exhaust, 3500 hp/2600 kW. *(Elliott Company, Jeannette, Pa.)*

BYRH-BYRHH

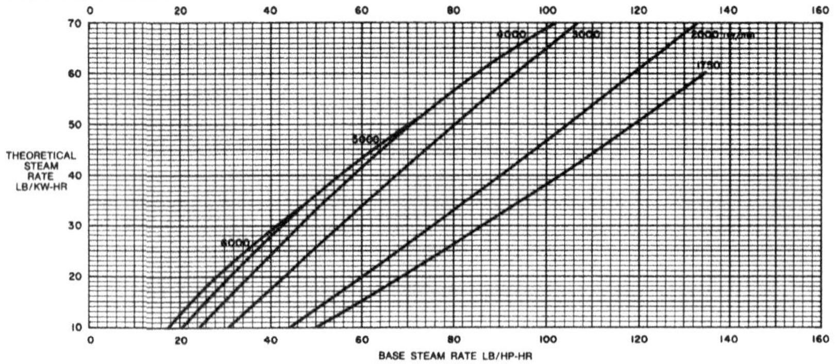

(a)

Figure 14.9 Base steam rates of single-valve, single-stage, high back-pressure steam turbines, 8 in/200 mm exhaust, 3000 hp/2250 kW. *(Elliott Company, Jeannette, Pa.)*

BYRH-BYRHH

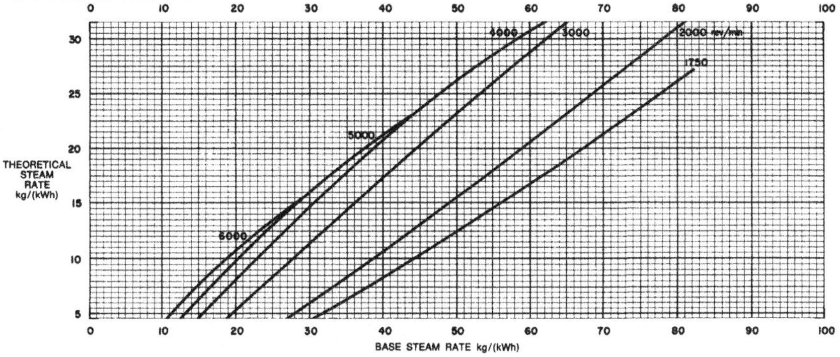

(b)

Figure 14.9 *(Continued)*

AYR

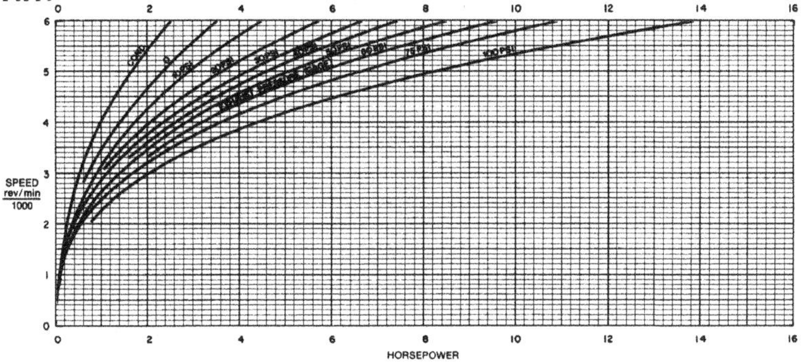

(a)

AYR

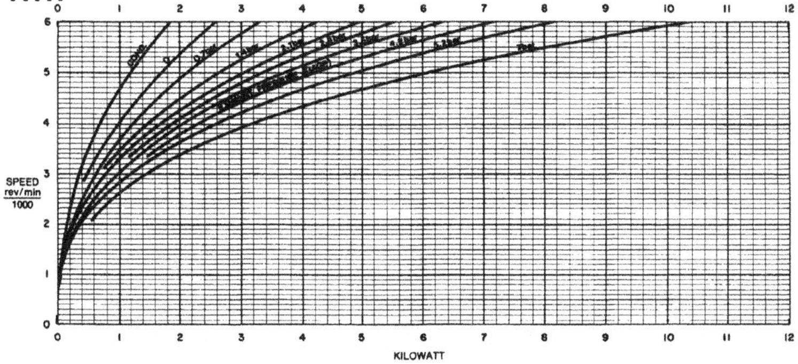

(b)

Figure 14.10 Mechanical losses of single-valve, single-stage steam turbines, 6 in/150 mm exhaust, 750 hp/560 kW. *(Elliott Company, Jeannette, Pa.)*

BYR-BYRIH

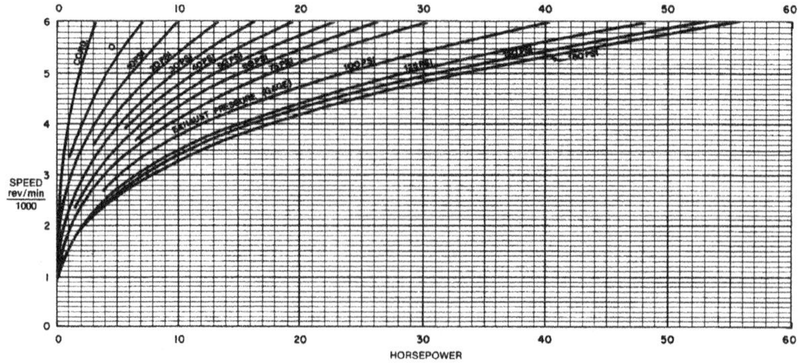

(a)

BYR-BYRIH

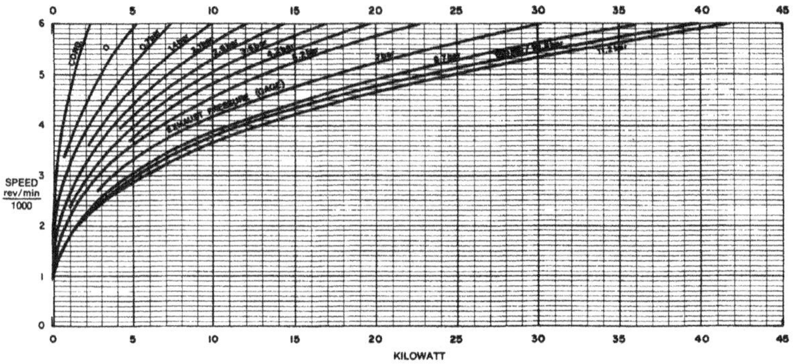

(b)

Figure 14.11 Mechanical losses of single-valve, single-stage steam turbines, 8 in/200 mm exhaust, 1400 hp/1050 kW. *(Elliott Company, Jeannette, Pa.)*

CYR

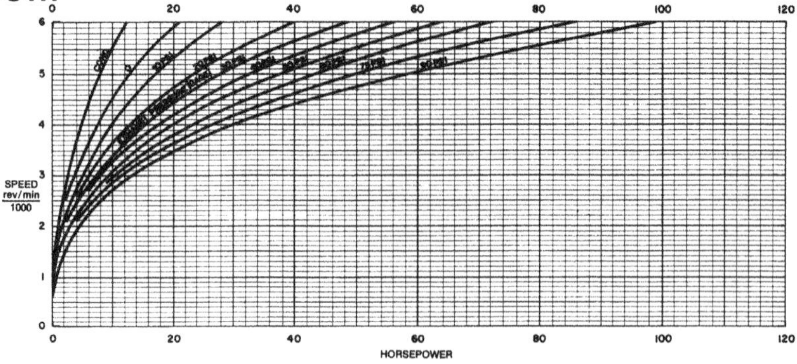

(a)

Figure 14.12 Mechanical losses of single-valve, single-stage steam turbines, 10 in/250 mm exhaust, 2500 hp/1850 kW. *(Elliott Company, Jeannette, Pa.)*

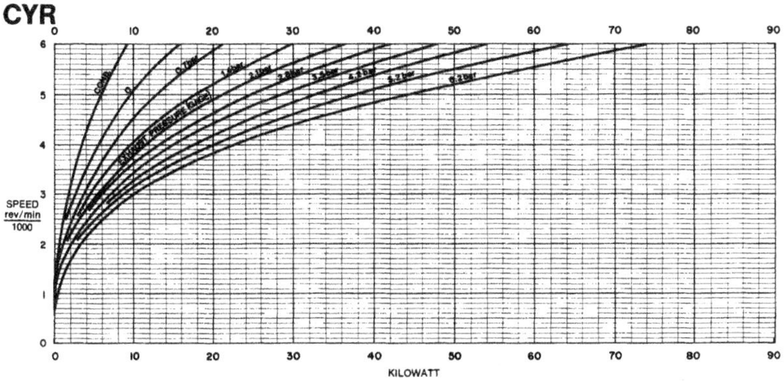

(b)

Figure 14.12 *(Continued)*

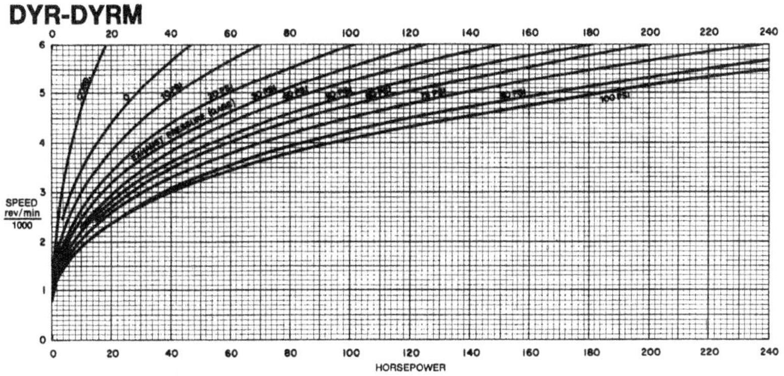

(a)

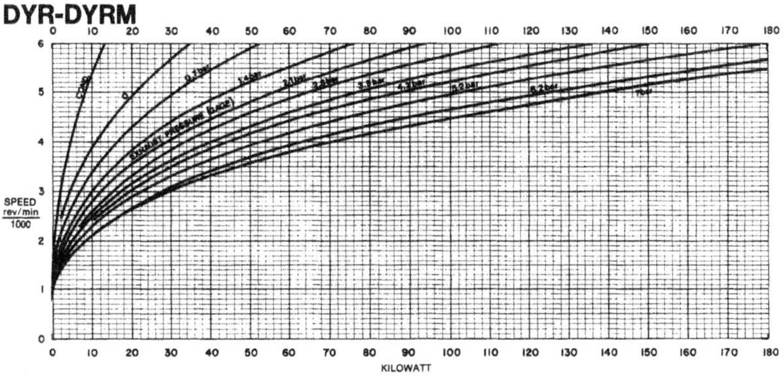

(b)

Figure 14.13 Mechanical losses of single-valve, single-stage steam turbines, 12 in/300 mm exhaust, 3500 hp/2600 kW. *(Elliott Company, Jeannette, Pa.)*

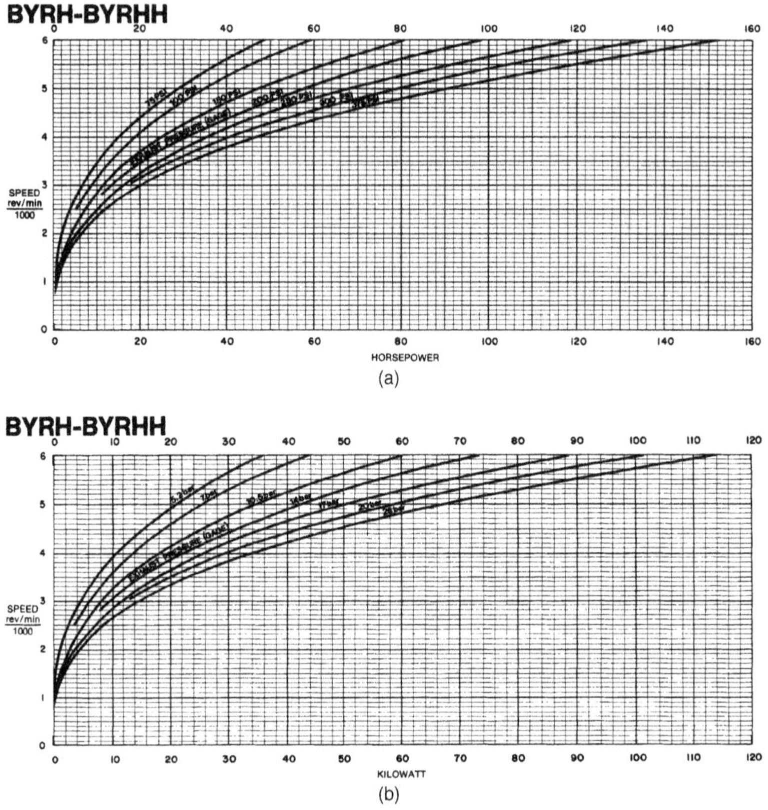

Figure 14.14 Mechanical losses of single-valve, single-stage, high back-pressure turbines, 8 in/200 mm exhaust, 3000 hp/2250 kW. *(Elliott Company, Jeannette, Pa.)*

Using metric dimensions,

$$\text{TSR} = \frac{3600}{h_1 - h_2} \text{ kg/(kWh)}$$

where h_1 is the enthalpy in kJ/kg of steam at the turbine inlet, and h_2 is the enthalpy in kJ/kg of steam at the exhaust pressure and at the inlet entropy.

Step 2: Determine *base steam rate* (BSR) using Figs. 14.5 through 14.9, based on TSR, turbine speed, and turbine frame.

Step 3: Determine the *mechanical loss* using Figs. 14.10 through 14.14, based on turbine speed, exhaust pressure, and turbine frame.

Step 4: Determine *superheat* (use Table 14.5 or 14.6).

Step 5: Determine the *superheat correction factor* (SCF) using Fig. 14.15, knowing superheat and TSR.

TABLE 14.5 Temperature of Dry and Saturated Steam (English)

To obtain superheat in degrees F, subtract temperature given in tabulation from total initial temperature.

psig	Saturation temp., °F	psig	Saturation temp., °F	psig	Saturation temp., °F	psig	Saturation temp., °F
0	212	150	366	300	422	450	460
5	228	155	368	305	423	455	461
10	240	160	371	310	425	460	462
15	250	165	373	315	426	465	463
20	259	170	375	320	428	470	464
25	267	175	378	325	429	475	465
30	274	180	380	330	431	480	466
35	281	185	382	335	432	485	467
40	287	190	384	340	433	490	468
45	293	195	386	345	434	495	469
50	298	200	388	350	436	500	470
55	303	205	390	355	437	510	472
60	308	210	392	360	438	520	474
65	312	215	394	365	440	530	476
70	316	220	396	370	441	540	478
75	320	225	397	375	442	550	480
80	324	230	399	380	444	560	482
85	328	235	401	385	445	570	483
90	331	240	403	390	446	580	485
95	335	245	404	395	447	590	487
100	338	250	406	400	448	600	489
105	341	255	408	405	449	610	491
110	344	260	410	410	451	620	492
115	347	265	411	415	452	630	494
120	350	270	413	420	453	640	496
125	353	275	414	425	454	650	497
130	356	280	416	430	455	660	499
135	358	285	417	435	456	670	501
140	361	290	419	440	457	680	502
145	364	295	420	445	458	690	504

TABLE 14.6 Temperature of Dry and Saturated Steam (Metric)

To obtain superheat in degrees Celsius, subtract temperature given in tabulation from total initial temperature.

p_g, bar	t_{sat}, °C	p_g, bar	t_{sat}, °C	p_g, bar	t_{sat}, °C
0	100	15.5	203	37	247
0.5	112	16.0	204	38	249
1.0	120	16.5	206	39	250
1.5	128	17.0	207	40	252
2.0	134	17.5	208	41	253
2.5	139	18.0	210	42	255
3.0	144	18.5	211	43	256
3.5	148	19.0	212	44	258
4.0	152	19.5	214	45	259
4.5	156	20.0	215	46	260
5.0	159	20.5	216	47	261
5.5	162	21.0	217	48	263
6.0	165	21.5	218	49	264

TABLE 14.6 Temperature of Dry and Saturated Steam (Metric) (*Continued*)

To obtain superheat in degrees Celsius, subtract temperature given in tabulation from total initial temperature.

p_g, bar	t_{sat}, °C	p_g, bar	t_{sat}, °C	p_g, bar	t_{sat}, °C
6.5	168	22.0	220	50	265
7.0	171	22.5	221	51	266
7.5	173	23.0	222	52	268
8.0	176	23.5	223	53	269
8.5	178	24.0	224	54	270
9.0	180	24.5	225	55	271
9.5	182	25	226	56	272
10.0	184	26	228	57	273
10.5	186	27	230	58	275
11.0	188	28	232	59	276
11.5	190	29	234	60	277
12.0	192	30	236	61	278
12.5	193	31	238	62	279
13.0	195	32	239	63	280
13.5	197	33	241	64	281
14.0	198	34	243	65	282
14.5	200	35	244		
15.0	201	36	246		

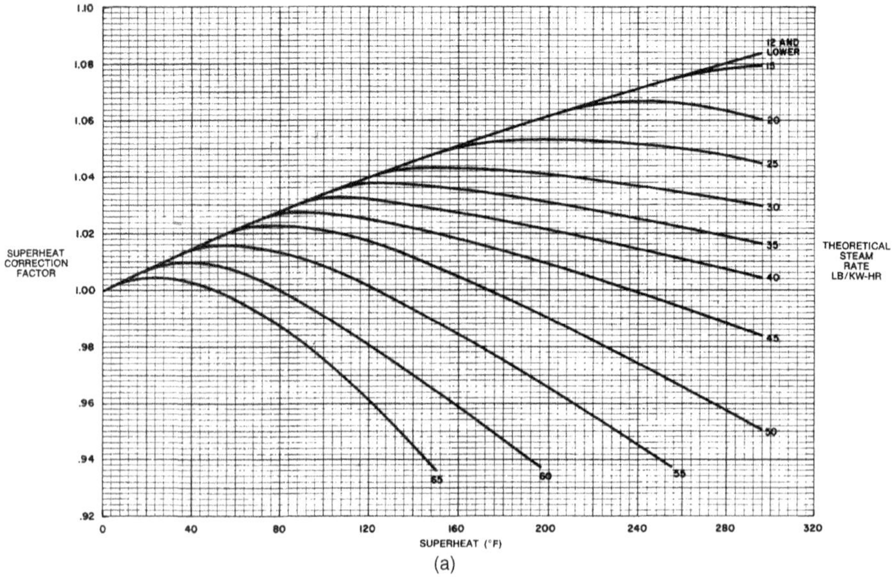

(a)

Figure 14.15 Superheat correction factors for single-valve, single-stage steam turbines. *(Elliott Company, Jeannette, Pa.)*

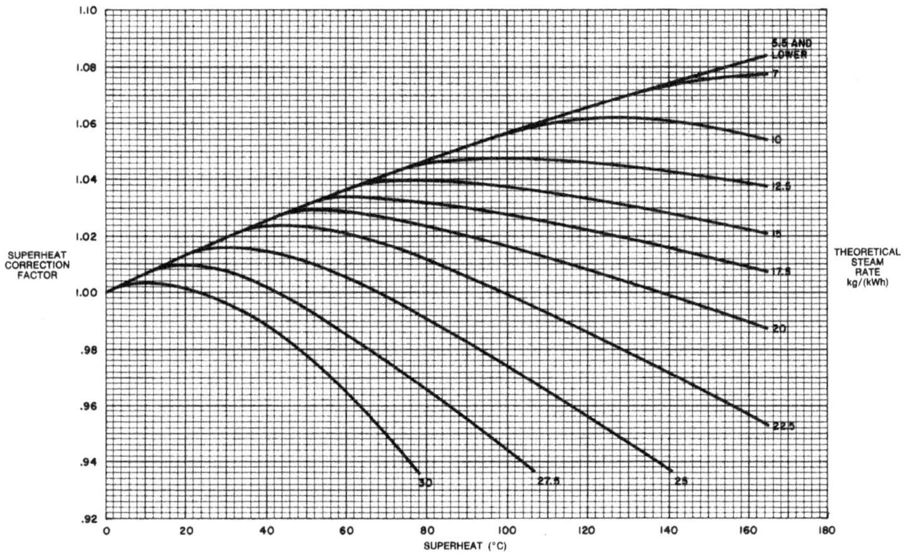

(b)

Figure 14.15 *(Continued)*

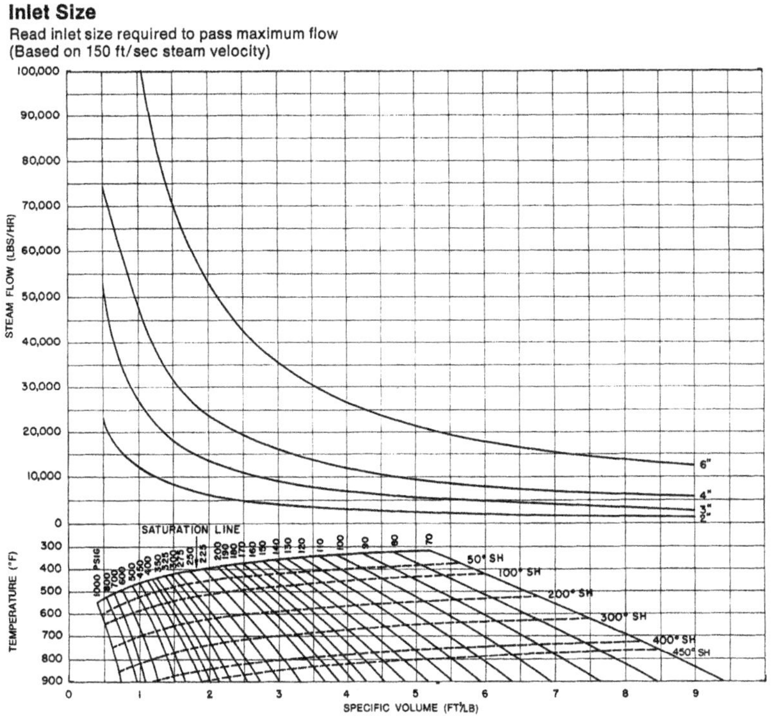

Figure 14.16 Inlet size requirements for single-valve, single-stage steam turbines, English units. *(Elliott Company, Jeannette, Pa.)*

Inlet Size

Read inlet size required to pass maximum flow
(Based on 46 m/s steam velocity)

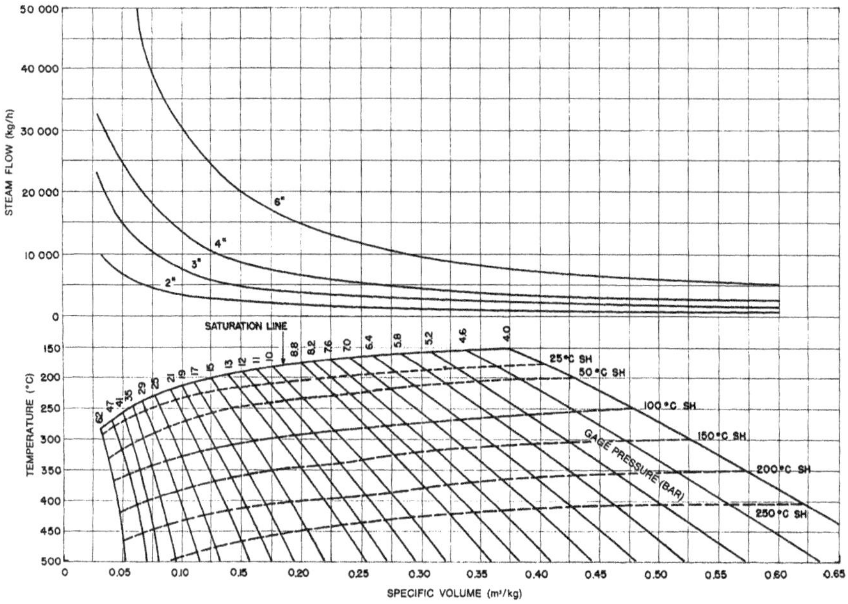

Figure 14.17 Inlet size requirements for single-valve, single-stage steam turbines, metric units. *(Elliott Company, Jeannette, Pa.)*

Step 6: Determine *corrected steam rate* by formula:

$$\text{Corrected steam rate} = \frac{\text{base steam rate}}{\text{superheat correction factor}} \times \frac{(\text{hp} + \text{hp loss})}{\text{hp}}$$

or

$$\frac{\text{Base steam rate}}{\text{SCF}} \times \frac{(\text{kW} + \text{kW loss})}{\text{kW}}$$

Step 7: Determine *steam flow* (SF) by the formula:

$$\text{Steam flow} = \text{corrected steam rate} \times \text{hp}$$

or

$$\text{Steam flow} = \text{corrected steam rate} \times \text{kW}$$

Step 8: Determine the required *inlet size* using Fig. 14.16 or 14.17, knowing SF and turbine inlet pressure and temperature.

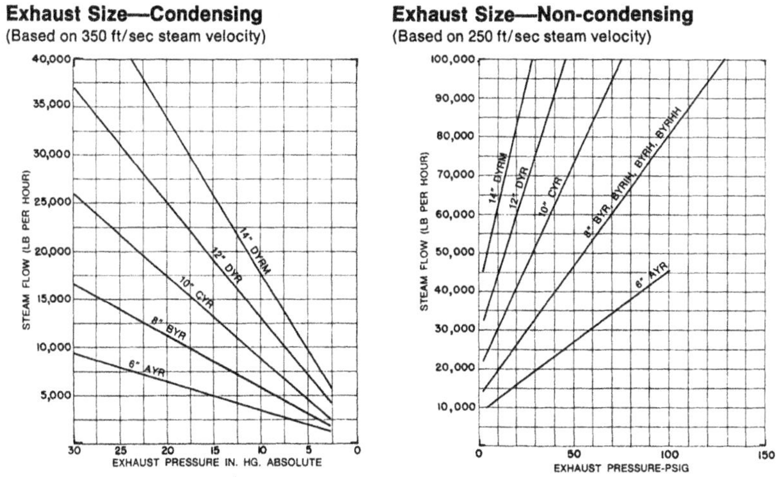

Figure 14.18 Exhaust size requirements for single-valve, single-stage steam turbines, English units. *(Elliott Company, Jeannette, Pa.)*

Step 9: Determine the required *exhaust size* using Fig. 14.18 or 14.19, knowing SF and exhaust pressure.

Sample calculations using first the English and then the metric system of measurement are shown next.

Example 14.1 (see Fig. 14.20)

Steam conditions: Inlet 250 psig, 575°F, Exhaust 50 psig,

Design conditions: Turbine to develop 500 hp at 4000 r/min

The simplified steam rate calculations establish that, for most applications, the DYR turbine offers the lowest corrected steam rate (see below), the AYR the highest. The DYR is the most expensive turbine. However, the larger and more efficient turbine will usually have a lower evaluated cost if steam cost is considered.

Step 1: TSR = 26.07 lb/kWh

Step 2: BSR = 36

Step 3: Mechanical loss = 55 hp

Step 4: Superheat = 169°F

Step 5: SCF = 1.050

Step 6: $\dfrac{36}{1.05} \times \dfrac{(500 + 55)}{500} = 38.0$ lb/hp·h

Step 7: SF = 38.0 × 500 = 19,000 lb/h

Step 8: Inlet size = 4 in ANSI

Step 9: Exhaust size = 6 in ANSI

(Similar calculation for AYR turbine results in corrected steam rate of 49.8 lb/hp·h)

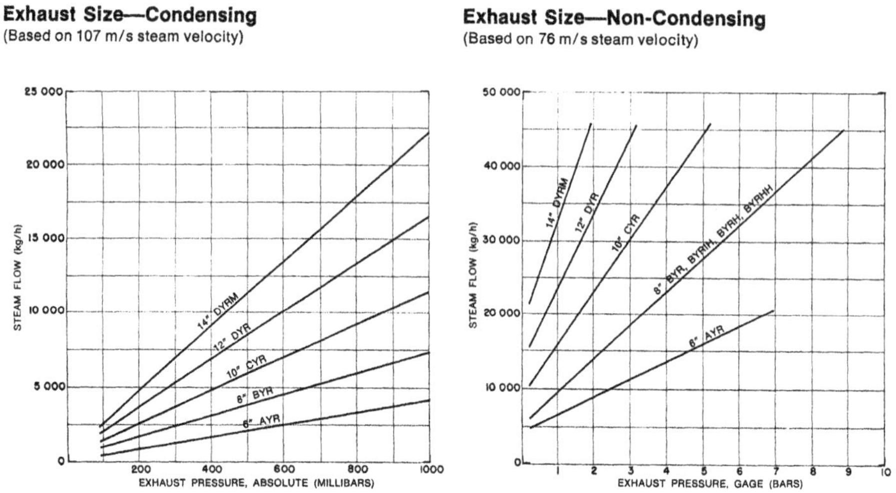

Figure 14.19 Exhaust size requirements for single-valve, single-stage steam turbines, metric units. *(Elliott Company, Jeannette, Pa.)*

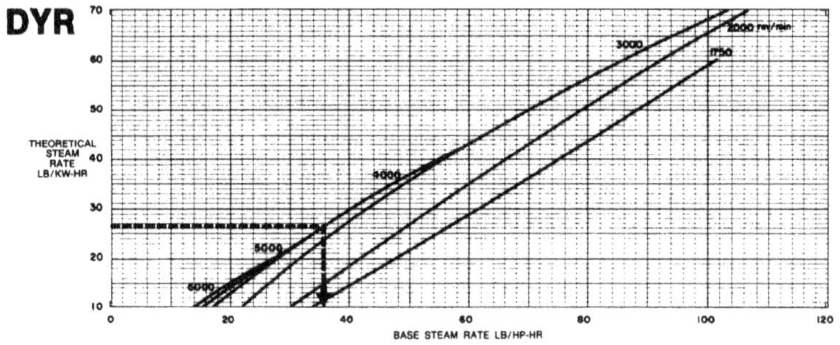

(a)

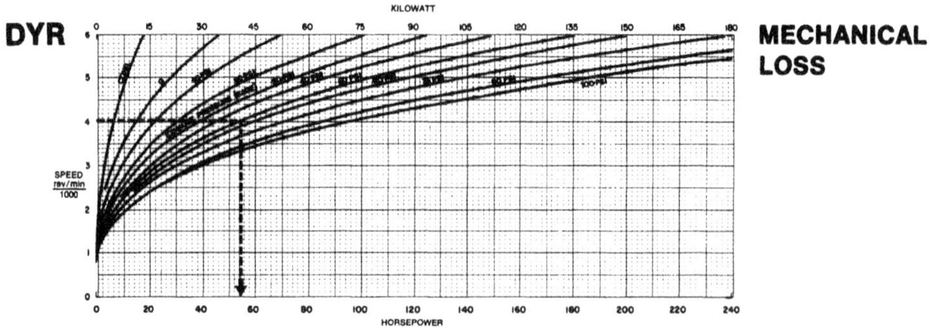

(b)

Figure 14.20 Approximate performance of single-valve, single-stage turbine, English units. Typical sizes are 12 in exhaust, 3500 hp/2600 kW. (*a*) Basic steam rate; (*b*) mechanical losses; (*c*) Superheat correction factor. *(Elliott Company, Jeannette, Pa.)*

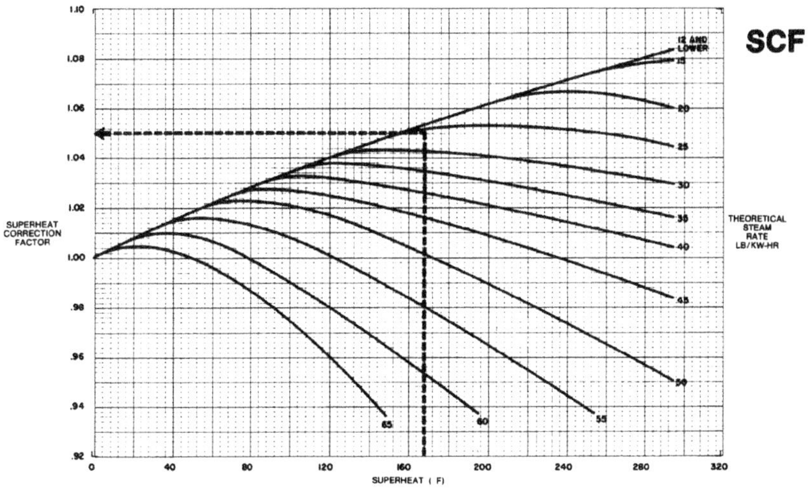

(c)

Figure 14.20 *(Continued)*

Example 14.2 (see Fig. 14.21)

Steam conditions: Inlet (gauge) 17.25 (Absolute) 18.25 bar, 300°C, Exhaust (gauge) 3.5 bar (Absolute) 4.5 bar;

Design conditions: Turbine to develop 375 kW at 4000 r/min

The simplified steam rate calculations establish that, for most applications, the DYR turbine offers the lowest corrected Steam Rate (see below), the AYR the highest. The DYR is the most expensive turbine. However, the larger and more efficient turbine will usually have a lower evaluated cost if steam cost is considered.

Step 1: $h_1 = 3027.8$ kJ/kg at $s_1 = 6.808$ kJ/(kg·K)

$h_2 = 2723.7$ kJ/kg at $s_2 = 6.808$ kJ/(kg·K)

$$\text{TSR} = \frac{3600}{3027.8 - 2723.7} = 11.84 \text{ kg/(kWh)}$$

Step 2: BSR = 21.9

Step 3: Mechanical loss = 41 kW

Step 4: Superheat 300 – 207.5 = 92.5°C

Step 5: SCF = 1.050

Step 6:

$$\frac{21.9}{1.05} \times \frac{(375 + 41)}{375} = 23.1 \text{ kg/(kWh)}$$

Step 7: SF = 23.1 × 375 = 8660 kg/h

Step 8: Inlet size = 4 in ANSI

Step 9: Exhaust size = 6 in ANSI

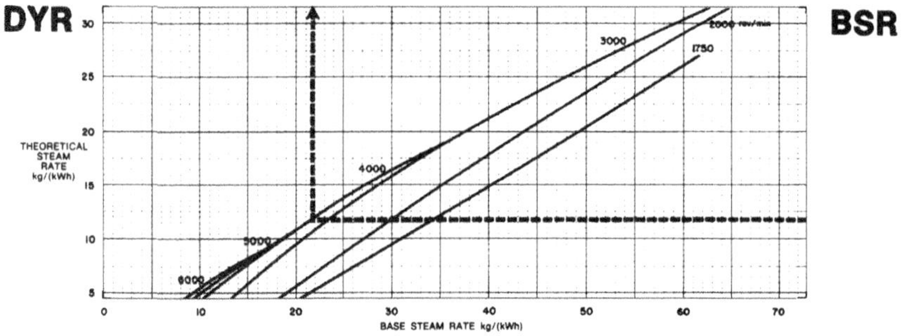

(a)

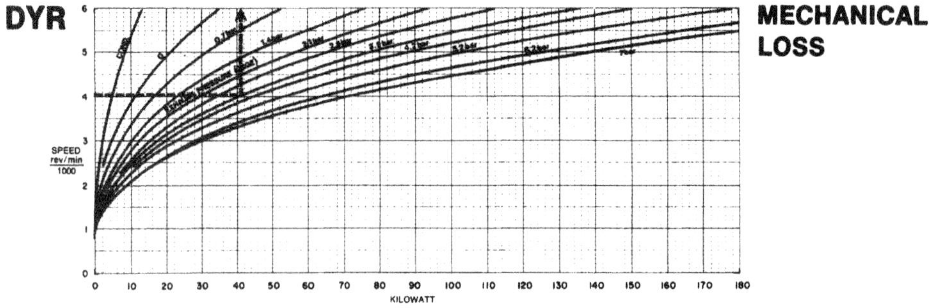

(b)

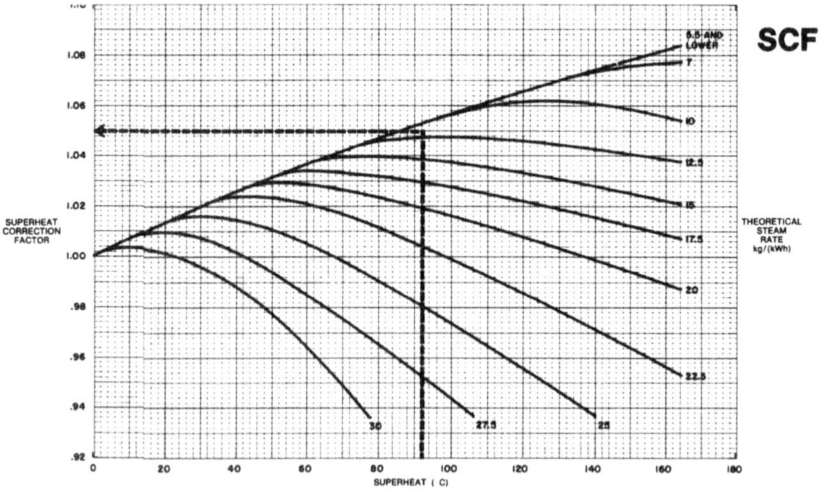

(c)

Figure 14.21 Approximate performance of single-valve, single-stage turbine, metric units. Typical sizes are 300-mm exhaust, 2600 kW. (*a*) Basic steam rate; (*b*) Mechanical losses; (*c*) Superheat correction factor. *(Elliott Company, Jeannette, Pa.)*

Similar calculation for AYR turbine results in corrected steam rate of 30.3 kg/(kW·h).

Preliminary selection of *single-valve, multistage steam turbines* is facilitated by similar sets of curves that are often available from the turbine manufacturer.

Using Figs. 14.22 through 14.33, we can ascertain, quickly and easily, *approximate* steam rates for selected single-valve, multistage impulse turbine frames that cover the more common horsepower/speed requirements. For guaranteed steam rates, prices, and other specific information, the manufacturer will have to be consulted. There are six steps in estimating performance of these turbines.

Step 1: Select the performance curve that is most applicable to the horsepower and speed requirements of the driven machine.

Step 2: Using your Theoretical Steam Rate Tables, determine the TSR in lb/kWh (or kg/kWh) for the given steam conditions.

Step 3: Locate this TSR on the horizontal axis of the appropriate curve selected in Step 1.

Step 4: Moving upward on the proper TSR line, locate the required speed. Interpolation may be necessary.

Step 5: At this point, read the overall efficiency (η) on the vertical axis.

Step 6: The TSR divided by this overall efficiency is equal to the approximate steam rate.

$$SR = \frac{TSR}{\eta}$$

Note: The calculated SR will be in the same units as the TSR from the steam tables. When TSR is in lb/kWh, and SR is desired in lb/hp·h, TSR must be multiplied by 0.746:

$$SR_{\text{lb/hp·h}} = \frac{TSR_{\text{lb/kWh}}\,(0.746)}{\eta}$$

Again, we can visualize the estimating procedure by two examples.

Example 14.3: English units What is the steam rate of a condensing turbine driving a compressor requiring 6000 hp at 5000 r/min? Steam inlet conditions are 600 psig, 650°F; exhaust pressure 4-in. Hga.

Step 1: The horsepower and r/min requirements suggest that turbine model 2E7 (Fig. 14.24) should be selected.

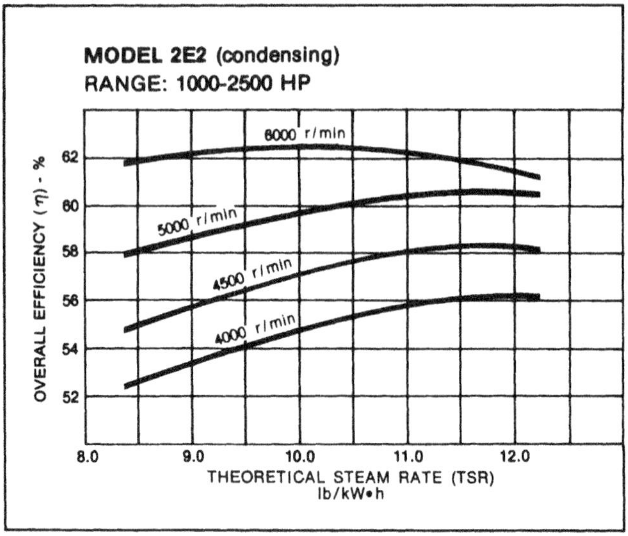

(a)

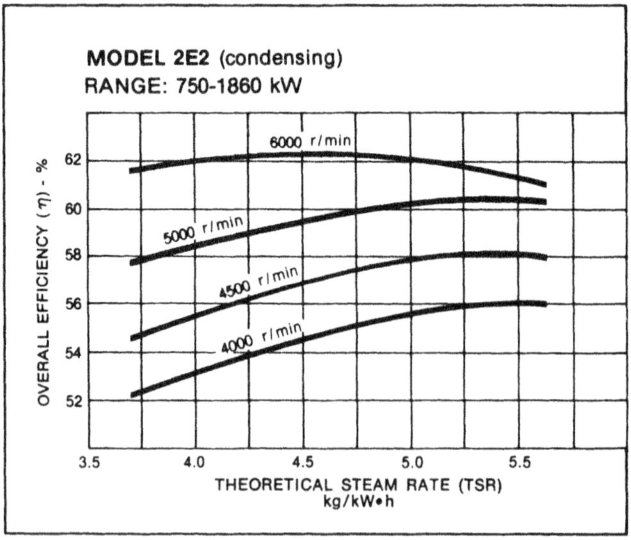

(b)

Figure 14.22 Selection charts, three-stage, single-valve turbines (condensing). *(Elliott Company, Jeannette, Pa.)*

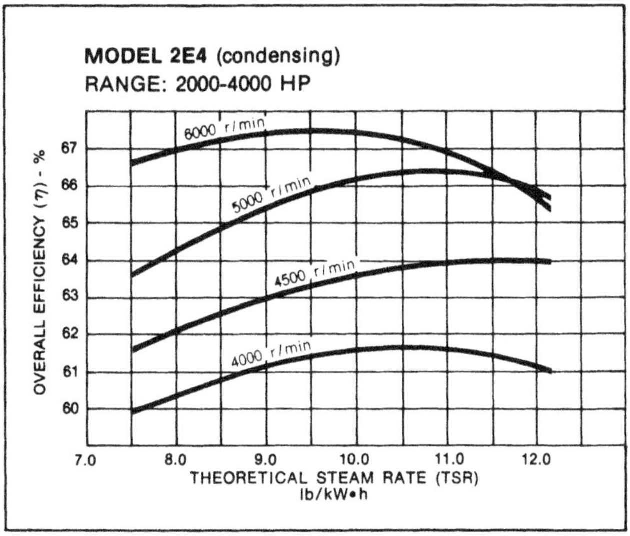

(a)

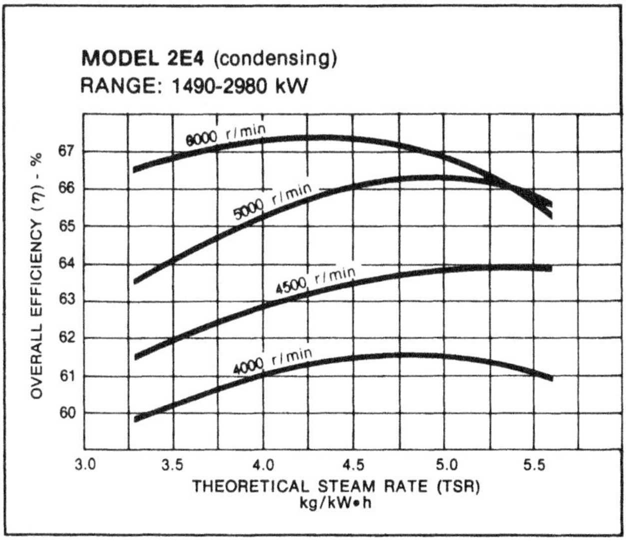

(b)

Figure 14.23 Selection charts, five-stage, single-valve turbines (condensing). *(Elliott Company, Jeannette, Pa.)*

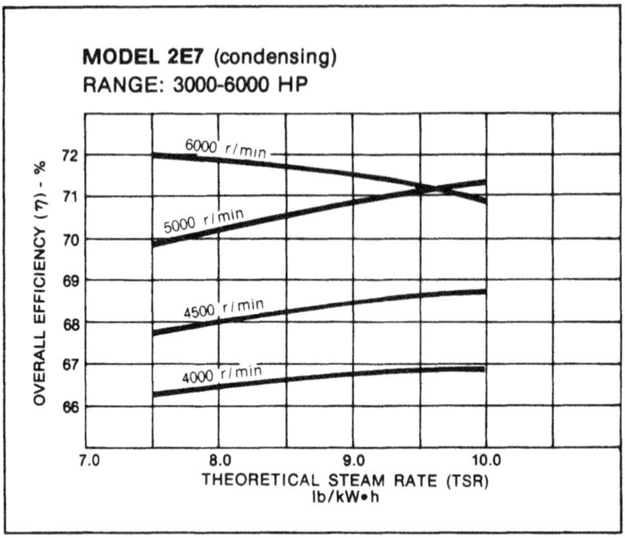

(a)

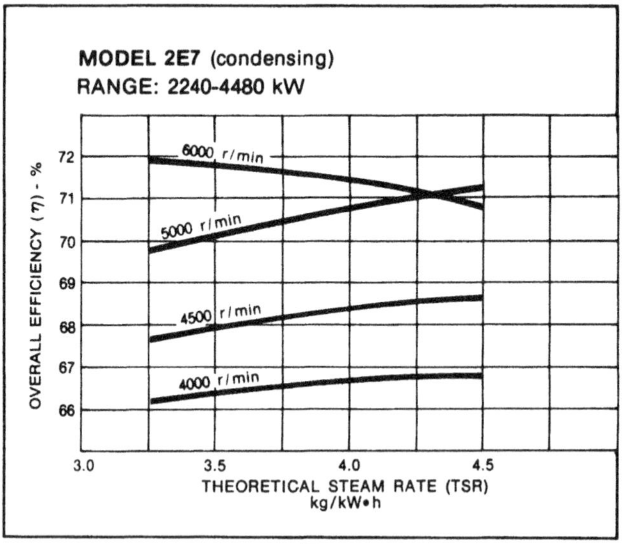

(b)

Figure 14.24 Selection charts, eight-stage, single-valve turbines (condensing). *(Elliott Company, Jeannette, Pa.)*

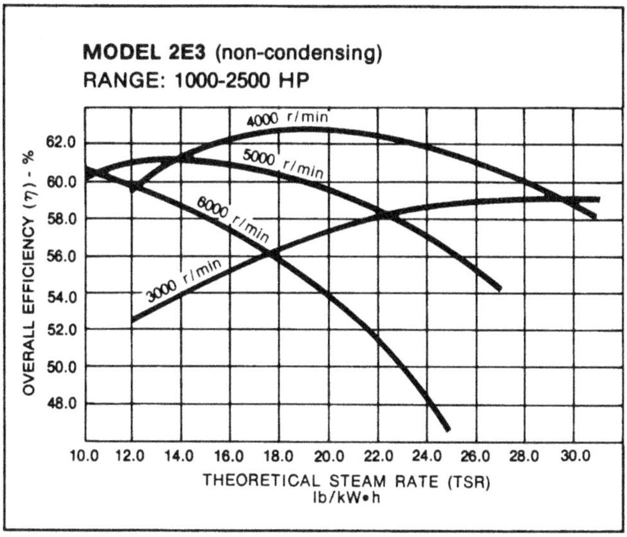

(a)

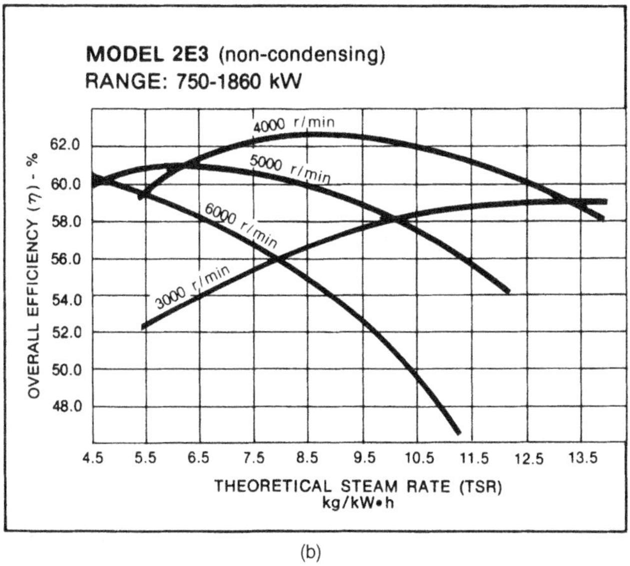

(b)

Figure 14.25 Selection charts, four-stage, single-valve turbines (noncondensing). *(Elliott Company, Jeannette, Pa.)*

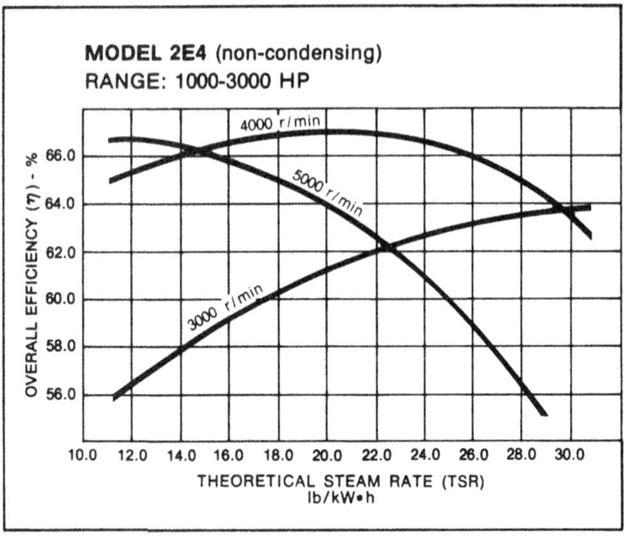

(a)

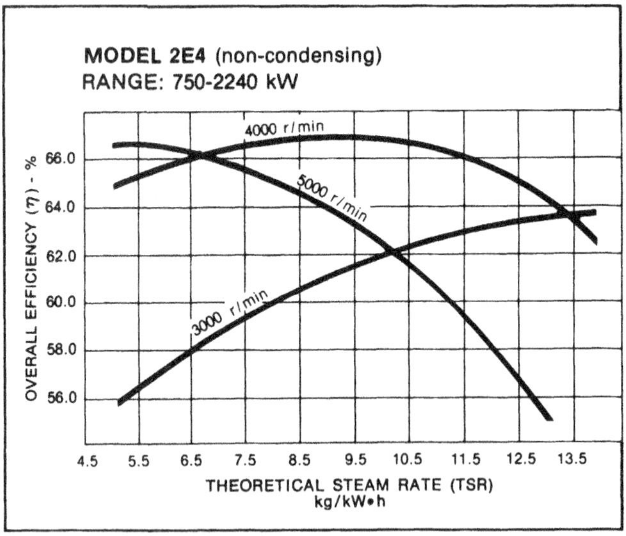

(b)

Figure 14.26 Selection charts, five-stage, single-valve turbines (noncondensing). *(Elliott Company, Jeannette, Pa.)*

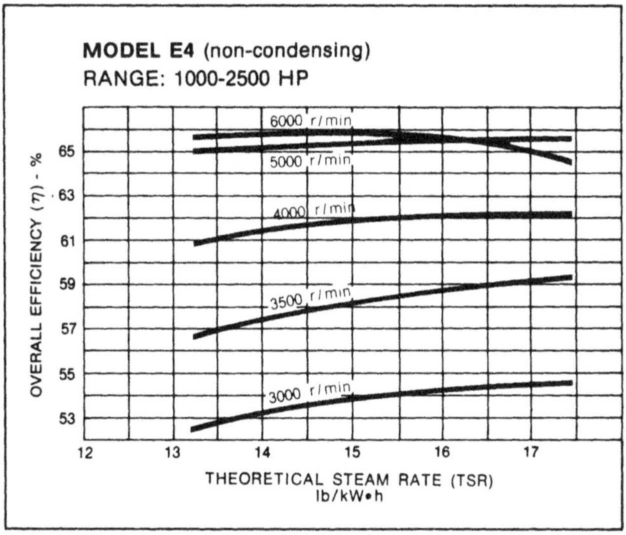

(a)

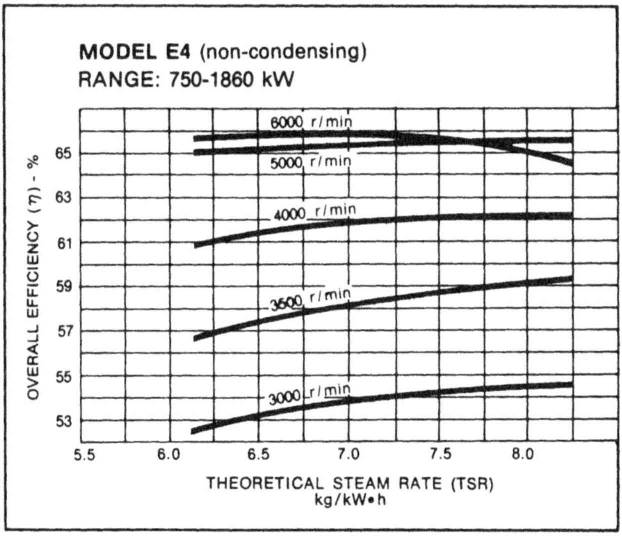

(b)

Figure 14.27 Selection charts, four-stage, single-valve turbines (noncondensing). *(Elliott Company, Jeannette, Pa.)*

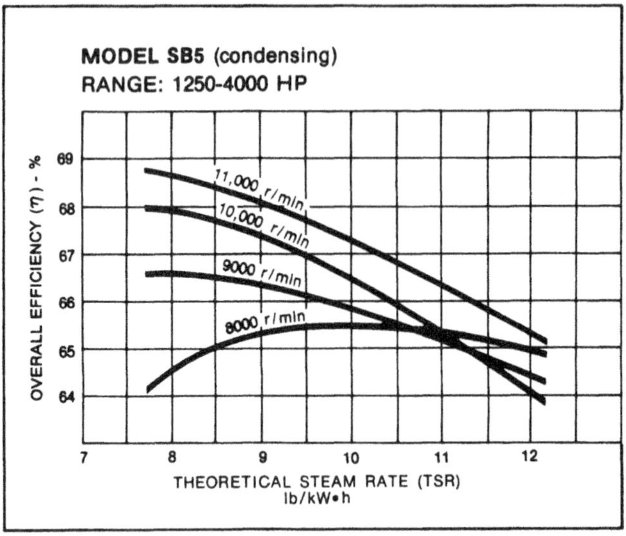

(a)

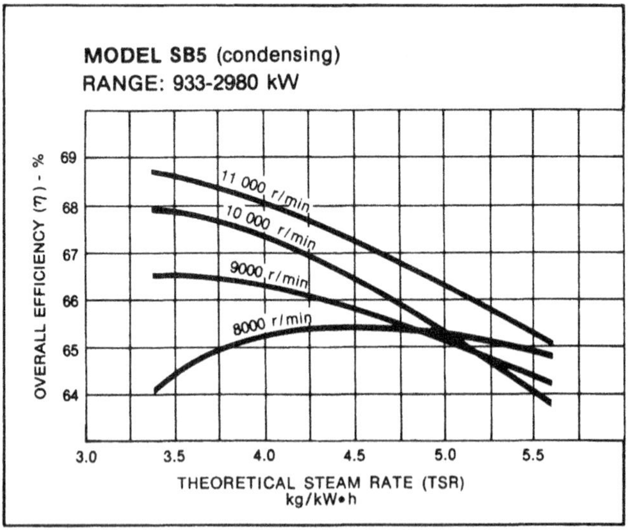

(b)

Figure 14.28 Selection charts, five-stage, single-valve turbines (condensing). *(Elliott Company, Jeannette, Pa.)*

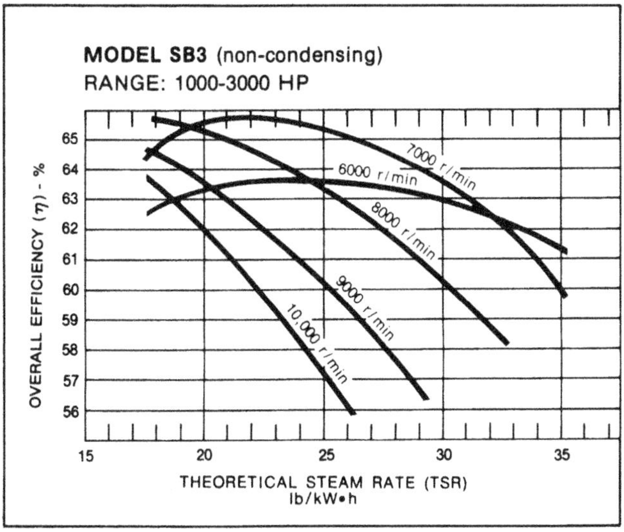

(a)

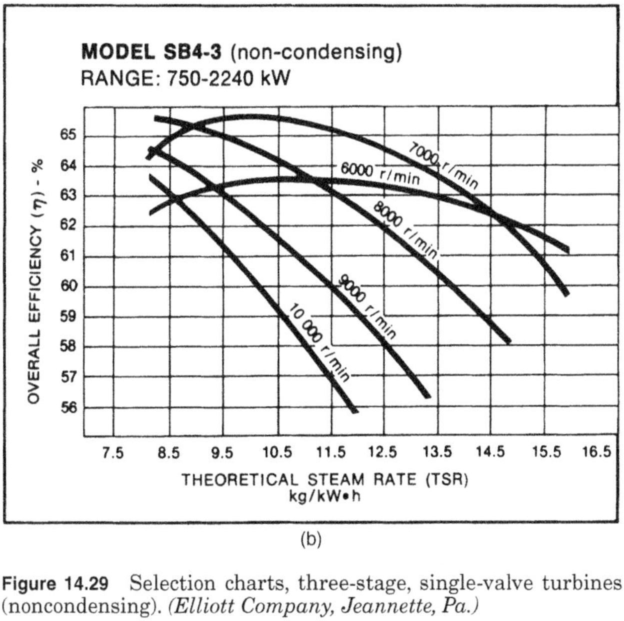

(b)

Figure 14.29 Selection charts, three-stage, single-valve turbines (noncondensing). *(Elliott Company, Jeannette, Pa.)*

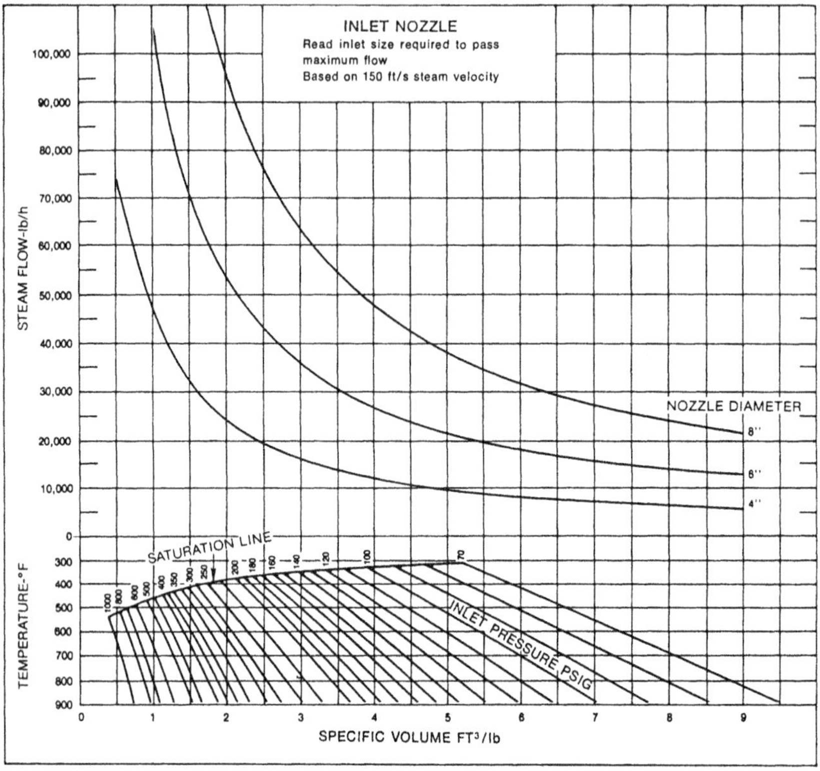

Figure 14.30 Inlet nozzle sizing for single-valve, multistage steam turbines, English units. *(Elliott Company, Jeannette, Pa.)*

Step 2: Using your Theoretical Steam Rate Tables, Tables 14.3 and 14.4, the Mollier diagrams reproduced in Figs. 14.1 or 14.2, or Polar Mollier diagrams available from some steam turbine manufacturers, the TSR will be found to be 8.18 lb/kWh.

Steps 3, 4, 5: Reading upward on the 8.18 line of the 2E7 approximate performance curve at 5000 r/min, the overall efficiency is found to be 70.4%.

Step 6: Applying these values to the equation:

$$SR_{\text{lb/hp}\cdot\text{h}} = \frac{TSR_{\text{lb/kWh}}\,(0.746)}{\eta} = \frac{8.18\,(0.746)}{0.704}$$

$$= 8.67 \text{ lb/hp}\cdot\text{h}$$

Total steam flow is then equal to 8.67 lb/hp·h × 6000 hp = 52020 lb/h.

Using the nozzle curves in Figs. 14.30 and 14.32, inlet and exhaust nozzles are found to be 6 and 36 in.

Example 14.4: Metric units What is the steam rate of a condensing turbine driving a compressor requiring 4475 kW at 5000 r/min? Steam inlet conditions are 41.4 bar, 340°C; exhaust 135 m bar.

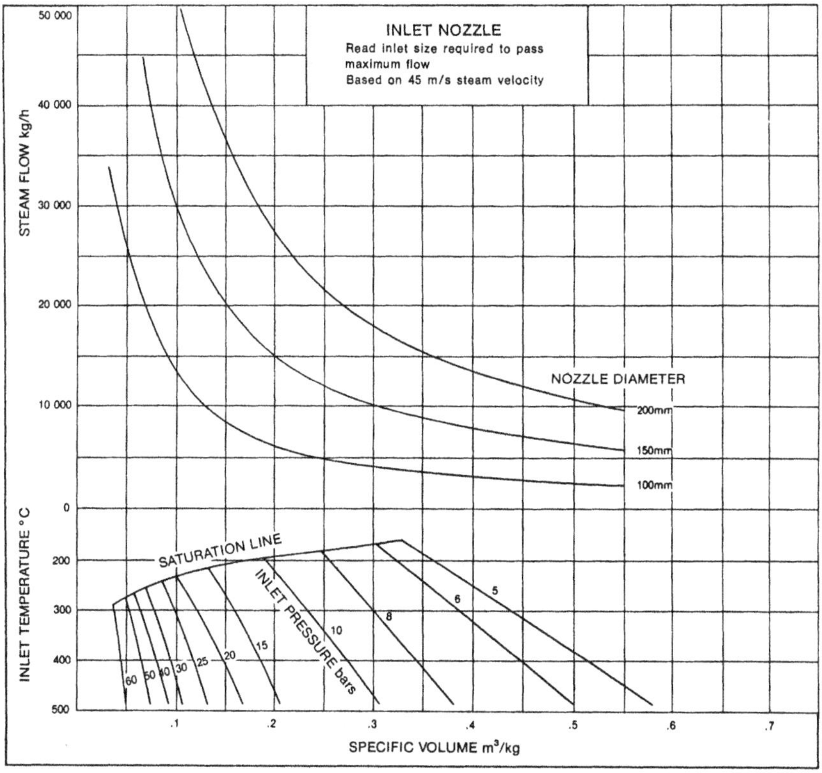

Figure 14.31 Inlet nozzle sizing for single-valve, multistage steam turbines, metric units. *(Elliott Company, Jeannette, Pa.)*

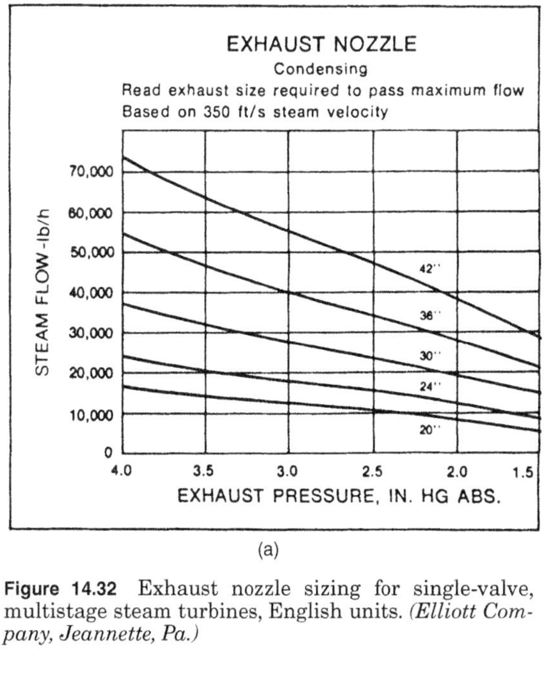

(a)

Figure 14.32 Exhaust nozzle sizing for single-valve, multistage steam turbines, English units. *(Elliott Company, Jeannette, Pa.)*

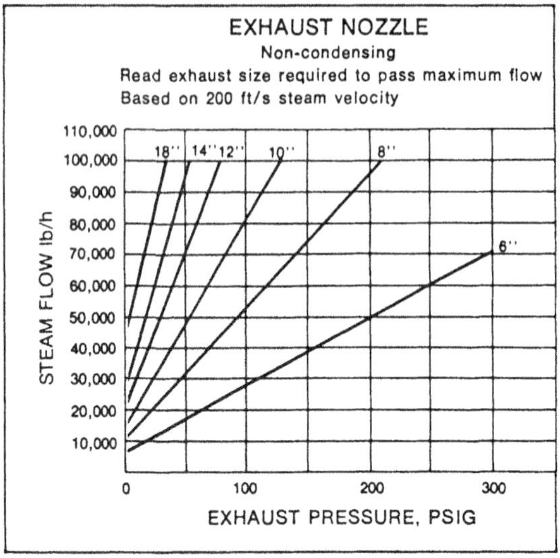

(b)

Figure 14.32 *(Continued)*

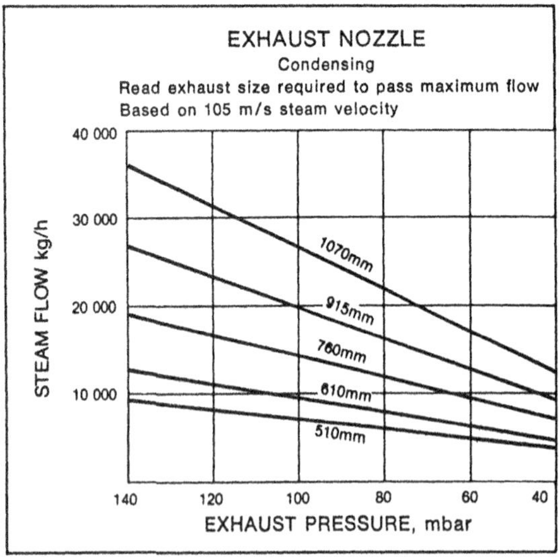

(a)

Figure 14.33 Exhaust nozzle sizing for single-valve, multistage steam turbines, metric units. *(Elliott Company, Jeannette, Pa.)*

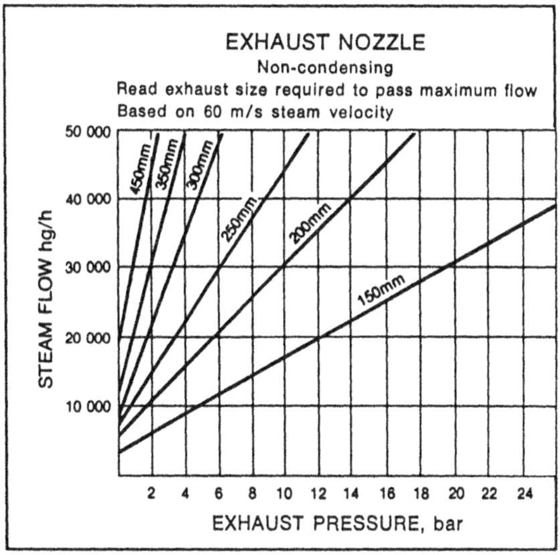

(b)

Figure 14.33 *(Continued)*

Step 1: The kW and r/min requirements suggest that the curves for turbine Model 2E7, Fig. 14.24, should be selected.

Step 2: Using your Theoretical Steam Rate Tables, the TSR will be found to be 3.71 kg/kW·h.

Steps 3, 4, 5: Reading upward on the 3.71 line of the 2E7 approximate performance curve at 5000 r/min, the overall efficiency is found to be 70.4%.

Step 6: Applying these values to the equation:

$$SR_{\text{kg/kW·h}} = \frac{TSR_{\text{kg/kW·h}}}{\eta} = \frac{3.71}{0.704} = 5.27 \text{ kg/kW·h}$$

Total steam flow is then equal to 5.27 kg/kW·h × 4475 kW = 23 583 kg/h.

Using the nozzle curves on Figs. 14.31 and 14.33, inlet and exhaust nozzles are found to be 150 and 900 mm.

Finally, we can look up weights and dimensions for these single-valve, multistage steam turbines in Figure 14.34.

14.3 Quick Reference Information to Estimate Steam Rates of Multivalve Multistage Steam Turbines

Example 14.5: Straight condensing turbine (General Electric Company) A rated load of 6000 hp (4474 kW), speed of 7000 r/min; throttle steam at 600 psig (42

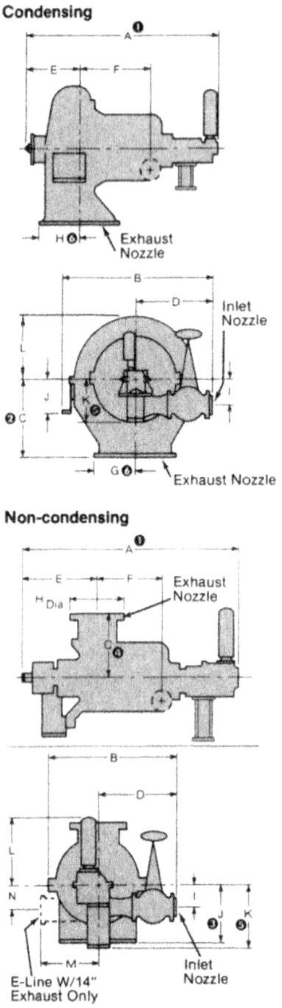

Frame	No. of Stages	Inlet Size	Exhaust Size	A	B	C	D	E	F	G	H	I	J	K	L	M	N	Net Wt.	Upper Half Wt.	Rotor Wt.
				INCHES/*CENTIMETERS*														LBS/*KG*		
CONDENSING																				
2E2	3	4 *10*	24 *61*	75 *190*	68 *173*	30 *76*	39 *99*	23 *58*	38 *97*	16 *41*	16 *41*	9 *23*	22 *56*	22 *56*	24 *61*	— —	— —	8000 *3640*	1650 *750*	1060 *480*
2E3 & 4E1	4	4 *10*	30 *76*	75 *190*	73 *185*	35 *89*	39 *99*	23 *58*	39 *99*	20 *51*	20 *51*	9 *23*	22 *56*	22 *56*	23 *58*	— —	— —	8600 *3910*	1800 *815*	1200 *545*
2E4	5	6 *15*	30 *76*	79 *201*	80 *203*	35 *89*	46 *117*	23 *58*	43 *109*	20 *51*	20 *51*	9 *23*	22 *56*	22 *56*	28 *71*	— —	— —	9300 *4230*	2100 *950*	1400 *635*
2E6	7	6 *15*	36 *91*	90 *229*	80 *203*	40 *102*	46 *117*	26 *66*	51 *130*	23 *58*	23 *58*	9 *23*	22 *56*	22 *56*	28 *71*	— —	— —	14000 *6365*	4000 *1810*	1815 *825*
2E7	8	6 *15*	42 *107*	100 *254*	86 *218*	40 *102*	46 *117*	26 *66*	61 *155*	27 *69*	27 *69*	9 *23*	22 *56*	22 *56*	30 *76*	— —	— —	16000 *7275*	5000 *2270*	2040 *925*
(S)E5	5	6 *15*	36 *91*	79 *201*	72 *183*	40 *102*	46 *117*	26 *66*	40 *102*	23 *58*	23 *58*	9 *23*	22 *56*	22 *56*	28 *71*	— —	— —	12000 *5455*	2100 *950*	1465 *665*
(S)E6	6	6 *15*	36 *91*	83 *211*	72 *183*	40 *102*	46 *117*	26 *66*	44 *112*	23 *58*	23 *58*	9 *23*	22 *56*	22 *56*	28 *71*	— —	— —	14000 *6365*	3000 *1360*	1680 *760*
(S)B4	4	4 *10*	24 *61*	67 *170*	66 *168*	27 *69*	39 *99*	20 *51*	34 *86*	16 *41*	16 *41*	9 *23*	22 *56*	22 *56*	20 *51*	— —	— —	8300 *3775*	2700 *1225*	870 *400*
(S)B5	5	6 *15*	30 *76*	72 *183*	79 *201*	34 *86*	46 *117*	19 *48*	40 *102*	20 *51*	20 *51*	9 *23*	22 *56*	22 *56*	23 *58*	— —	— —	9500 *4320*	3000 *1360*	1050 *475*
SB5DF	5	6 *15*	42DF *107*	100 *254*	97 *246*	40 *102*	46 *117*	35 *89*	48 *122*	29 *74*	22 *56*	9 *23*	22 *56*	22 *56*	32 *81*	— —	— —	15000 *6820*	5000 *2270*	1560 *710*
NON-CONDENSING																				
E3	3	6 *15*	14 *36*	80 *203*	72 *183*	— —	46 *117*	22 *56*	45 *114*	— —	21 *53*	9 *23*	22 *56*	22 *56*	29 *74*	25 *64*	11 *28*	6000 *2730*	1800 *815*	960 *435*
E4	4	6 *15*	14 *36*	80 *203*	72 *183*	— —	46 *117*	22 *56*	45 *114*	— —	21 *53*	9 *23*	22 *56*	22 *56*	29 *74*	25 *64*	11 *28*	6500 *2955*	2100 *950*	1100 *500*
2E2	3	6 *15*	14 *36*	80 *203*	72 *183*	— —	46 *117*	22 *56*	45 *114*	— —	21 *53*	9 *23*	22 *56*	22 *56*	29 *74*	25 *64*	11 *28*	6000 *2730*	1800 *815*	960 *435*
2E3	4	6 *15*	14 *36*	80 *203*	72 *183*	— —	46 *117*	22 *56*	45 *114*	— —	21 *53*	9 *23*	22 *56*	22 *56*	29 *74*	25 *64*	11 *28*	6500 *2955*	2100 *950*	1100 *500*
2E4	5	6 *15*	14 *36*	80 *203*	72 *183*	— —	46 *117*	22 *56*	45 *114*	— —	21 *53*	9 *23*	22 *56*	22 *56*	29 *74*	25 *64*	11 *28*	7000 *3180*	2400 *1090*	1260 *570*
E2	2	8 *20*	14 *36*	84 *213*	83 *211*	— —	58 *147*	22 *56*	31 *79*	— —	21 *53*	8 *20*	22 *56*	22 *56*	34 *86*	25 *64*	11 *28*	5500 *2500*	1500 *680*	700 *320*
E3	3	8 *20*	14 *36*	84 *213*	83 *211*	— —	58 *147*	22 *56*	31 *79*	— —	21 *53*	8 *20*	22 *56*	22 *56*	34 *86*	25 *64*	11 *28*	6000 *2730*	1800 *815*	835 *380*
E4	4	8 *20*	14 *36*	84 *213*	83 *211*	— —	58 *147*	22 *56*	31 *79*	— —	21 *53*	8 *20*	22 *56*	22 *56*	34 *86*	25 *64*	11 *28*	6500 *2955*	2100 *950*	960 *435*
E3	3	6 *15*	18 *46*	78 *198*	82 *208*	30 *76*	46 *117*	26 *66*	38 *97*	— —	25 *64*	9 *23*	22 *56*	22 *56*	29 *74*	— —	— —	7000 *3138*	2300 *1040*	960 *435*
E4	4	6 *15*	18 *46*	78 *198*	82 *208*	30 *76*	46 *117*	26 *66*	38 *97*	— —	25 *64*	9 *23*	22 *56*	22 *56*	29 *74*	— —	— —	7500 *3410*	2500 *1140*	1100 *500*
2E2	3	6 *15*	18 *46*	78 *198*	82 *208*	30 *76*	46 *117*	26 *66*	38 *97*	— —	25 *64*	9 *23*	22 *56*	22 *56*	29 *74*	— —	— —	7000 *3180*	2300 *1045*	960 *435*
2E3	4	6 *15*	18 *46*	78 *198*	82 *208*	30 *76*	46 *117*	26 *66*	38 *97*	— —	25 *64*	9 *23*	22 *56*	22 *56*	29 *74*	— —	— —	7500 *3410*	2500 *1140*	1100 *500*
2E4	5	6 *15*	18 *46*	78 *198*	82 *208*	30 *76*	46 *117*	26 *66*	38 *97*	— —	25 *64*	9 *23*	22 *56*	22 *56*	29 *74*	— —	— —	8000 *3640*	2900 *1320*	1260 *570*
(S)B4-4	4	6 *15*	12 *31*	91 *231*	65 *165*	24 *61*	46 *117*	29 *74*	39 *99*	— —	21 *53*	9 *23*	22 *56*	22 *56*	29 *74*	— —	— —	14500 *6590*	4000 *1810*	950 *430*
(S)B4-3	3	6 *15*	12 *31*	91 *231*	65 *165*	24 *61*	46 *117*	29 *74*	39 *99*	— —	21 *53*	9 *23*	22 *56*	22 *56*	29 *74*	— —	— —	14000 *6365*	3800 *1730*	800 *360*
(S)B4-2	2	6 *15*	12 *31*	91 *231*	65 *165*	24 *61*	46 *117*	29 *74*	39 *99*	— —	21 *53*	9 *23*	22 *56*	22 *56*	29 *74*	— —	— —	13500 *6140*	3600 *1630*	700 *310*
(S)B4-1	1	6 *15*	12 *31*	91 *231*	65 *165*	24 *61*	46 *117*	29 *74*	39 *99*	— —	21 *53*	9 *23*	22 *56*	22 *56*	29 *74*	— —	— —	13000 *5910*	3400 *1540*	550 *250*

❶ Add 6 inches (15 cm) to remove governor valve.
❷ "UP" nozzles also available.
❸ 18-inch (46 cm) exhaust supports are positioned on the sides of the unit the same as the condensing models shown above.
❹ "DOWN" nozzle also available except for 14-inch (36 cm) size which is right or left side orientation.
❺ Add 4 inches (10 cm) to lowest point of T & T valve support.
❻ If G = H nozzle is circular; if G ≠ H nozzle is rectangular.

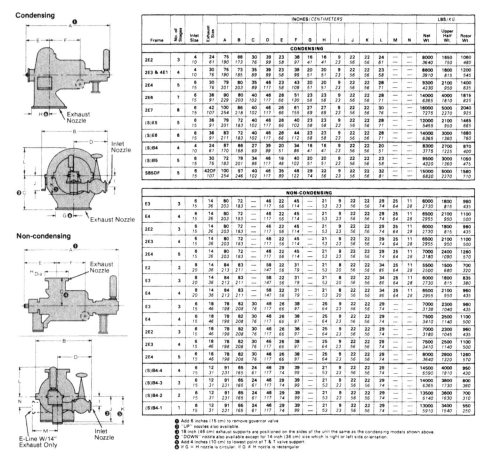

Figure 14.34 Dimensions and weights of typical single-valve, multistage steam turbines. *(Elliott Company, Jeannette, Pa.)*

kg/cm²), 750°F (399°C) (261°F) (127°C) superheat; exhaust at 3 in. Hg abs (0.1 atm abs)

1. Theoretical steam rate, Table 14.3 at 600 psig, 750°F, 3 in. Hg abs . . . 7.4 lb/kWh (3.36 kg/kWh)
2. Basic efficiency, Fig. 14.35 at 6,000 hp and 600 psig . . . 0.729
3. Superheat factor, Fig. 14.36 at 261°F superheat . . . 1.03
4. Speed factor, Fig. 14.37 at 7000 r/min and (6000 hp/3 in. Hg abs = 2000) . . . 0.989

$$\text{Estimated steam rate} = \frac{\text{TSR}}{\text{efficiency}}$$

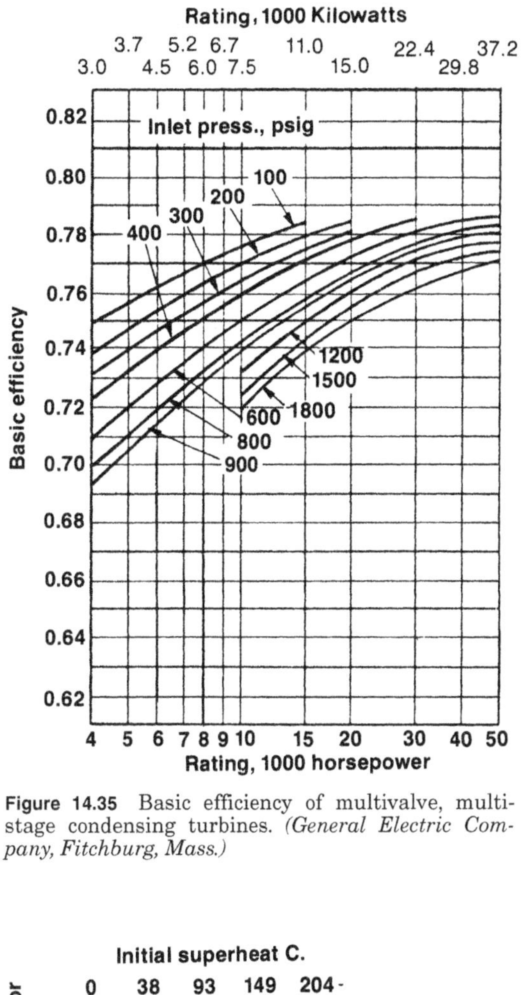

Figure 14.35 Basic efficiency of multivalve, multistage condensing turbines. *(General Electric Company, Fitchburg, Mass.)*

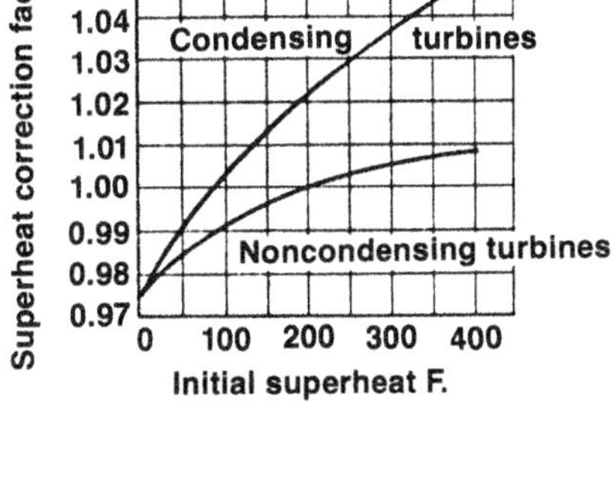

Figure 14.36 Superheat correction factors for multivalve, multistage steam turbines. *(General Electric Company, Fitchburg, Mass.)*

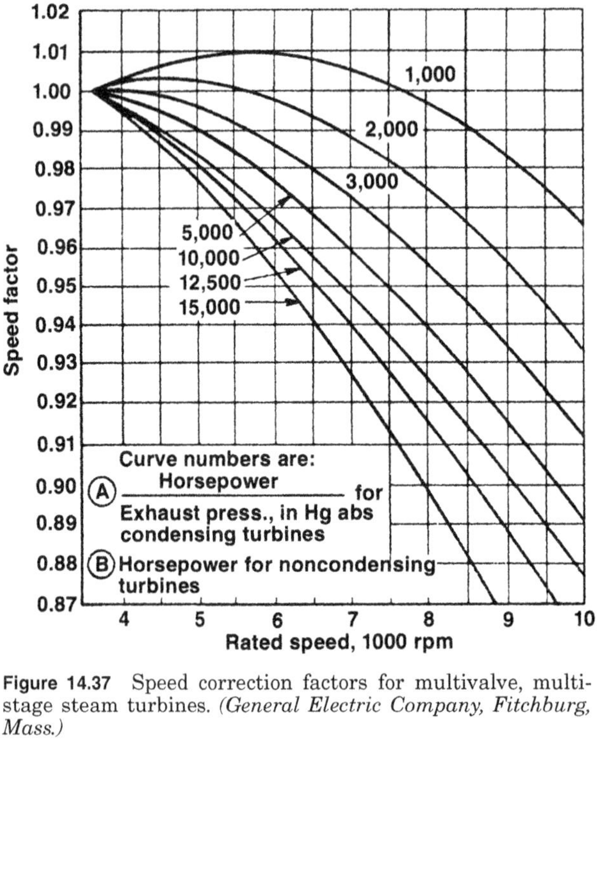

Figure 14.37 Speed correction factors for multivalve, multistage steam turbines. *(General Electric Company, Fitchburg, Mass.)*

$$= \frac{0.746 \times 7.4}{0.729 \times 1.03 \times 0.989}$$

= 7.43 lb/hph (4.54 kg/kWh)

Example 14.6: Straight noncondensing turbine (General Electric Company) At rated load of 6000 hp, speed of 7000 r/min; throttle steam at 600 psig, 750°F (261°F superheat); exhaust at 100 psig (7 kg/cm^2)

1. Theoretical steam rate, Table 14.3, at 600 psig, 750°F, 100 psig . . . 19.44 lb/kWh (8.82 kg/kWh)
2. Basic efficiency, Fig. 14.38 at 6000 hp and 600 psig . . . 0.737
3. Superheat factor, Fig. 14.36 at 261°F superheat . . . 1.003
4. Speed factor, Fig. 14.37 at 7000 r/min and 6000 hp . . . 0.957
5. Pressure-ratio factor, Fig. 14.39, at P_2/P_1 = 100 + 15/600 + 15 = 0.187 . . . 1.013

$$\text{Estimated steam rate} = \frac{\text{TSR}}{\text{efficiency}}$$

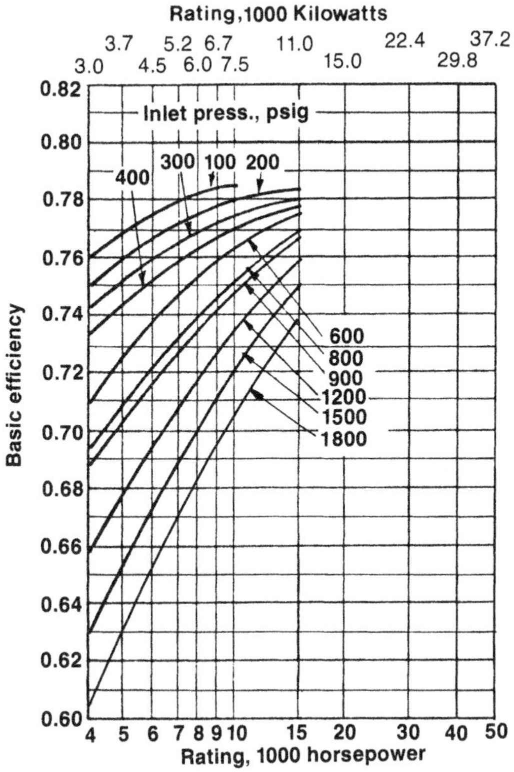

Figure 14.38 Basic efficiency of multivalve, multistage noncondensing steam turbines. *(General Electric Company, Fitchburg, Mass.)*

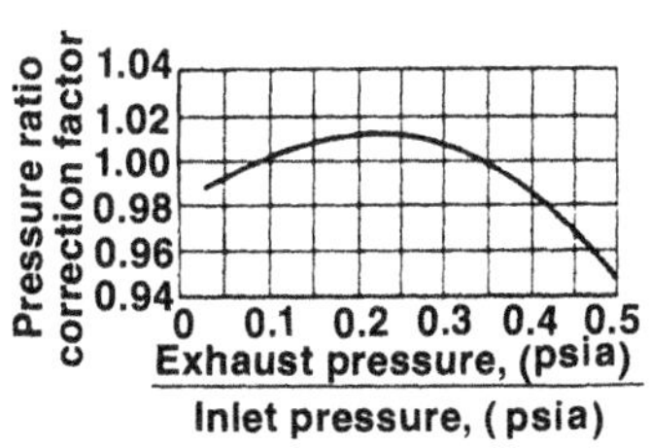

Figure 14.39 Pressure ratio correction for noncondensing steam turbines. *(General Electric Company, Fitchburg, Mass.)*

$$= \frac{0.746 \times 19.44}{0.737 \times 1.003 \times 0.957 \times 1.013}$$

$$= 20.2 \text{ lb/hph } (12.4 \text{ kg/kWh})$$

Chapter

15

Elliott Shortcut Selection Method for Multivalve, Multistage Steam Turbines

Manufacturers and their computers are usually available to assist the user with turbine selection. Using these resources, the purchaser should be able to evaluate the options available for each application. Blade stresses, thrust loads, stage pressures, temperatures, and other pertinent information are speedily determined with the computer's help (see later). Many manufacturers, in fact, use the computer to calculate the efficiency of each stage and of the entire turbine. In other words, every stage of every machine is completely defined before a proposal is made. The user is thus assured of getting the most reliable turbine at the highest possible efficiency.

In the early stages of planning, the engineer may want to do some approximate turbine selecting of his own. A Mollier diagram and some examples are included to enable you to:

1. Estimate full and part load steam rates
2. Find stage pressures and temperatures
3. Estimate an extraction turbine

15.1 Approximate Steam Rates

Figures 15.1 and 15.2 can be used to determine an approximate turbine efficiency when horsepower, speed, and steam conditions are known. In many instances an *approximate* efficiency may well serve the purpose since the parameters mentioned may change.

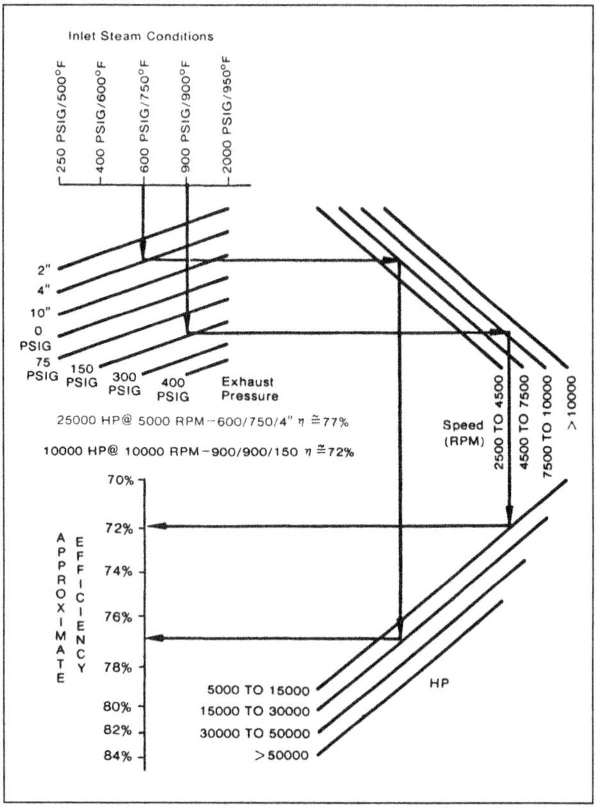

Figure 15.1 Determination of approximate efficiency for multivalve, multistage turbines, English units. *(Elliott Company, Jeannette, Pa.)*

As can be seen in the example, a 25,000 hp, 5000 r/min turbine using steam at 600 psig/750°F/4 in Hga has an approximate efficiency of 77 percent.

Applying this approximate efficiency to the theoretical steam rate (TSR), Table 14.3, results in a steam rate and steam flow as follows:

$$\text{TSR 600 psig/750°F/4 in Hga} = 7.65 \text{ lb/kWh}$$

$$\text{Approximate efficiency} = \eta_a = 77\%$$

$$\text{Approximate steam rate (ASR)} = \text{TSR}/\eta_a$$
$$= (7.65 \text{ lb/kWh})\,(.746 \text{ kW/hp})/0.77 = 7.40 \text{ lb/hph}$$

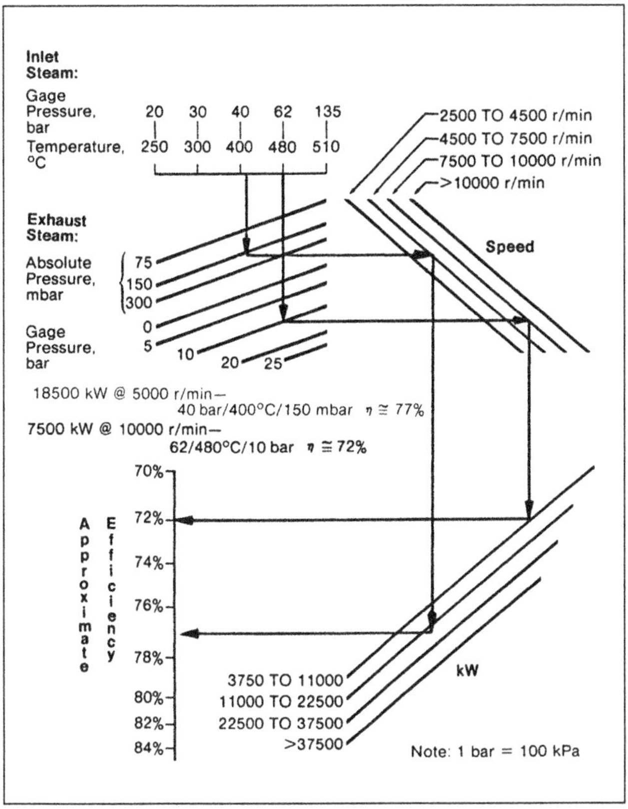

Figure 15.2 Determination of approximate efficiency for multivalve, multistage turbines, metric units. *(Elliott Company, Jeannette, Pa.)*

$$\text{Approximate steam flow} = \text{ASR} \times \text{hp}$$
$$= (7.40 \text{ lb/hph})\ (25{,}000 \text{ hp}) = 185{,}000 \text{ lb/h}$$

The Mollier diagram, Fig. 14.1 or 14.2, can be used to determine the TSR for steam conditions not tabulated.

Figure 15.3 can be used to find approximate steam rates for turbines operating at part load and speed. For example, find the approximate steam rate when the 25,000 hp, 5000 r/min turbine we've discussed is operated at 20,000 hp and 4500 r/min.

$$\%\ \text{hp} = 20{,}000 \text{ hp}/25{,}000 \text{ hp} = 80\%$$

$$\%\ \text{r/min} = 4500 \text{ r/min}/5000 \text{ r/min} = 90\%$$

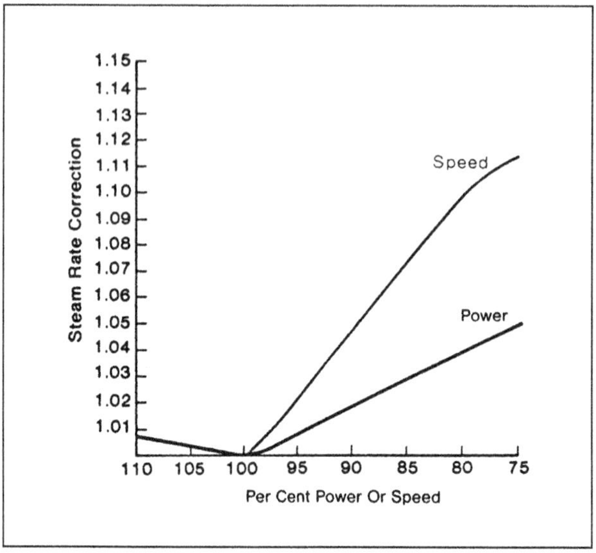

Figure 15.3 Part load steam rate correction for multivalve, multistage turbines. *(Elliott Company, Jeannette, Pa.)*

From the curve, the hp correction is 1.04 and the r/min correction 1.05. Total correction is 1.04 × 1.05 = 1.09. The part load steam rate is therefore 7.40 × 1.09 = 8.06 lb/hph.

Alternatively, the same procedure in metric units: In this example, Fig. 15.2, an 18,500 kW, 5000 r/min turbine using steam at 40 bar/400°C exhausting to 150 mbar has an approximate efficiency of 77 percent.

Applying this approximate efficiency to the theoretical steam rate (TSR), Table 14.4, results in a steam rate and steam flow as follows:

$$\text{TSR} = 3.52 \text{ kg/(kWh)}$$

$$\text{Approximate efficiency} = \eta_a = 77\%$$

$$\text{Approximate steam rate (ASR)} = \text{TSR}/\eta_a = 3.52/.77 = 4.57 \text{ kg/(kWh)}$$

$$\text{Approximate steam flow} = \text{ASR} \times \text{kW} = 4.57 \text{ kg/(kWh)} \times 18{,}500 \text{ kW} = 85{,}550 \text{ kg/h}$$

The Mollier diagram, Fig. 14.2, can be used to determine the TSR for steam conditions not tabulated. Figure 15.3 can again be used to find approximate steam rates for turbines operating at part load and speed.

For example, find the approximate steam rate when the 18,500 kW, 5000 r/min turbine we've discussed is operated at 14,800 kW and 4500 r/min.

$$\% \text{ Power} = 14{,}800 \text{ kW}/18{,}500 \text{ kW} = 80\%$$

$$\% \text{ Speed} = 4500/5000 = 90\%$$

From Fig. 15.3, the power correction is 1.04 and the speed correction 1.05. Total correction is $1.04 \times 1.05 = 1.09$. The part load steam rate is therefore $4.57 \times 1.09 = 4.98$ kg/(kWh).

15.2 Stage Performance Determination

The following examples illustrate the performance of various combinations of impulse staging. It should be understood, of course, that stage selection and overall turbine efficiency are affected by many important considerations other than stage efficiency. Speed limitations, mechanical stresses, leakage and throttling losses, windage, bearing friction, and reheat must be factored into the ultimate turbine design. That's the job of the vendor's turbine specialist.

The approximate efficiencies of a Curtis stage and a Rateau stage are shown in Fig. 15.4.

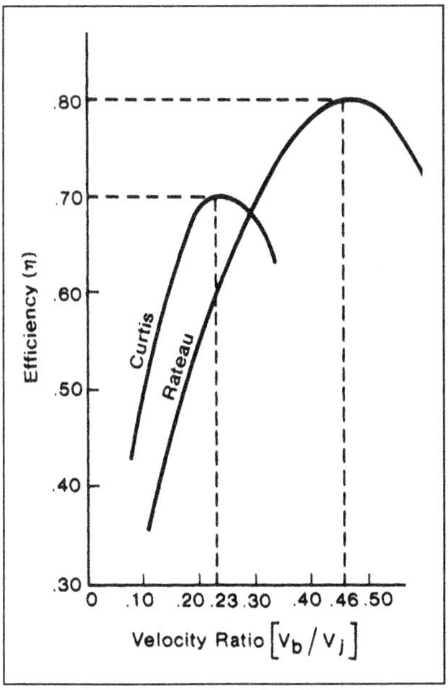

Figure 15.4 Approximate efficiencies of Curtis and Rateau stages. *(Elliott Company, Jeannette, Pa.)*

Basic formulas:

$$V_b = \frac{\pi DN}{720} \qquad V_j = 224\sqrt{h_1 - h_2} = 224\sqrt{\Delta H}$$

where V_b = pitch line (blade) velocity, ft/s
D = pitch diameter of wheel, inches (base diameter plus height of blade)
N = rotative speed, r/min
V_j = steam jet velocity, ft/s
h_1 = inlet steam enthalpy, Btu/lb
h_2 = isentropic exhaust steam enthalpy, Btu/lb
h_{2e} = stage exit steam enthalpy, Btu/lb
ΔH = isentropic heat drop, Btu/lb $(h_1 - h_2)$
V_b/V_j = velocity ratio, dimensionless

Example 15.1: Curtis stage performance *Conditions:* 1500 psig (1515 psia), 950°F, 5000 r/min, 25-in wheel diameter. Assume 1-in blade height. Find isentropic heat drop and end point. (See Fig. 15.5.)

$$V_b = \frac{\pi DN}{720} = \frac{(3.14)(25+1)(5000)}{720} = 568 \text{ ft/s}$$

From Figure 15.4, velocity ratio for optimum Curtis stage efficiency = 0.23

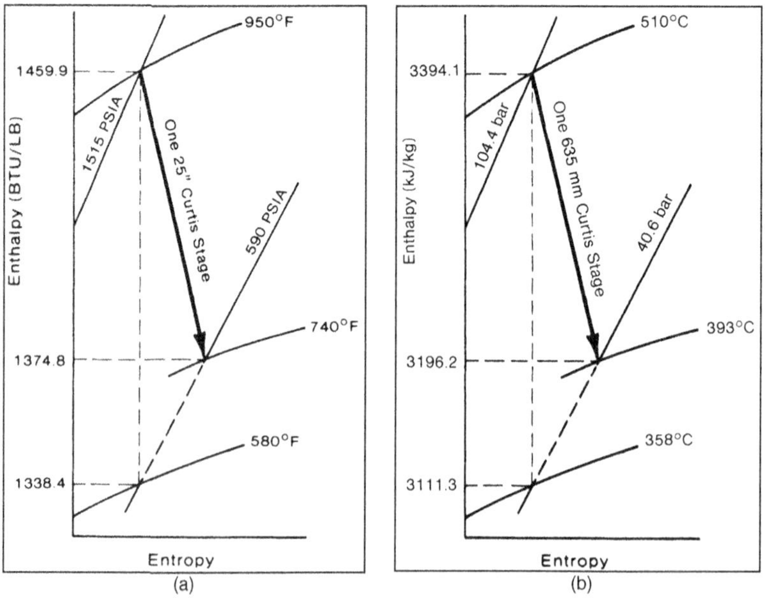

Figure 15.5 Isentropic heat drop and end points for a Curtis stage. *(Elliott Company, Jeannette, Pa.)*

$$V_b/V_j = 0.23; V_j = \frac{V_b}{.23} = \frac{568}{.23} = 2470 \text{ ft/s}$$

$$V_j = 224\sqrt{\Delta H} = 2470 \text{ ft/s}$$

$$\Delta H = 121.5 \text{ Btu/lb}$$

h_1 (from Mollier chart or steam table) = 1459.9 Btu/lb
h_2 = 1459.9 – 121.5 = 1338.4 Btu/lb
Exhaust pressure (from Mollier chart) = 590 psia

Assuming a stage efficiency of 70 percent, the stage exit conditions are:

exhaust pressure 590 psia
h_{2e} = 1459.9 – (.7) (121.5) 1374.8 Btu/lb.

Example 15.2: Rateau stage performance *Conditions:* 400 psig (415 psia), 600°F, 5000 r/min, 35-in wheel diameter. Assume 1-in blade height. Find isentropic heat drop and end point. (See Fig. 15.6.)

$$V_b = \frac{\pi DN}{720} = \frac{(3.14)(35+1)(5000)}{720} = 785 \text{ ft/s}$$

From Fig. 15.4, velocity ratio for optimum Rateau stage efficiency = 0.46

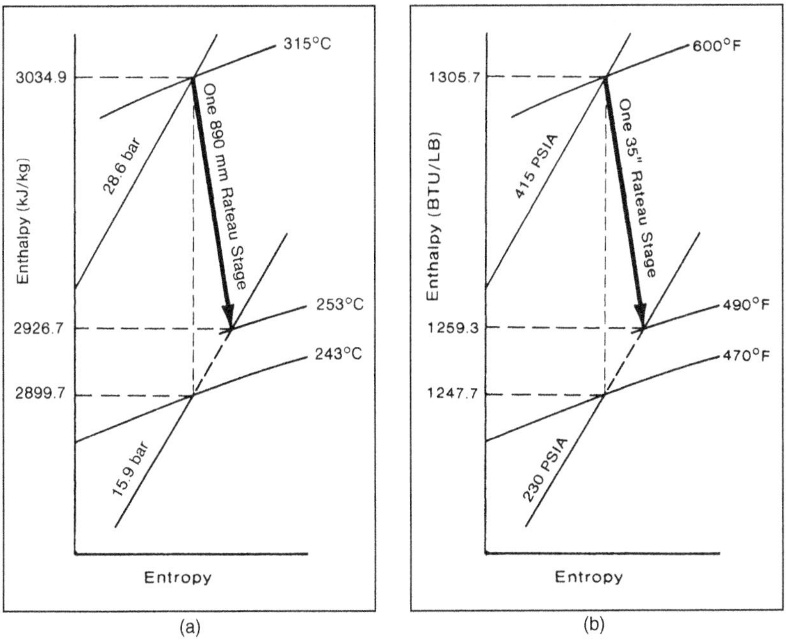

Figure 15.6 Isentropic heat drop and end points for a Rateau stage. *(Elliott Company, Jeannette, Pa.)*

$$V_b/V_j = 0.46;\ V_j = V_b/0.46 = \frac{785}{.46} = 1705 \text{ ft/s}$$

$$V_j = 224\sqrt{\Delta H} = 1705 \text{ ft/s}$$

$$\Delta H = (V_j/224)^2 = (1705/224)^2 = 58.0 \text{ Btu/lb}$$

h_1 (from Mollier chart or steam tables) = 1305.7 Btu/lb
h_2 = 1305.7 − 58.0 = 1247.7 Btu/lb
Exhaust pressure (from Mollier chart) = 230 psia

Assuming a stage efficiency of 80 percent, the stage exit conditions are:

exhaust pressure 230 psia;
h_{2e} = 1305.7 − (.8) (58.0) 1259.3 Btu/lb.

Example 15.3: Straight Rateau staging *Conditions:* 400 psig (415 psia), 600°F, exhaust pressure 100 psig (115 psia). Find the number of 35-in diameter Rateau stages required, at 5000 r/min, assuming optimum stage efficiency. (See Fig. 15.7.)

From Mollier chart, the isentropic heat available is:

$$1305.7 - 1184.2 = 121.5 \text{ Btu/lb}$$

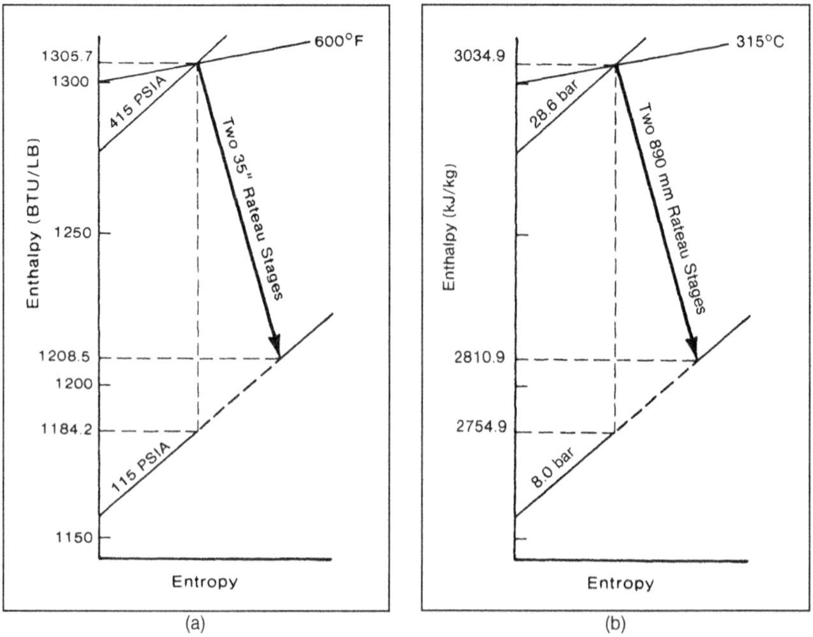

Figure 15.7 Isentropic heat drop and end points for straight Rateau staging. *(Elliott Company, Jeannette, Pa.)*

Alternatively, using a TSR table:

$$\text{TSR} = 28.08 \text{ lb/kWh}$$

Using the basic definition of TSR:

$$\Delta H = h_1 - h_2 = \frac{3413}{\text{TSR}} = \frac{3413}{28.08} = 121.5 \text{ Btu/lb}$$

From Example 15.2, the optimum isentropic heat drop per Rateau stage = 58.0 Btu/lb

$$\text{Approximate stages required} = \frac{121.5}{58.0} = 2.1 \text{ (thermodynamically)}$$

The turbine will require two Rateau stages.

Assuming 80 percent stage efficiency:

$$121.5 \times 0.80 = 97.2 \text{ Btu/lb}$$

$$1305.7 - 97.2 = 1208.5 \text{ Btu/lb}$$

Example 15.4: Curtis and Rateau staging *Conditions:* 1500 psig (1515 psia), 950°F, exhaust pressure 150 psig (165 psia). Find the number of 35-in-diameter Rateau stages required, at 5000 r/min, when using one 25-in-diameter Curtis stage, assuming optimum stage efficiencies. (See Fig. 15.8.)

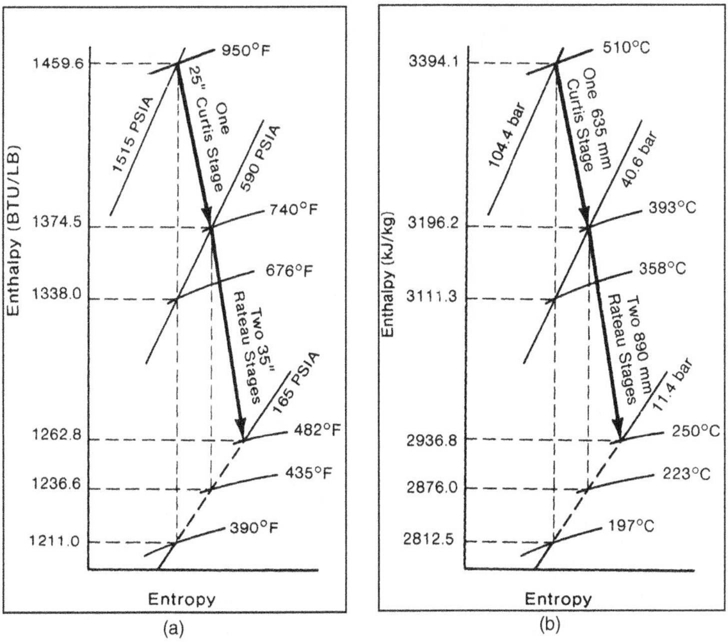

Figure 15.8 Isentropic heat drop and end points for Curtis and Rateau staging. *(Elliott Company, Jeannette, Pa.)*

From Example 15.1, we found that a Curtis stage removes 121.6 Btu/lb

$$1459.6 - 121.6 = 1338.0 \text{ Btu/lb}$$

$$\text{Pressure} = 590 \text{ psia}$$

Assuming 70 percent stage efficiency:

$$121.6 \times 0.70 = 85.1 \text{ Btu/lb}$$

$$1459.6 - 85.1 = 1374.5 \text{ Btu/lb}$$

From this end point to 165 psia, the isentropic heat drop for the Rateau stages is:

$$1374.5 - 1236.6 = 137.9 \text{ Btu/lb}$$

From Example 15.2, the optimum isentropic heat drop per Rateau stage = 58.0 Btu/lb

$$\text{Approximate stages required} = \frac{137.9}{58.0} = 2.4 \text{ (thermodynamically)}$$

The turbine will therefore require one 25-in Curtis stage and two 35-in Rateau stages.

Assuming an overall Rateau efficiency of 81 percent, the end point will be:

$$1374.5 - .81\,(137.9) = 1262.8 \text{ Btu/lb}$$

Here are the basic relationships in *metric units:*

$$V_b = \frac{\pi DN}{60{,}000} \quad V_j = \sqrt{2000\,(h_1 - h_2)} = 44.72\,\sqrt{\Delta H_i}$$

where V_b = pitch line (blade) velocity, m/s

D = pitch diameter of wheel, mm (base diameter plus height of blade)
n = rotative speed, r/min
V_j = steam jet velocity, m/s
h_1 = inlet steam enthalpy, kJ/kg
h_2 = isentropic exhaust steam enthalpy, kJ/kg
h_{2e} = stage exit steam enthalpy, kJ/kg

ΔH_i = isentropic heat drop, kJ/kg $(h_1 - h_2)$

V_b/V_j = velocity ratio, dimensionless
p_a = absolute pressure, bar
p_g = gauge pressure, bar = $p_a - 1.01$
s = specific entropy, kJ/(kg K)

Example 15.5: Curtis stage performance *Conditions:* $p_g = 103.4$ bar ($p_a = 104.4$ bar), 510°C, 5000 r/min, 635-mm wheel diameter. Assume 25-mm blade height. Find isentropic heat drop and end point. (See Fig. 15.5.)

$$V_b = \frac{\pi DN}{60{,}000} = \frac{(3.14)(635 + 25)(5000)}{60{,}000} = 173 \text{ m/s}$$

From Fig. 15.4, velocity ratio for optimum Curtis stage efficiency = 0.23

$$V_b/V_j = 0.23;\ V_j = \frac{V_b}{0.23} = \frac{173}{0.23} = 752 \text{ m/s}$$

$$V_j = 44.72\sqrt{\Delta H_j} = 752 \text{ m/s}$$

$$\Delta H_i = 282.8 \text{ kJ/kg}$$

h_1 (from Mollier chart or steam table) = 3394.1 kJ/kg
h_2 = 3394.1 – 282.8 = 3111.3 kJ/kg
Exhaust pressure (from Mollier chart) = 40.6 bar

Assuming a stage efficiency of 70 percent, the stage exit conditions are:

exhaust pressure p_a 40.6 bar
h_{2e} = 3394.1 – (.7) (282.8) 3196.2 kJ/kg

Example 15.6: Rateau stage performance *Conditions:* p_g = 27.6 bar (p_a = 28.61 bar), 315°C, 5000 r/min, 890-mm wheel diameter. Assume 25-mm blade height. Find isentropic heat drop and end point. (See Fig. 15.6.)

$$V_b = \frac{\pi DN}{60{,}000} = \frac{(3.14)(890 + 25)(5000)}{60{,}000} = 239 \text{ m/s}$$

From Fig. 15.6, velocity ratio for optimum Rateau stage efficiency = 0.46

$$V_b/V_j = 0.46;\ V_j = V_b/.46 = \frac{239}{0.46} = 520 \text{ m/s}$$

$$V_j = 44.72\sqrt{\Delta H_i} = 520 \text{ m/s}$$

$$\Delta H_i = (V_j^2/2000) = (520^2/2000) = 135.2 \text{ kJ/kg}$$

h_1 (from Mollier chart or steam tables) = 3034.9 kJ/kg
h_2 = 3034.9 – 135.2 = 2899.7 kJ/kg
Exhaust pressure (from Mollier chart) = 15.9 bar

Assuming a stage efficiency of 80%, the stage exit conditions are:

exhaust pressure 15.9 bar;
h_{2e} = 3034.9 – (.8) (135.2) 2926.7 kJ/kg.

Example 15.7: Straight Rateau staging *Conditions:* p_g = 27.6 bar (p_a = 28.61 bar), 315°C, exhaust p_g = 7 bar (p_a = 8.01 bar). Find the number of 890-mm-diameter Rateau stages required, at 5000 r/min, assuming optimum stage efficiency. (See Fig. 15.7.)

From Mollier chart, or steam table, the isentropic heat available is:

$$3034.9 - 2754.9 = 280.0 \text{ kJ/kg}$$

Alternately, using a TSR table:

$$\text{TSR} = 12.8 \text{ kg/(kWh)}$$

Using the basic definition of TSR:

$$\Delta H_j = h_1 - h_2 = \frac{3600}{\text{TSR}} = \frac{3600}{12.8} = 281 \text{ kJ/kg}$$

From Example 15.6, the optimum isentropic heat drop per Rateau stage = 135.2 kJ/kg.

$$\text{Approximate stages required} = \frac{280}{135.2} = 2.1 \text{ (thermodynamically)}$$

The turbine will require two Rateau stages.

Assuming 80 percent stage efficiency:

$280 \times 0.80 = 224$ kJ/kg

$3034.9 - 224 = 2810.9$ kJ/kg

Example 15.8: Curtis and Rateau staging *Conditions:* $p_g = 103.4$ bar ($p_a = 104.4$ bar), 510°C, exhaust $p_g = 10.4$ bar ($p_a = 11.4$ bar). Find the number of 890-mm-diameter Rateau stages required, at 5000 r/min, when using one 635-mm diameter Curtis stage, assuming optimum stage efficiencies. (See Fig. 15.8.)

From Example 15.5 we found that a Curtis stage removes 282.8 kJ/kg

$3394.1 - 282.8 = 3111.3$ kJ/kg

Exhaust pressure $p_a = 40.6$ bar

Assuming 70 percent stage efficiency:

$$282.8 \times .70 = 197.9 \text{ kJ/kg}$$

$$3394.1 - 197.9 = 3196.2 \text{ kJ/kg}$$

From this end point to 11.4 bar, the isentropic heat drop for the Rateau stages is:

$$3196.2 - 2876.0 = 320.2 \text{ kJ/kg}$$

From Example 15.6, the optimum isentropic heat drop per Rateau stage = 135.2 kJ/kg.

$$\text{Approximate stages required} = \frac{320.2}{135.2} = 2.4 \text{ (thermodynamically)}$$

The turbine will therefore require one 635-mm Curtis stage and two 890-mm Rateau stages.

Assuming an overall Rateau efficiency of 81 percent, the end point will be:

$$3196.2 - .81\,(320.2) = 2936.8 \text{ kJ/kg}$$

15.3 Extraction Turbine Performance

Today's fuel costs demand that the maximum amount of energy be squeezed from each pound of steam generated. To help in this effort, when both process steam and shaft power are required, many plant designers are turning to extraction turbines.

The extraction turbine can substantially reduce the energy charged to the driven machine if process steam would otherwise be supplied through pressure-reducing valves. Even though back-pressure turbines can be used to supply process steam, they are rather inflexible since the shaft power and process steam requirements must be closely matched. An extraction turbine, however, can cope with changes in these variables and satisfy the requirements of each over a broad range.

Figures 15.9 and 15.10 show performance maps for a typical extraction turbine. Determining the shape of this diagram is a problem that often arises. Here is an example that demonstrates the procedure to follow in drawing an approximate extraction diagram, Fig. 15.9:

Assume: Shaft hp and speed: 25,000 hp at 4500 r/min
Steam conditions: 600 psig/750°F/4 Hga
Extraction requirements: 150,000 lb/h at 250 psig

First tabulate the TSR.

TSR (inlet-to-extraction),

600 psig/750°F/250 psig = 35.4 lb/kWh

TSR (inlet-to-exhaust),

600 psig/750°F/4-in HGA = 7.64 lb/kWh)

Now assume that the efficiency of the entire turbine (inlet-to-exhaust) is 75 percent and the efficiency of the inlet-to-extraction section is 70 percent. Therefore:

Approximate SR (inlet-to-extraction) will be:

$$(35.4 \text{ lb/kWh} \times 0.746 \text{ kW/hp}) \div 0.70 = 37.8 \text{ lb/hph}$$

Approximate SR (inlet-to-exhaust) will be:

$$(7.64 \text{ lb/kWh} \times 0.746 \text{ kW/hp}) \div 0.75 = 7.60 \text{ lb/hph}$$

Point A is the first point to be located on the diagram by multiplying:

$$\text{Total hp} \times \text{approximate SR (inlet-to-exhaust)} = 25{,}000 \text{ hp} \times 7.60 \text{ lb/kWh} = 190{,}000 \text{ lb/h}$$

Locate Point A at 25,000 hp and 190,000 lb/h throttle flow.

Point B is located at zero hp and a throttle flow of 5% of A. This 5% flow is cooling steam going to the extraction section.

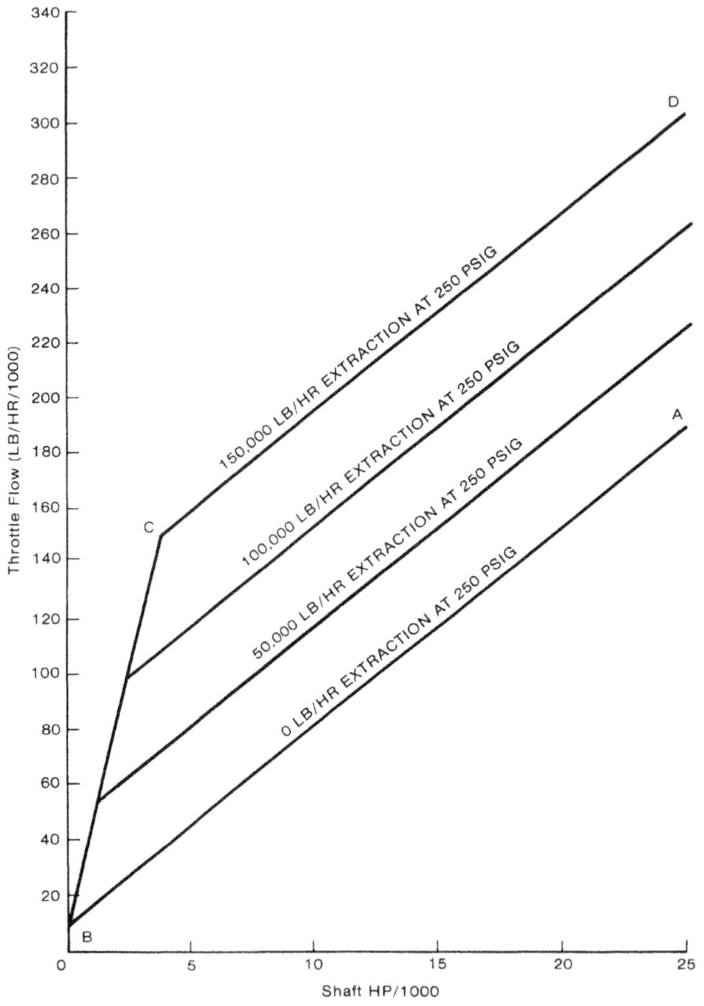

Figure 15.9 Performance map for typical extraction turbine, English units. *(Elliott Company, Jeannette, Pa.)*

Thus,

$$B = .05 \times 190{,}000 \text{ lb/h} = 9500 \text{ lb/h}$$

The zero extraction line results from connecting Points A and B.

Point C is located by dividing:

Extraction flow requirement/Approximate steam rate (inlet-to-extraction):

$$(150{,}000 \text{ lb/h}) \div (37.8 \text{ lb/hp-h}) = 3970 \text{ hp}$$

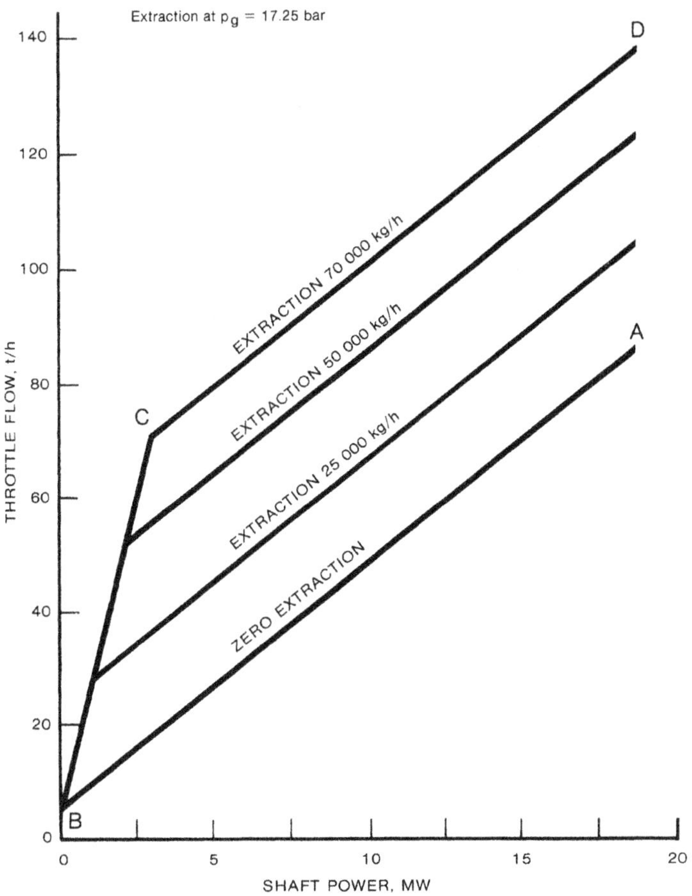

Figure 15.10 Performance map for typical extraction turbine, metric units. *(Elliott Company, Jeannette, Pa.)*

Locate Point C at 3970 hp and 150,000 lb/h throttle flow. Now draw line C-D parallel to A-B and another line C to B. Label A-B "Zero lb/h extraction at 250 psig" and C-D "150,000 lb/h extraction at 250 psig."

Notice that the general shape of the diagram and the slopes of the lines are determined mainly by the steam conditions used. The turbine investigated here could be built, but one must remember that we have dealt only with the thermodynamic aspects of this application. The mechanical aspects, such as blade stresses, nozzle flow limits, cooling steam, and other factors must also be checked.

This example can also be carried further to determine the number of stages in each section using the procedures outlined earlier. To do so requires finding the energy available in each portion of the turbine.

The energy available (ΔH_i) to the inlet-to-extraction section is:

$$\Delta H = 3413 \text{ Btu/kWh} \div \text{TSR (inlet-to-extraction)}$$

$$= 3413 \text{ Btu/kWh} \div 35.4 \text{ lb/kWh} = 96.4 \text{ Btu/lb}$$

$$V_b = \pi(25 + 1)(4500)/720 = 511 \text{ ft/s}$$

V_j for an ideal Curtis stage would therefore be:

$$V_j = \frac{V_b}{0.23} = \frac{511}{0.23} = 2220 \text{ ft/s}$$

V_j for an ideal Rateau stage would be:

$$V_j = \frac{V_b}{0.46} = \frac{511}{0.46} = 1110 \text{ ft/s}$$

With a ΔH_i of 96.5 Btu/lb, V_j for the inlet-to-extraction section with one stage would be:

$$V_j = 224\sqrt{\Delta H} = \sqrt{224\ 96.5} = 2200 \text{ ft/s}$$

This is seen to be very close to the 2220 ft/s for an ideal Curtis stage. We will therefore assume that the inlet-to-extraction section will contain one 25-in Curtis stage.

Now for the extraction-to-exhaust section. To find the energy available to this section we need the temperature of the steam entering this portion of the turbine (extraction steam temperature). The enthalpy of this steam will be:

$$\text{Inlet steam enthalpy} - \Delta H_i \text{ (inlet-to-extraction)} \times \eta \text{ (inlet-to-extraction)}$$

$$= 1378 \text{ Btu/lb} - 96.4 \text{ Btu/lb} \times 0.70 = 1378 \text{ Btu/lb} - 67.5 \text{ Btu/lb} = 1310.5 \text{ Btu/lb}$$

From the Mollier diagram, Fig. 14.1 at 250 psig and 1310.5 Btu/lb the extraction steam temperature is found to be close to 590°F (say 600°F).

The extraction-to-exhaust portion of this turbine therefore operates on steam conditions of 250 psig/600°F/4-in Hga. The TSR (extraction-to-exhaust) is 9.35 lb/kWh. The energy available to the extraction-to-exhaust section is therefore:

$$\Delta H_i = (3413 \text{ Btu/kWh}) \div (9.35 \text{ lb/kWh}) = 365 \text{ Btu/lb}$$

The blade velocity for 35-in nominal diameter staging with a 1-in blade height will be:

$$V_b = \pi(35 + 1)(4500)/720 = 706 \text{ ft/s}$$

If all staging is of the Rateau type in this portion of the turbine:

$$V_j = 706/.46 = 1535 \text{ ft/s}$$

ΔH_i per stage is therefore:

$$\Delta H_i = \left(\frac{V_j}{224}\right)^2 = \left(\frac{1535}{224}\right)^2 = (6.85)^2 = 47.0 \text{ Btu/lb}$$

The number of Rateau stages in this section would therefore be:

$$\frac{\text{Total energy available}}{\text{Energy removed per stage}} = \frac{365}{47.0} = 7.7 \text{ (say 8)}$$

Our turbine will, therefore, contain one 25-in diameter Curtis stage followed by eight 35-in diameter Rateau stages with the extraction opening after the Curtis stage.

Using metric data, we would observe Fig. 15.10 and proceed as follows:

Assume: Shaft power and speed: 18,500 kW at 4500 r/min
Steam conditions: 40 bar (gage)/400°C/150 mbar (abs.)
Extraction requirements:
70,000 kg/h at 17.25 bar (gauge)

First tabulate the TSR for an absolute inlet pressure of 41.0 bar at 400°C:

TSR to extraction (p_g = 17.25 bar, p_a = 18.26 bar):
h_1 = 3211.8 kJ/kg at s = 6.756 kJ/(kg K)

From steam table, at the same entropy,

at p_a = 18.0 bar, h = 2991.5 kJ/kg
at p_a = 18.5 bar, h = 2998.2 kJ/kg

Therefore, at 18.26 bar, h_2 = 2995.0 kJ/kg

$$\text{TSR} = \frac{3600}{3211.8 - 2995.0} = 16.6 \text{ kg/(kWh)}$$

TSR to exhaust (p_a = 150 mbar): From steam table or TSR tabulation (Table 14.4), TSR = 3.52 kg/(kWh).

Now assume that the efficiency of the entire turbine (inlet-to-exhaust) is 75 percent and the efficiency of the inlet-to-extraction section is 70 percent. Therefore:

Approximate SR (inlet-to-extraction) will be:

$$\frac{16.6}{0.70} = 23.7 \text{ kg/(kW·h)}$$

Approximate SR (inlet-to-exhaust) will be:

$$\frac{3.52}{0.75} = 4.69 \text{ kg/(kW·h)}$$

Point A is the first point to be located on the diagram by multiplying:

$$\text{Total power} \times \text{approximate SR (inlet-to-exhaust)} = 18{,}500 \text{ kW} \times 4.69 \text{ kg/(kW·h)} = 86{,}750 \text{ kg/h}$$

Locate Point A at 18,500 kW and 86,750 kg/h throttle flow.

Point B is located at zero kW and a throttle flow of 5 percent of A. This 5 percent flow is cooling steam going to the extraction section. Thus,

$$B = 0.05 \times 86{,}750 \text{ kg/h} = 4338 \text{ kg/h}$$

The zero extraction line results from connecting Points A and B.

Point C is located by dividing:

Extraction flow requirement/approximate steam rate (inlet-to-extraction):

$$70{,}000 \text{ kg/h} \div 23.7 \text{ kg/(kW·h)} = 2955 \text{ kW}$$

Locate Point C at 2955 kW and 70,000 kg/h throttle flow. Now draw line C-D parallel to A-B and another line C to B. Label A-B *Zero Extraction* and C-D *70,000 kg / h Extraction.*

Notice that the general shape of the diagram and the slopes of the lines are determined mainly by the steam conditions used. The turbine investigated here could be built, but one must remember that we have dealt only with the thermodynamic aspects of this application. The mechanical aspects, such as blade stresses, nozzle flow limits, cooling steam and other factors must also be checked.

This example can also be carried further to determine the number of stages in each section using the procedures outlined earlier. To do so requires finding the energy available in each portion of the turbine.

The energy available (ΔH_i) to the inlet-to-extraction section is:

$$\Delta H_i = 3211.8 - 2995.0 = 216.8 \text{ kJ/kg}$$

$$V_b = \pi(635 + 25)\ (4500)/60{,}000 = 155.5 \text{ m/s}$$

V_j for an ideal Curtis stage would therefore be:

$$V_j = \frac{V_b}{0.23} = \frac{155.5}{0.23} = 676 \text{ m/s}$$

V_j for an ideal Rateau stage would be:

$$V_j = \frac{V_b}{0.46} = \frac{155.5}{0.46} = 338 \text{ m/s}$$

With a ΔH_i of 216.8 kJ/kg, V_j for the inlet-to-extraction section with one-stage would be:

$$V_j = \sqrt{2000\ \Delta H_i} = 44.72\sqrt{216.8} = 658 \text{ m/s}$$

This is seen to be very close to the 676 m/s for an ideal Curtis stage. We will therefore assume that the inlet-to-extraction section will contain one 635-mm Curtis stage.

Now for the extraction-to-exhaust section. To find the energy available to this section we need the temperature of the steam entering this portion of the turbine (extraction steam temperature). The enthalpy of this steam will be:

inlet steam enthalpy – ΔH_i (inlet-to-extraction) $\times$ η (inlet-to-extraction) = 3211.8 – 216.8 $\times$ 0.70 = 3211.8 – 151.8 = 3060.0 kJ/kg

From the Mollier diagram, Fig. 14.2, at 18.26 bar and 3060 kJ/kg the extraction steam temperature is found to be close to 315°C.

The extraction-to-exhaust portion of this turbine therefore operates on steam conditions of 17.25 bar (gauge)/315°C/150 mbar (abs.). The energy available to the extraction-to-exhaust section is therefore (from steam table):

$$\Delta H_i = 3060.0 - 2228.0 = 832.0 \text{ kJ/kg}$$

The blade velocity for 890-mm nominal diameter staging with a 25-mm blade height will be:

$$V_b = \pi\ (890 + 25)\ (4500)/60{,}000 = 215 \text{ m/s}$$

If all staging is of the Rateau type in this portion of the turbine:

$$V_j = 215/0.46 = 467 \text{ m/s}$$

ΔH_i per stage is therefore:

$$\Delta H_i = \frac{V_j^2}{2000} = \frac{467^2}{2000} = 109.0 \text{ kJ/kg}$$

The number of Rateau stages in this section would therefore be:

$$\frac{\text{Total energy available}}{\text{Energy removed per stage}} = \frac{832}{109.0} = 7.7 \text{ (say 8)}$$

The turbine will, therefore, contain one 635-mm diameter Curtis stage followed by eight 890-mm-diameter Rateau stages with the extraction opening after the Curtis stage.

Chapter

16

Rerates, Upgrades, and Modifications

Typically, steam turbines are designed for long, useful lives of 30 to 50 years. Periodic replacement of steam turbines' primary working parts may be required, due to the damaging effect of the environment in which they operate. Many parts are custom designed for each particular application, but standardized components are also used to facilitate delivery and minimize cost. Therefore, a steam turbine can be effectively redesigned several times during its useful life. These redesigns are usually referred to as rerates and upgrades, depending on the reasons for doing them. A typical cutaway view of a turbine is shown in Fig. 16.1. For example, the flow path components can be freely redesigned while the basic structure of these turbines can remain essentially unchanged. Steam turbines, because of the unique design, manufacturing, and operating environment in which they exist, can be reworked at regular intervals to match the unit's performance with the owner's changing needs.

Changes in hardware in an existing turbine may be required for many reasons, which can be categorized into four distinct needs: efficiency upgrade, reliability upgrade (including life extension), rerating due to a change in process, and reapplication and modification of existing turbines for a use different from that of its original design.

Efficiency upgrades might be required and desirable for older turbines. As a turbine is used and ages, it experiences loss of performance as a result of change in the nozzle and bucket throat areas and blade profile. Deterioration of finishes always happens, and clearances may also increase.

Depending on a user's specific needs, a steam turbine can be rerated, upgraded, or modified. Possible reasons for these changes include

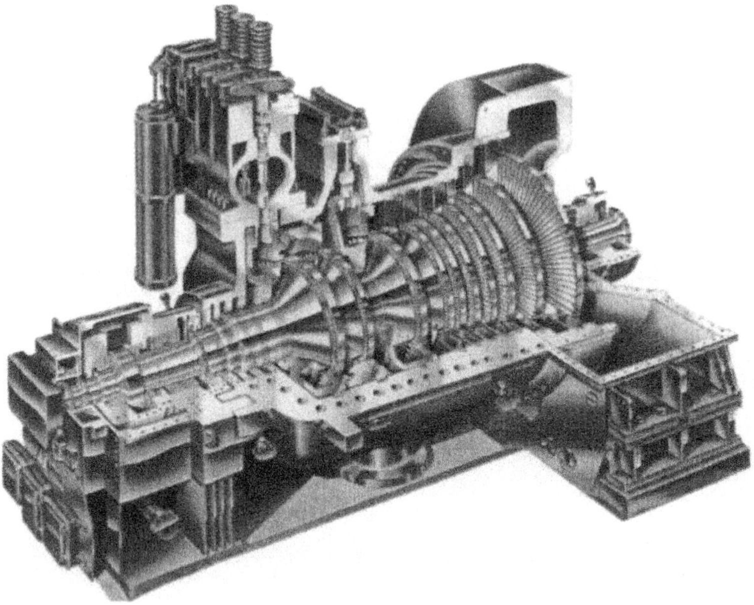

Figure 16.1 Cutaway view of a steam turbine. *(GE Energy)*

increasing power output, optimizing performance, improving reliability, reducing maintenance requirements, solving operating problems, extending equipment life, and, finally, replacing the turbine, either in whole or in part, as a result of catastrophic failure or normal wear, or to rectify problems found during an inspection. A decision can be made regarding possible action, bearing in mind clearly defined goals and taking into account any constraints of the existing equipment and economics. Needless to say, the success of upgrades and rerates depends on effective coordination between the user and the manufacturer.

Experience has shown that reliability may decrease after many years of service. For a typical unit, the forced, unplanned outage rate can increase 3 to 4 percent every 10 years. To arrest this trend, reliability improvements can be made by utilizing new and improved components that may also improve performance. Utilization of new-generation components, such as buckets and nozzles, will aid in an upgrade.

Process change may also require the turbine to be modified, redesigned, or rerated to accommodate changes in process parameters such as new throttle, extraction admission, and/or exhaust conditions. An existing turbine can also be reapplied at a different location or for a different application. Many reviews are required to assess whether a given turbine may be reapplied and what modifications are necessary. (See Table 16.1.)

TABLE 16.1 Reasons and Modifications for Turbine Redesign

Reasons for modification	Possibly due to	Candidates for modification
Performance and efficiency upgrade	New technologies	Buckets/blades
Reliability upgrade	New materials	Rotor
Life extension	Improved manufacturing tools	Control system
Modification and reapplication rather than buying new turbine	Improved manufacturing processes	Thrust bearing
Process change		Journal bearing
Maintenance		Brush seals
Problem solving		Improved leakage control
		Nozzle/diaphragm
		Lube and hydraulic system
		Auxiliaries
		Casing modification
		Exhaust end
		Condensing bucket (L-0)
		Valves—bar lift
		Valves—steam lift
		Tip seals
		Coatings

16.1 Performance and Efficiency Upgrade

The basic power and/or speed requirements of a steam turbine may change after commissioning, for various reasons. The most common reason is an increase (or decrease) in the power required by the driven machine due to a plant expansion or de-bottlenecking. Other reasons include a search for increased efficiency, a change in the plant steam balance, or a change in steam pressure or temperature. Because steam turbines are periodically refurbished, an opportunity exists to update the design for the current operating environment.

In many cases, the desired objective of a rerate is an increase in output power. An increase in power usually requires more steam flow area inside the turbine, which may or may not be possible within the physical limits of the existing casing. A decrease in power is nearly always possible simply by blocking off some flow area, but maintaining efficiency usually requires a more sophisticated solution. Sometimes increased flow is accomplished by removing the second stage of the turbine. It may seem counterintuitive that removing this stage results in additional power, but this modification increases the flow capacity of the stages downstream of the control stage. Increased flow capacity in the downstream stages allows the turbine to pass more flow as a whole (the control-stage nozzle may also need to be modified). More flow translates into more power, even though the turbine efficiency will be reduced.

For example, consider a turbine with a nominal efficiency of 80 percent. If the flow capacity is increased by 10 percent, the power will be increased by 10 percent (at constant efficiency). Even if the turbine efficiency were reduced by 5 percent (0.5 percent to 1.0 percent would be more common), the overall increase in power would still be about 4.5 percent.

More complex design changes can be employed when necessary. If the flow path components need to be replaced, the entire flow path may be redesigned for the higher power and flow. This allows the rerate to be accomplished while maintaining the best possible turbine efficiency.

Rerates are also done when the steam conditions vary from the original design values. For example, it might be determined that the savings in maintenance costs for dropping boiler pressure and temperature outweigh the slight reduction in plant output. The turbine can be operated at the reduced conditions without modification, but it may be desirable to redesign the unit for the new inlet conditions during the next outage. Thus, the maximum amount of turbine efficiency is retained and the reduction in plant output is minimized.

Turbines can be refurbished every 5 to 15 years, depending on the severity of the operating environment. During this time interval, since the turbine was designed and commissioned or since the last outage, technological development will have taken place. These advancements may be in the areas of new technologies, new materials, improved manufacturing tools, and/or improved manufacturing processes. This enables users to decide whether to replace worn components in kind or take advantage of the technological development and incorporate state-of-the-art components. An updated flow path will increase turbine performance, which will result in more power, reduced fuel consumption, or some combination thereof. Typical areas of technological development include seals, airfoil design, stage design (reaction, area ratios, or other flow path refinements), and increased-area final stages (taller blades in the last row).

16.1.1 Brush seals and labyrinth seals

Steam turbines require a method of preventing steam from escaping from the casing at the region of the rotating shaft. Due to severe temperature and pressure conditions, many of the sealing methods used for pumps and compressors have not been practical for steam turbines. Until recently, the most common methods employed labyrinths or close-clearance carbon bushings. In reality, however, these components do not serve as true seals; they act only as throttling devices to minimize leakage. A typical carbon ring seal is illustrated in Fig. 16.2. Depending on design and operating conditions, substantial steam losses can be expected. Even a small steam path leads to *steam-cutting action*. Steam cutting describes an escape-jet action whereby high-velocity steam rapidly and exponentially causes erosive wear of the bushing bore.

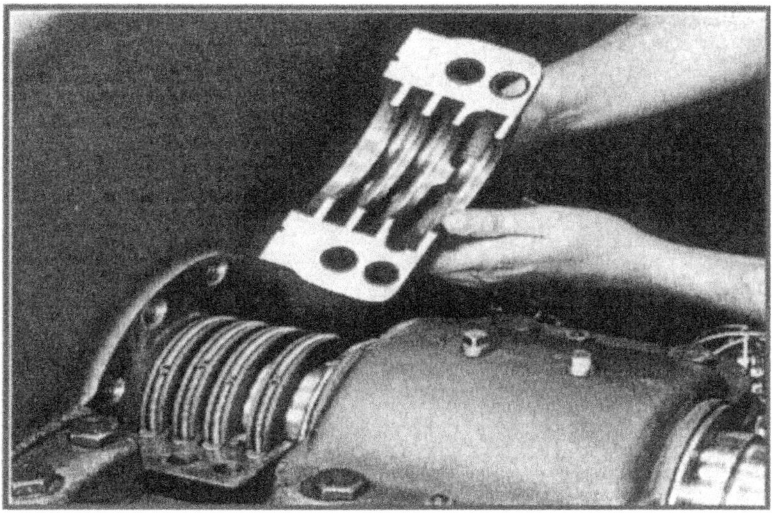

Figure 16.2 Conventional carbon ring seal. *(Saudi Aramco)*

Over the past decades, the industry has witnessed development of brush seals and retractable seals in addition to increased use of spring-back seals. Better seal performance increases turbine performance by ensuring that more steam passes through the flow path, generating useful work. New end gland seal systems return steam to the turbine rather than sending it directly to a gland condenser. In most cases, sealing improvements can be retrofitted into an older design.

Leakage through labyrinth seals can be reduced by up to 80 percent by integrating brush seals with the usual stationary labyrinth seals, as shown in Fig. 16.3. These types of seals reduce gaps, thereby decreasing interstage leakage without rotor rubbing. If not properly selected, however, unwarranted rotor vibration can occur.

The brush seal consists of bristles that are angled slightly with shaft rotation. They can tolerate some deflection and still spring back to their original position. Brush seals can be incorporated between the labyrinth teeth as well as at the ends. Figure 16.4 shows brush seals in a heavy metal retainer. For higher pressures, brush seals may require pressure balancing to avoid excessive downstream deflection. Note that because of the angled bristles, some brush seals may not tolerate reverse rotation. General Electric's advanced compliant brush seal design is based on gas turbine and aviation engine technology. Experience shows that this design, when properly installed, reduces leakage by about 70 percent. See Fig. 16.5 for variation of brush seal designs.

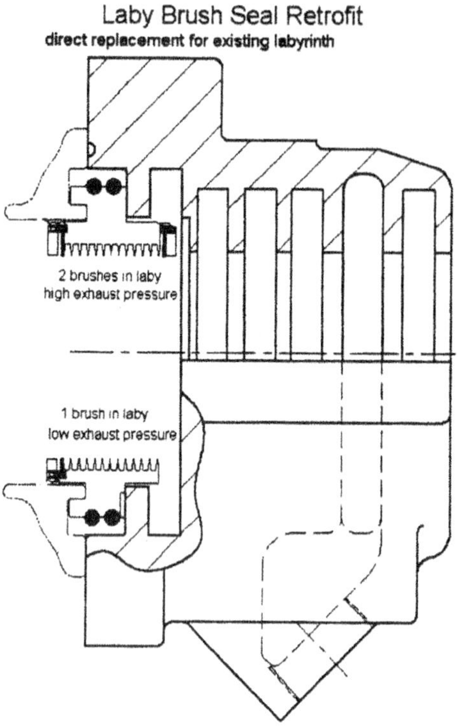

Figure 16.3 Labyrinth-brush seal. *(Elliott Company, Jeannette, Pa.)*

Figure 16.4 A brush seal in a metal retainer. *(Elliott Company, Jeannette, Pa.)*

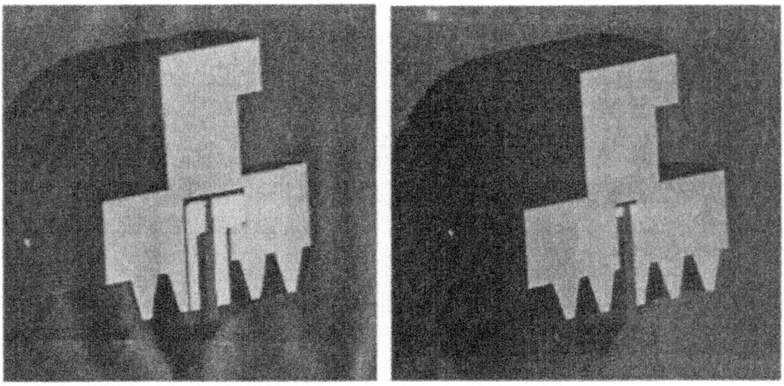

Figure 16.5 Two variations of a brush seal installation. *(Advanced Turbomachine, LLC)*

Labyrinth teeth can be damaged by rubs, especially during start-up or coast-down when the turbine rotor passes through a lateral critical speed. Leakage may also occur as a result of rubs. Rubs may also occur at start-up due to different rates of thermal expansion between the seal and the rotor. The rub opens the clearance, results in leakage, and reduces efficiency. Retractable packing, as illustrated in Fig. 16.6, is a possible solution to this problem. The labyrinth ring is circumferentially divided into segments that are spring-loaded to hold them apart and give a very generous clearance. Once the turbine starts and steam pressure builds up on the outside diameter of the seal, it overcomes the spring pressure and closes the seal to normal clearance. During turbine trips, steam pressure is reduced and the spring again opens the clearance for coast-down. Reduction in leakage flow as a result of improved seal design is illustrated in Fig. 16.7.

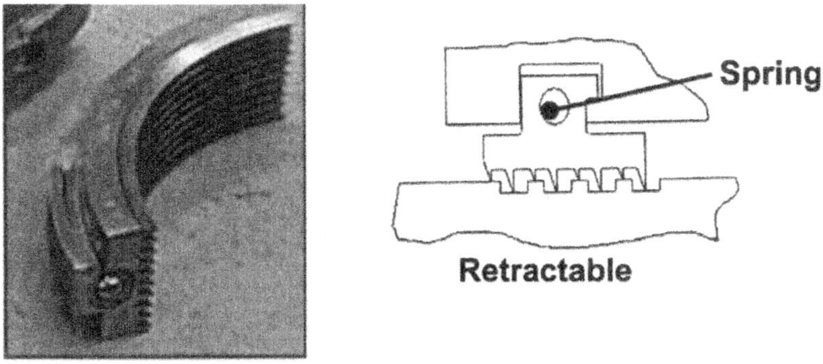

Figure 16.6 An example of a retractable seal. *(Advanced Turbomachine, LLC)*

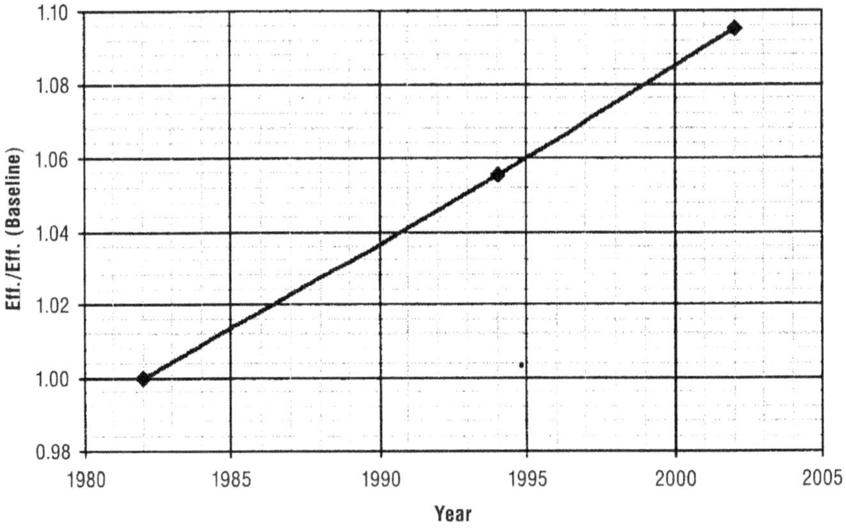

Figure 16.7 Reduction in leakage flow due to improved seal designs. *(Advanced Turbomachine, LLC)*

16.1.2 Wavy face dry seals

Carbon rings are currently the most common way of sealing steam turbines in the size range up to approximately 1500 kW. They are simple in design, but the sealing capability of these devices is very low. This results in high steam losses, lower efficiency, and decreased equipment reliability. New developments since 1980 have superseded the old seal design and are worth considering in both new and retrofit situations.

The design, selection, and installation of a modern noncontacting seal for operation in general-purpose steam turbines is discussed as follows. The goals, objectives, and results of a specific project conducted by a single user from 2000 through 2005 are referenced. The overall project goal aimed to improve turbine efficiency, enhance the reliability of the equipment, and reduce maintenance cost. The design has features that include proper application along with shop fabrication, field installation, and subsequent operation requirements. Details from field application at the user's plants and guidelines for installation are included. The operation of steam turbine seals is discussed, along with the cost justification for the use of dry seals in steam turbines.

Conventional carbon ring seals are made of a special form of graphite. Special graphites have good self-lubricating properties that allow for close operating clearances. Ideally, at final operating temperature, the diametrical clearance between the shaft and the carbon ring should not exceed 0.002 in (0.05 mm). Problems often develop when the turbine is brought up to operating conditions too rapidly. When this happens, the shaft expands faster than the carbon and seizure can occur. Seizure generally causes severe vibration and carbon ring breakage. Hence, closely following a proven run-in procedure is recommended to reduce the risk of carbon ring seizure and to ascertain steam leakage in the more reasonable range.

Because steam leakage is not normally considered an environmental problem, it has not undergone the same scrutiny as other process leakages. Controlling steam leakage, though, would be beneficial for a number of reasons, including environmental reasons. Steam released to the atmosphere represents a loss in revenue, since the user has invested money in water, its preparation and chemical treatment, and energy to create the steam. Since 1985 the significant potential cost savings achievable using gas seal technology in lieu of carbon rings was reported by reliability engineers who had successfully retrofitted high-temperature metal bellow seals to medium-size steam turbines.

Figure 16.8 shows a cloud of steam around the turbine and adjacent areas that was generated as a result of steam leakage. Clouds of steam and condensed water promote corrosion and in turn make working conditions for operators unpleasant and potentially unsafe. Unless advanced bearing housing protector seals are used, the escaping steam will find its way into the bearing lubrication and associated systems.

Figure 16.8 Steam clouds due to an ineffective sealing device. *(Saudi Aramco)*

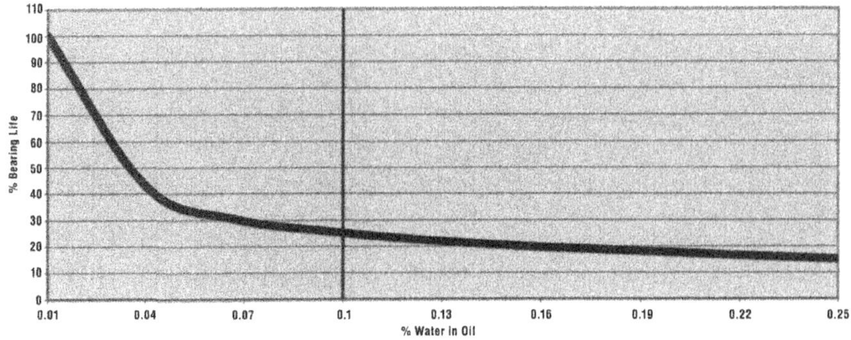

Figure 16.9 The effect of water contamination on the life of oil-lubricated bearings. *(Saudi Aramco)*

The condensed water vapors, an undesirable contaminant, considerably reduce the lubricating properties of the oil and therefore reduce the overall reliability of the equipment. Figure 16.9 illustrates the effect of water content in lube oil on the bearing life. For example, standard rolling element bearings, exposed to 3000 ppm of water in the oil, can typically lose 50 percent of their design L_{10} life. The *L_{10} life* of a bearing, also called *rated life* or sometimes *minimum life,* is the expected life of 90 percent of a group of bearings that meet or exceed the established failure criterion.

Replacing carbon rings with wavy dry seals. A user has 15 multistage pumps driven by 1500-hp medium-speed, single-stage, and noncondensing turbines, which are used to transfer crude oil to critically important stabilizer columns. The turbines operate at 4200 r/min with 625 psig/610°F inlet steam. Steam is discharged into a 60-psig header. The turbines had conventional carbon ring seals with a leak-off line to a gland condenser. Typically, carbon ring seal life is short in high-backpressure service, and excessive leakage was reported after only one week of operation. The most likely reason for this appears to be the worn condition of the gland condensers. Every time the unit went down, there was a standing repair order to change the carbon ring seal.

Chronic maintenance problems began in 1970 and extended to early 2000, including:

- Safety hazards from steam leakage causing potential personnel injury
- Steam seal leakage, producing water, which contaminated the oil system

- Shaft damage from carbon rings rubbing on the shaft
- Vibration problems from poorly fitted carbon rings

Chronic rotor vibration and lube contamination problems are not unusual on conventionally sealed steam turbines in the size range up to 1000 kW. In an effort to find a solution to these problems, different seal designs were tested. As illustrated in Fig. 16.10, the application of noncontacting wavy face dry gas seals to steam turbines has the potential to solve many of these problems. Technologies used to design these seals have been proven in turbomachinery applications. It was established that wavy face dry gas seals in steam service represented one of the latest developments for steam turbine shaft sealing. Similar to the noncontacting wavy face dry gas seals used in centrifugal natural gas liquid (NGL) pump equipment, the steam seal uses its process gas (steam) to separate the seal faces by an amount in the micron range, thereby effectively sealing the turbine steam.

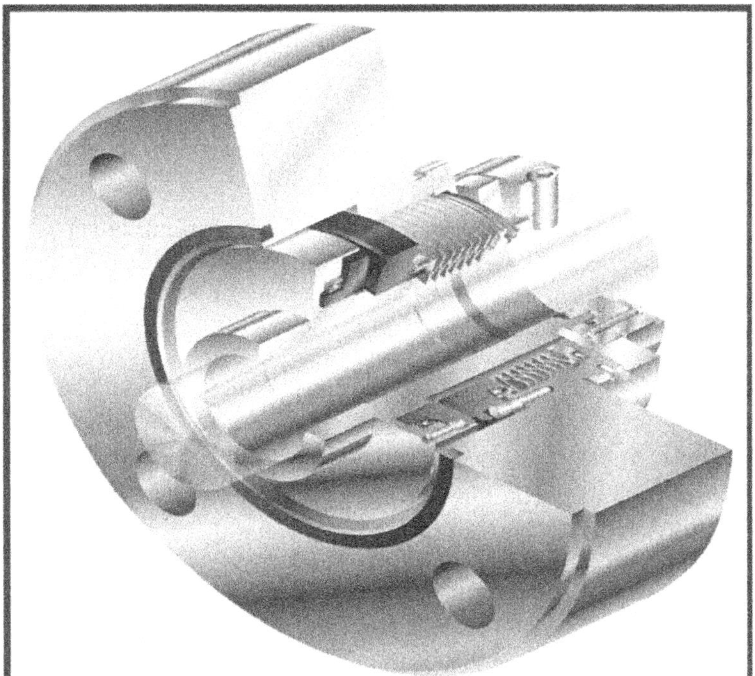

Figure 16.10 Cutaway view of wavy face steam turbine seal. *(Flowserve Corporation)*

Some of the advantages anticipated and subsequently ascertained in the dry gas seal over the more traditional, conventional carbon ring bushing include:

- A safer sealing system, reducing potential for injury from leaking steam.
- More reliable, longer sealing component life expectancy, typically exceeding five years. The average life of carbon seals varies from one to three years.
- Yearly steam leakage reduced by 96 to 99 percent, thus improving the efficiency of the turbine.
- Reduced bearing oil contamination due to steam leakage, thus decreasing the potential for bearing failure and downtime.
- Elimination of the need for gland condensers, steam extraction, and associated piping, reducing maintenance costs.
- Elimination of shaft wear caused by rubbing contact with carbon rings, thereby reducing the need for overhaul.
- Elimination of the requirement for warm-up and run-in cycles.
- Avoidance of an investment in and maintenance of oil purification (reclaiming and/or dehydration unit) for the oil system.

Design goals and available technologies. The following design goals, including the technologies to achieve these goals, were considered:

- Steam is considered to be a poor lubricant for traditional mechanical seal designs. With very light face loading, it may be possible to use contacting faces in low-pressure, low-speed applications. Due to the construction of API-style general-purpose steam turbines, the seals will only be exposed to exhaust pressure in operation, usually not exceeding 100 psig. The exhaust pressure still requires the use of liftoff faces similar to those used in compressors and dry gas pump seals.
- Noncontacting gas seals operate on a very thin gas/steam film to both minimize leakage and prevent face contact. While steam turbines operate on steam, they are also exposed to condensate, a variety of water treatment chemicals, and various forms of pipe scale and deposits. The presence of foulants in this thin film can cause damage to the seals. In addition, a buildup of debris due to condensate flashing across the seal faces can increase both the film thickness and seal leakage. In its various manifestations, this presents a challenge to seal designers.

- Material selection is also a key consideration. Steam conditions are relatively aggressive, and components must withstand long exposure to high temperatures. Elastomeric materials have not proven reliable in high-temperature steam applications. Metallic materials also require close examination. Although most turbines are constructed from steel, the combination of steam, high temperature, and air on the atmospheric side of the seal may cause corrosion. At high temperatures, it is also beneficial to match thermal expansion of critical surfaces to assist in piloting the rotating equipment components and providing sleeve sealing.
- The new seal must be adaptable to existing equipment and must be installed in place of the existing carbon rings. It should be noted that carbon ring seal boxes generally fall into two categories: detachable and integral.

  A *detachable* carbon ring seal box is bolted into place on the side of the turbine. This allows for easier maintenance and replacement. Steam seals for these turbines are mounted externally and replace the carbon ring seal box.

  An *integral* carbon ring seal box is part of the casing or the case end covers; thus, an integral carbon ring box cannot be removed from the turbine. The selected steam seal is then mounted internally and must fit within the dimensional envelope of the existing box. Typically, and for existing installations, the turbine casing will have to be machined in the shop to accept the new cartridge seal modification.

Close attention should be paid to design details when choosing a dry gas/steam seal. From a reliability point of view, the components in any single mechanical seal are not independent and are basically functioning in series. Therefore, the failure of a single component due to design error is likely to lead to an overall failure of the seal assembly.

Seal design specifics. When high temperature prevents the use of elastomeric materials, a metal bellows seal should be selected. This eliminates the problem associated with hang-up due to debris building up on the dynamic seal interface. Static gaskets made of flexible graphite or high-temperature composites should be chosen. Using floating faces would eliminate problems associated with shrink fits or face-clamping designs. Wavy face technology is used for face liftoff, since it provides the highest degree of water and debris tolerance.

Figure 16.11 illustrates the main features of such a wavy face technology steam turbine seal. The bellows core of this type of seal is made of Inconel 718 alloy, which has been used successfully for many years in high-temperature pump applications. Inconel 718 alloy has excellent

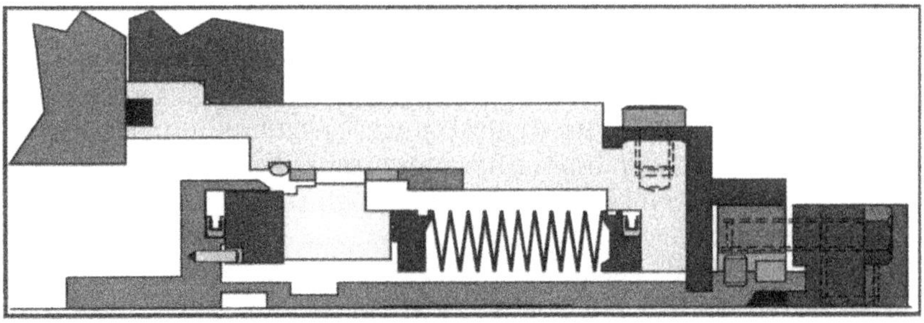

Figure 16.11 Cross section of wavy face shape (WFS) technology steam turbine seal. *(Flowserve Corporation)*

high-temperature properties and good corrosion resistance to chemicals present in steam. In this design, the bellows core provides only an axial load to the seal faces. It does not provide any drive function. This eliminates stick-slip, fatigue, and associated dynamic problems. The flange at the end of the bellows is lapped flat and it seals against the carbon seal face. This lapped support surface allows free thermal expansion between the bellows and the face, which helps maintain face flatness.

In order to design stationary carbon faces to eliminate pressure distortion, finite element analysis (FEA) has been used. Since the face is not clamped into place, it remains flat during operation. The face is driven by two drive lugs at the outer diameter (OD) and piloted to the inner diameter (ID) of the flange.

The rotating face is made of silicon carbide and is manufactured using wavy face technology for face liftoff. The wavy face shape (WFS) common to all such seal designs is shown in Fig. 16.12, which depicts the key features associated with this shape. These features are waviness, tilt and the valley of wave, and the seal dam. When mated against a flat seal face, the waviness forms circumferentially converging and diverging regions. For the seal pressurized at the OD under dynamic operation, the circumferentially converging regions hydrodynamically compress the gas to develop a pressure at the wave peaks. This pressure is considerably higher than the surrounding bulk gas and results in a hydrodynamic load to permit noncontact operation. The WFS is nonpumping, meaning that gas will not be forced across the face during dynamic operation.

Near the ID is a sealing dam that minimizes leakage past the seal faces. When the face is rotated, gas is drawn into the valleys of the waves and is compressed as the fluid film decreases at the peaks. This process creates a high-pressure region that is capable of separating the two faces, which allows the seals to operate without face contact. The second aspect of this design provides for the cleaning action of the

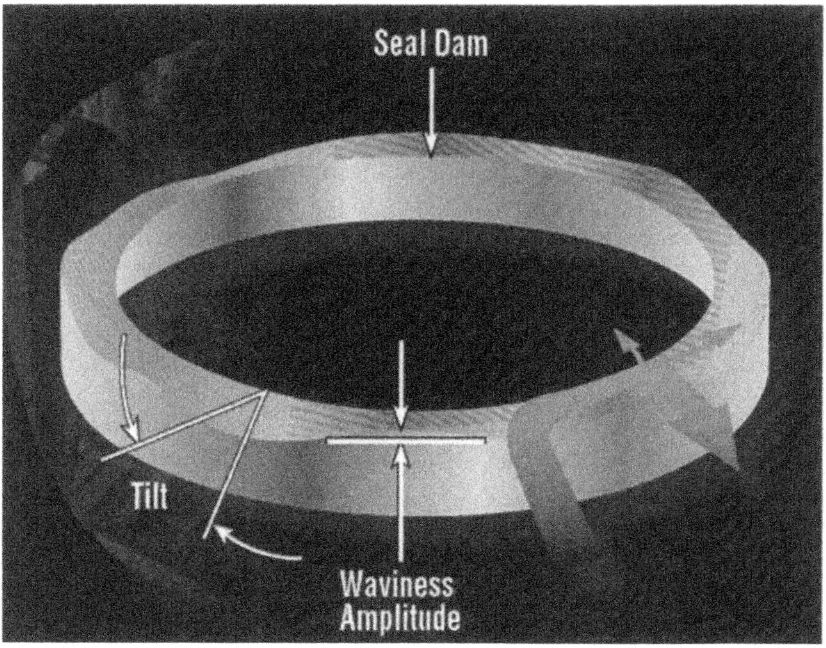

Figure 16.12 Wavy face technology features. *(Saudi Aramco)*

faces. As the gas is compressed near the peaks of the waves, it can relieve pressure in three directions, as shown in Fig. 16.13. Some fluid will go across the seal dam as leakage and some will flow over the peak to create lift. The majority of the fluid will flow toward the OD of the seal and go back into the seal chamber, which is at a lower pressure. This continual flow of gas across the faces creates a self-cleaning action. Since the face is free from grooves or slots, there is no place for

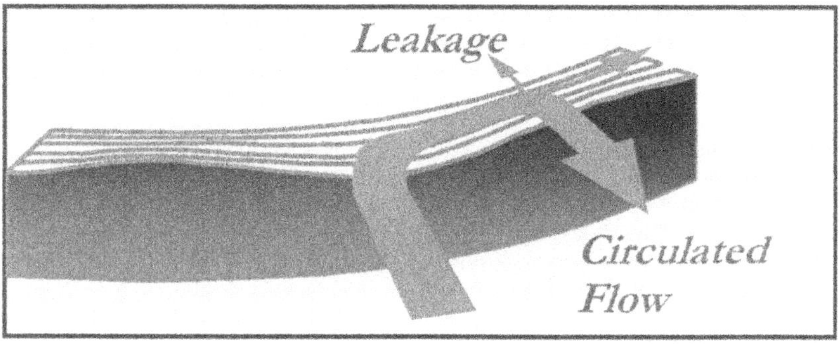

Figure 16.13 Circulation effect of wave shape. *(Saudi Aramco)*

the contamination to collect. This helps make the wavy face more tolerant of contamination than grooved face designs.

The rotating face, like the stationary face, is free floating and is not clamped into place, which minimizes clamping stresses and stresses imposed by differential thermal expansion between the face and the sleeve. The sleeve is also designed to minimize distortion. Due to the fact that part of the sleeve is exposed to hot steam conditions and part is exposed to cooler atmospheric conditions, the sleeve will distort as a result of thermal expansion. The sleeve was designed to decouple its hot and cold ends through a series of grooves, which allows the sleeve to be exposed to high temperature differences and still maintain a flat support surface for the face. The radial support surface for the face is kept small to minimize the effect of any residual distortion. (See Fig. 16.14.)

The design of wavy face seals was considered to be acceptable after a proper design review. The decision was made to move forward with the design and its implementation. The design had the potential for success, but field validation was desirable. A number of trial installations were run to evaluate the performance of the seals in operating turbines. These installations exposed the seals to a variety of the tasks and services that can be expected in industry. One of the conditions for selecting these field sites was that the units were operated for extended periods of time. Another requirement was that the turbine be examined to ensure that sealing and piloting surfaces were suitable for seal installations.

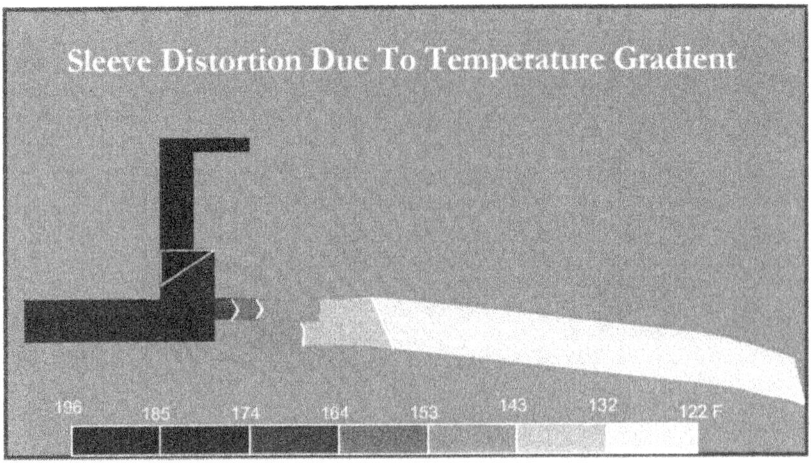

Figure 16.14 Sleeve distortion due to temperature. *(Flowserve Corporation)*

Figure 16.15 Eroded carbon ring boxes. *(Saudi Aramco)*

Installation of seals. Prior to assembly, the turbine case assembly was removed and sent to the user's shop for repairs and modification to accept the new dry steam seals. When the machine was opened for inspection, the following deficiencies were found:

- Several oil passages in the case were about 50 percent blocked by buildup of debris.
- Varnish deposits were found on parts caused by oil breakdown from high temperatures, most likely due to steam impingement on the bearing housing.
- Numerous carbon rings in the original gland seal packing boxes were broken.
- The carbon ring boxes were eroded on the downstream side of the sealing area. (See Fig. 16.15).
- The trailing edges of the nozzle block vanes were thin from erosion, with some sections missing.

These deficiencies came as no great surprise; after all, 11 years had elapsed since the machine had last been subjected to a case inspection.

Typical restorative work was done to return the turbine to original design specifications and some additional special machining was performed to accommodate the WFS seals, such as the following:

- The majority of sealing faces were refinished and the gland boxes were bored to fit the new seal cartridges (inside seal chosen).
- The case horizontal split surfaces were grounded.
- Both inlet and discharge faces were machined.
- Trip and throttle valve faces and valve seats were machined.
- Any eroded nozzle block was corrected.
- Other small deficiencies were corrected, as needed.

Progressive balancing was performed on the spare rotor before the seals were installed. Once the seals were installed on the shaft, no further balance checks could be performed because some parts of the cartridge were stationary in service.

Installation sequence

- To mount the seal cartridges on the shaft, all rotating flinger rings, thrust discs, the coupling hub, and worn gears were removed.
- The bare rotor was balanced; the thrust disc and the remaining accessories were added and balanced with the assembly.
- The upper case was set and the seal gland faces were checked relative to the shaft at that time and the total indicated run-out (TIR) was found to be 0.004 in. The seal manufacturer required them to be within 0.020 in of being square with the shaft, so no further corrections were made.
- The split seal housings were installed and swept with an indicator and were within 0.003 in TIR. The seal manufacturer asked for this to be within 0.005 in TIR.
- The upper case was then removed. All accessories were removed from the rotor, and the seals were installed on the rotor. The rotor was set in the case, the seals were located and secured, and the turbine assembly was completed. (See Fig. 16.16).

The turbine was reinstalled on schedule in December 2002 and the unit was started. As of mid-2007, the new seals had operated virtually trouble-free since plant start-up. In five years there was zero annual seal maintenance, and no change in seal leakage rates was experienced. Based on these highly favorable results, the user has now performed in-house retrofits on a total of 15 units. Various steam turbine manufacturers are represented, and the same type of liftoff seal was installed in each.

Cost vs. benefit analysis. Business requires steam turbines to achieve minimum runs of three years without downtime maintenance. This dry

Figure 16.16 A steam turbine fitted with the new seal. *(Saudi Aramco)*

gas seal retrofit project was initiated because it was less costly than other options. Traditional cost outlays were avoided because vacuum gland condensers and lube oil dehydrators were no longer needed.

Table 16.2 lists the estimated economic benefits of the dry gas seal. The cost of the seal was taken from the quotation provided by the manufacturer for one pair of dry seals.

The value of yearly steam leakage loss avoidance took into account ring leakage rates as a function of the backpressure of the unit and the information provided by the original equipment manufacturer (OEM). A total steam leakage rate of 200 lb/h was used. Leakage of noncontacting seals was based on 0.4 lb/h/seal and was estimated by the seal manufacturer. Total steam costs were then calculated from the unit steam cost times total leakage. Again, the cost will vary, depending on the availability of fuel and water at each plant.

The average expected seal life for the previously used carbon rings was based on plant history; the facility's records showed carbon ring lives to range from one to two years. Life expectancy depended on the operating conditions and how the carbon rings were broken in. The estimated seal life of a noncontacting wavy face steam seal is a minimum of five years. This is not unrealistic, because there are seals that have been operating for over six years without changeout.

Operating hours per year for the steam turbine are projected as 7560, which takes into account operation at full load, operation in slow-

TABLE 16.2 Cost and Payback Analysis

Steam seal type	Carbon ring	WFS
Cost of seals (Both INB/OUTB)	$1000	$20,000
Initial retrofit cost	0	$5000
Labor for seal repairs per year	$2160	0
Parts/labor on repairs (bearings, etc.) per year	$3000	0
Preventive maintenance costs per year	$1500	$1500
Maintenance on auxiliary equipment per year	$1120	0
Steam losses per year	$18,000	$200
Total operating cost per year	$26,780	$1700
Payback period		12 Months

roll condition, and operation in standby mode. This number is identical for both the carbon ring seal and the dry gas seal.

Yearly operating costs, including parts, labor, and steam loss, were based on actual and projected values. The cost of repairing a steam turbine would be applied when it occurred. For this case, the cost was distributed over the life of the sealing device. Simple expressions were used to calculate the three components of operating cost:

1. Cost of labor/year = cost of labor/expected seal life
2. Cost of parts/year = cost of parts/expected seal life
3. Steam losses/year = (leakage rate/side) (number of sides) (cost of steam) (operating h/year)

These costs were calculated for both carbon rings and dry gas seals and were totaled separately.

The payback period was calculated by adding the cost of the noncontacting wavy face steam seal and the operating costs per year for the seal, and then dividing by the operating costs per year for the carbon ring seal. The actual payback period, based on field results, was 12 months.

Table 16.2 indicates that current standards and the best available technology improve upon the basic design deficiencies that led to the problems that were originally experienced with conventional carbon ring seals.

16.1.3 Buckets

New design and analysis techniques allow more sophisticated airfoil sections to be generated. These new airfoil shapes have lower loss, increased incidence range, and better accommodation of supersonic flows. Such design enhancements reduce airfoil losses and increase the useful work extracted from each pound of steam. Also, the stacking of

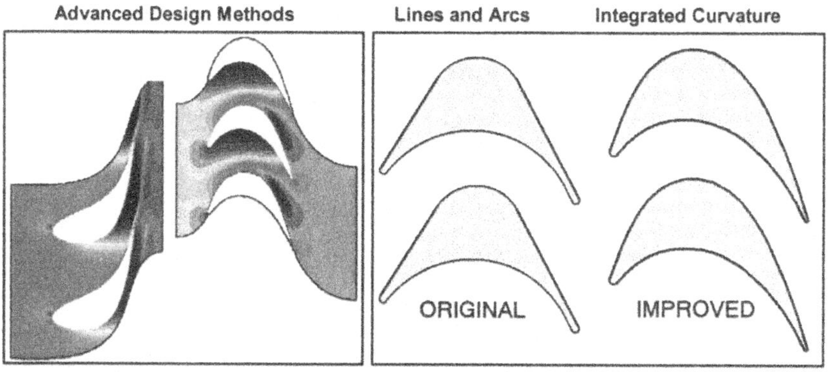

Figure 16.17 Improved numerical airfoil design methods, and resulting improved airfoil section design. *(Advanced Turbomachine, LLC)*

these sections can be used to improve turbine performance. Compound leaned, or bowed, vanes have higher losses than conventional vanes, but they reduce the secondary flow losses on the downstream blade enough to obtain a net increase in stage performance. Improvements in airfoil sections, and the numerical methods used to generate such designs, are presented in Fig. 16.17.

Finally, the basic design of the stage may be improved. Changes in the stage reaction level, nozzle–to–blade area ratio, and end wall shape can all contribute to increased performance. In general, increased reaction results in increased efficiency due to a more balanced sharing of the aerodynamic load between the stator vane and the rotor blade. End wall curvatures and slopes can be fine-tuned to direct the steam flow in directions more advantageous to overall performance, resulting in reduced stage losses. Figure 16.18 illustrates the improvements in stage efficiency through increased technology over a period of 20 years.

More efficient buckets may be available for older turbines. Considering the cost of energy, it may make sense to refurbish older turbines with new blades (both rotating and stationary) if the gain in efficiency is large enough. However, higher-efficiency buckets often require more axial spacing along the rotor, and older turbines may not provide enough room in the existing casing. In response, manufacturers have offered advanced-design blades. For example, Fig. 16.19 shows a collection of blades from control stage to back end, offered by General Electric. These buckets have a cylindrical tip profile for reliability and performance improvements. Flat skirts are provided for strength, while round skirts are offered for efficiency improvement. Plasma coating is offered for protection, and condensing end buckets are twisted and tapered. This free vortex design also provides improved section reliability and enhanced performance. For variable-speed,

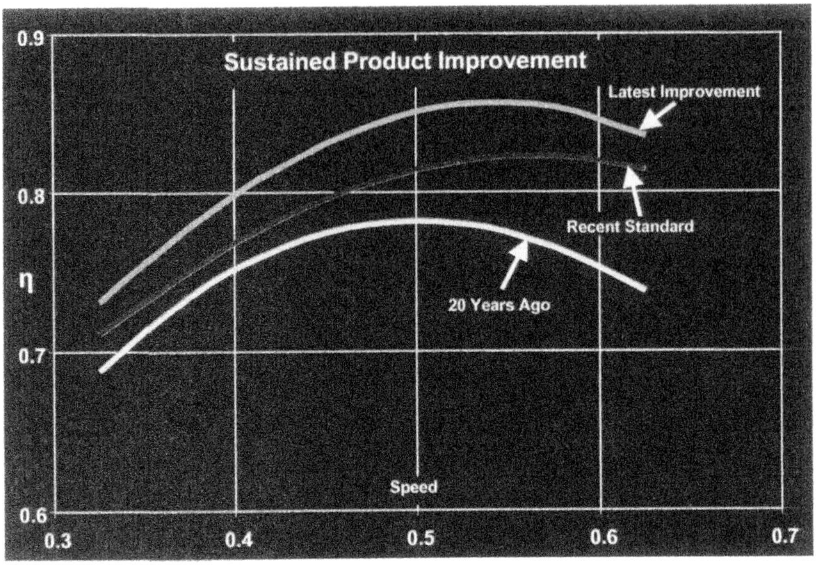

Figure 16.18 Example of stage efficiency improvements over a 20-year period. *(Advanced Turbomachine, LLC)*

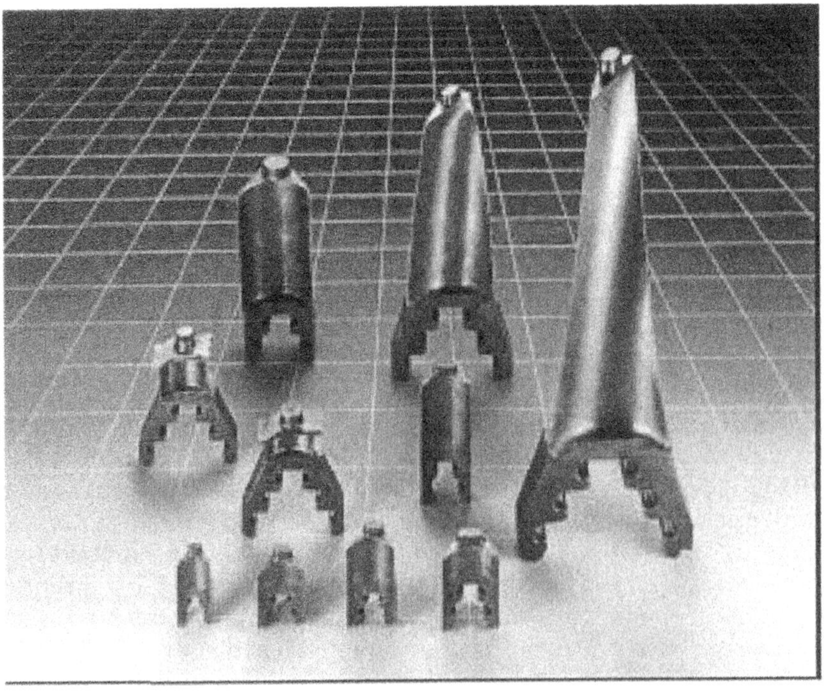

Figure 16.19 Improved blades for front end to back end. *(GE Energy)*

mechanical-drive turbines, buckets are provided with harmonically tuned nozzles and larger setbacks. Buckets are banded with long-arc shrouds to reduce stresses arising from low-frequency vibrations. Flame-hardening or solid material tails in buckets provide improvement in erosion resistance.

The overall effect of all turbine upgrades can be quite dramatic, as noted in Fig. 16.18. For example, in the unit represented in Fig. 16.20, improvements in sealing, airfoil design, and flow path refinement combined to produce an 11.4 percent measured increase in unit efficiency.

In many cases, rerating and upgrades are combined in one operation to increase efficiency. By combining an upgrade with a rerate, an existing turbine is brought up to current technological standards and, at the same time, fine-tuned for its current operating environment. Thus, the effectiveness of better matching the turbine to its current operating condition and the addition of improved technology designs typically result in significant increases in operating efficiency. In some cases, rerating is not possible without upgrading the turbine design as a result of power or stress limitations on the original components.

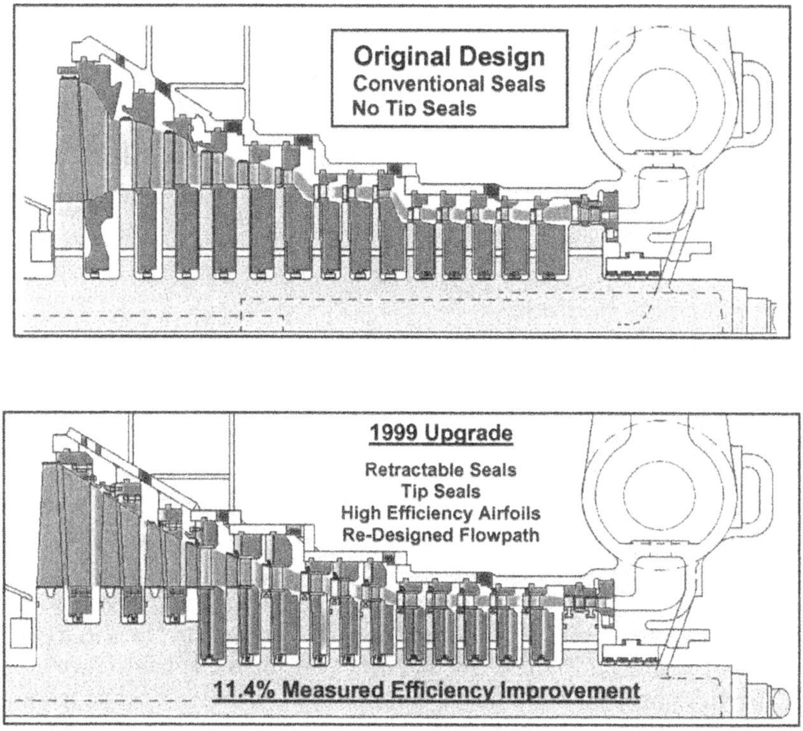

Figure 16.20 Performance improvement due to a comprehensive upgrade. *(Advanced Turbomachine, LLC)*

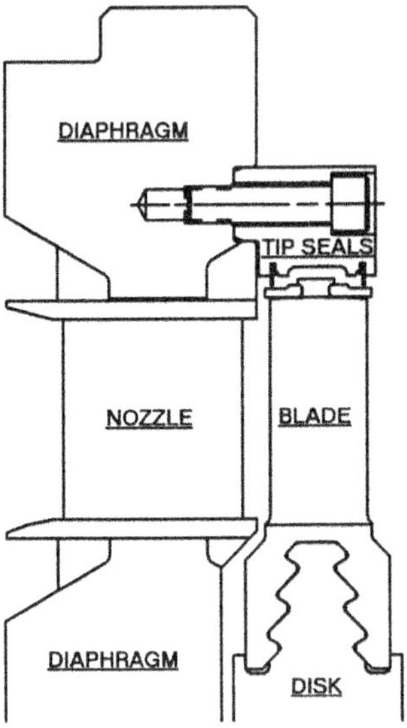

Figure 16.21 Tip seals. *(Elliott Company, Jeannette, Pa.)*

Almost all turbine stages have seals between each diaphragm and the shaft, but many impulse turbines do not have tip seals between the rotor blade tips and the casing/diaphragm. If room permits, tip seals can be added to increase efficiency. Tip seals, as shown in Fig. 16.21, are more effective when used in higher pressure stages and stages with greater reaction.

Any of the seals mentioned here can combine the features of the brush seal and the retractable seal.

16.2 Reliability Upgrade

The trend for longer runs and shorter turnarounds creates pressure to eliminate unplanned shutdowns and reduce required maintenance.

16.2.1 Electronic controls

One of the biggest changes in large steam turbines in the past two decades has been the conversion of the speed control and trip systems from wholly mechanical to wholly electronic.

Figure 16.22 illustrates a typical mechanical governor system for a straight-through turbine. The governor is powered off of the turbine shaft by a worm and wheel drive. The governor itself has many internal moving parts, including flyweights, springs, an oil pump, accumulator pistons, and several valves. Governors are reliable, but the moving, contacting parts result in inevitable wear, and maintenance is required.

Speed control for extraction turbines.

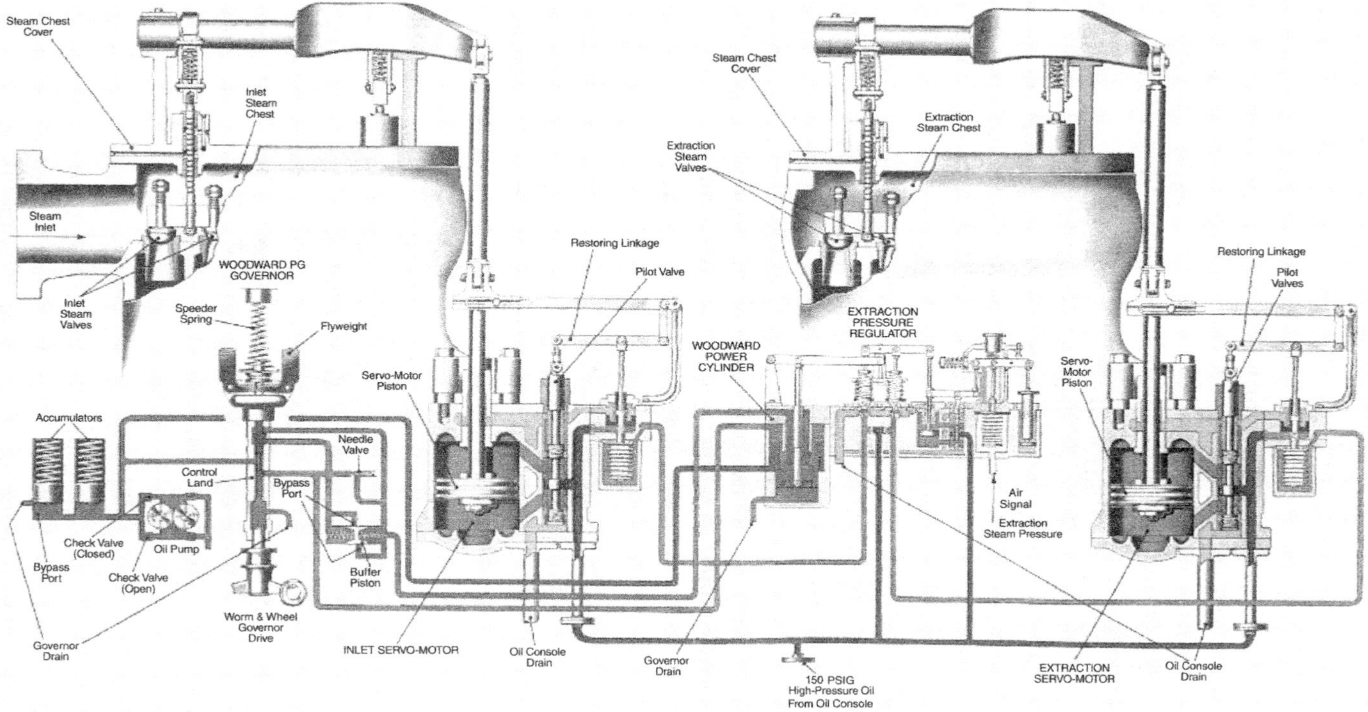

Figure 16.22 Mechanical governor controls for an extraction turbine. *(Elliott Company, Jeannette, Pa.)*

An electronic governor replaces the worm gear with a toothed wheel. Multiple noncontacting magnetic speed pickups read off the toothed wheel and provide the signal to the electronic governor. The power cylinder in Fig. 8.8 is replaced with an electronic or pneumatic actuator. The pilot valve and restoring linkage remain intact. Many electronic governors are also more versatile because they can be programmed to use parameters other than only speed to control the turbine. Most electronic governors can also interface with the user's DCS (distributed control system). A system gets more complicated when mechanical controls are applied to a controlled extraction turbine, as shown in Fig. 16.22. In Fig. 16.22, the power cylinder controls the added extraction pressure regulator, which includes even more linkages and springs as well as a pneumatic signal to monitor extraction steam pressure. The extraction pressure regulator sends signals to the prepilot cylinder on either or both servomotors. This system can maintain a set turbine speed and an extraction pressure level. However, the linkages and springs wear, and need adjustment, if there is a change to extraction pressure.

An electronic governor designed for extraction turbines works as described here for straight-through turbines, but it also receives an electronic signal for extraction steam pressure and outputs signals to the electronic (or pneumatic) actuators that replace the prepilot cylinders on each servomotor. Other signals (such as a compressor discharge pressure signal) can be input and processed by the electronic governor. Here a change to the extraction pressure becomes a simple set-point change.

Electronic governors are not failproof, but they are available in redundant and triple modular redundant formats, so failure of electronic components will cause the governor to switch to backup components while the failed components are replaced. The turbine continues to operate normally during the component replacement.

Most multivalve turbines still require hydraulic servomotors, since electronic ones are not powerful enough to move the valve racks. The prepilot and pilot valves may be eliminated, in addition to the linkages, cams, and rollers shown as the restoring linkage in Figs. 8.8 and 16.22. Figure 16.23 illustrates that the servomotor can be fed by a way valve. The way valve takes the oil directly from the control oil header and directs it to the proper side of the servomotor. The electronic governor sends a control signal to the actuator coil that is an integral part of the way valve. One or more linear variable differential transformers (LVDTs) replace the restoring linkage and provide feedback to the control system. The way valve may have dual actuator coils for redundancy.

Turbines are still tripped by dumping the oil that holds the trip and throttle valve open. Figure 7.8 illustrates the arrangement formerly used on most turbines. The overspeed trip was initiated by a trip pin loaded with a spring or a weighted Belleville spring, which struck a

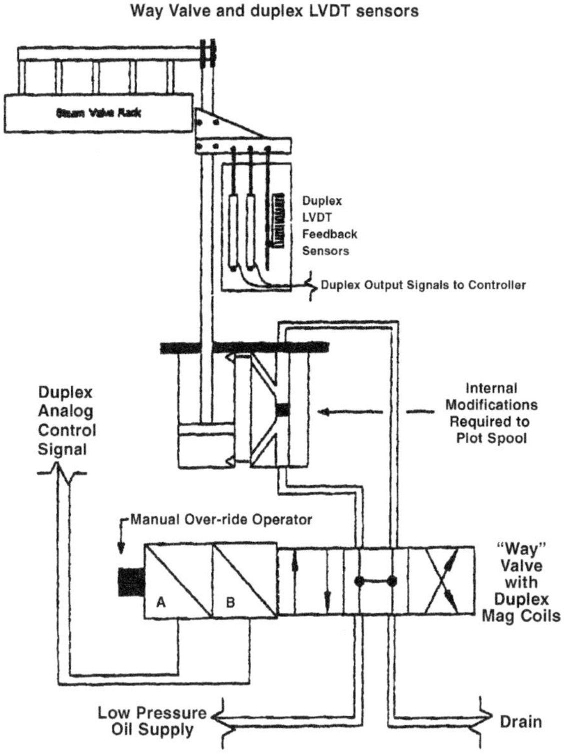

Figure 16.23 Way valve retrofit. *(Elliott Company, Jeannette, Pa.)*

mechanical lever that actuated a dump valve. The solenoid dump valve shown was for remote tripping purposes.

Electronic trips, which are standard on most new turbines, use three noncontacting magnetic speed pickups that read off a toothed wheel. This is the same method that sends the electronic governor its signal. The electronic overspeed trip control box uses two-out-of-three voting logic to trip the turbine and eliminate spurious trips. Two solenoid dump valves are piped in parallel to provide redundancy. It is sometimes difficult to find enough room on an existing turbine to mount the necessary speed pickups, but beyond that, it is fairly simple to add an electronic overspeed trip. It is important to be sure that the speed pickup wheel is attached to the shaft in a manner that will not allow it to come loose during operation. If the wheel should come loose, the trip function will be lost and the turbine will overspeed until something fails, unless it has a backup mechanical trip.

The solenoid dump valves that are activated by the electronic trip box are the only part of the system that cannot normally be tested without either tripping the turbine or defeating the overspeed trip function. If the turbine is scheduled to run for several years between shutdowns, this could be a reliability problem. A trip block system is available that can connect easily into existing trip systems and can constantly monitor and test the trip solenoids for function without accidentally tripping the turbine or temporarily defeating the trip function. This trip block is shown in Figs. 16.24 and 16.25.

16.2.2 Monitoring systems

Many steam turbines that have been in operation for at least 30 years do not have the level of instrumentation that is considered standard today. Radial and axial vibration probes, as well as bearing temperature instrumentation, may be fitted to existing turbines. Readouts from these instruments can be fed to sophisticated monitoring equipment that displays the current readings and also determines trends that may uncover a potential problem before it causes an unexpected outage. The instruments could indicate that a change to a different bearing type (tilt shoe, spherical seat) or orientation, such as load between pads, would be beneficial.

The main stumbling block to retrofitting vibration probes is finding the space in the bearing housings to mount them with a corresponding free area on the shaft for the probes to observe. The shaft area may also have to be treated to reduce electrical run-out to an acceptable level to avoid false readings. Both mechanical and electrical shaft run-out should be measured so they can be vectorially subtracted by the monitoring system to yield true vibration readings. Adding bearing temperature instrumentation normally requires some machining of the bearing retainers and housings to allow the wires to exit the turbine.

It is important to note that it is fairly common for machines that have been operating satisfactorily for many years to show high vibration readings once probes are installed. The excessive vibration may have always been there, but hasn't yet risen to the point where it has caused a problem. In such cases, the vibration level is normally still not a problem. Although it exceeds current standards, if the vibration has not caused any operating or maintenance problems in the past, there is no reason to believe that it will be a problem in the future. With the probes installed, however, it is now possible to monitor and trend the vibration levels and to record and observe indications of potential future problems.

16.3 Life Extension

There are several options available to extend the life of an existing turbine. Machining off the damaged area, rebuilding it with a weld, and remachining the rotor can repair most types of rotor damage.

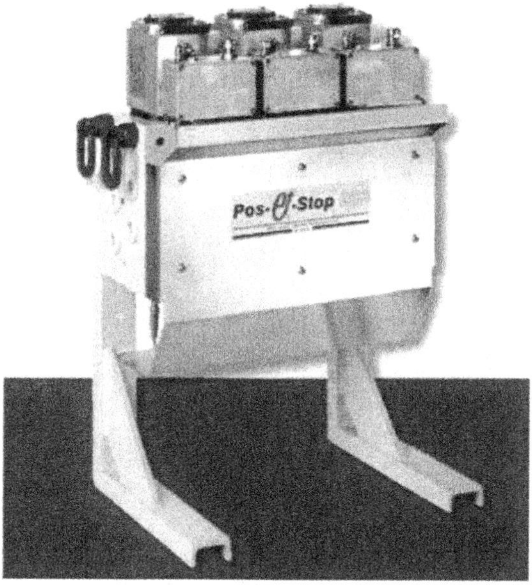

Figure 16.24 Solenoid trip block. *(Elliott Company, Jeannette, Pa.)*

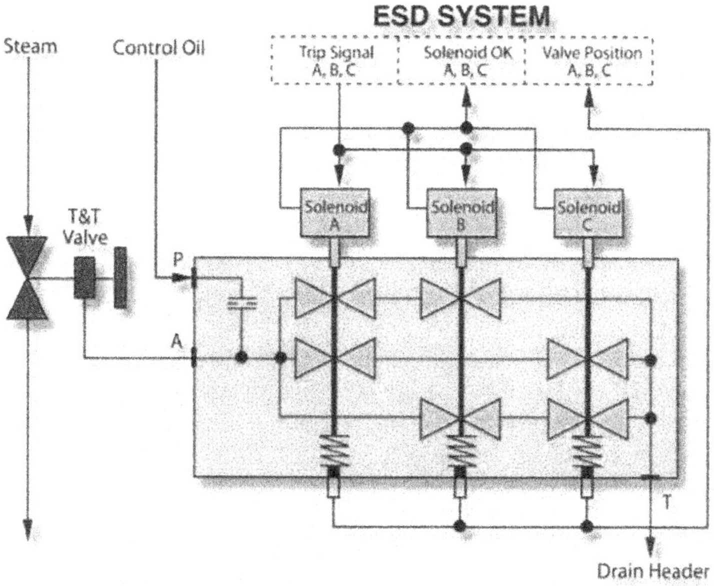

Figure 16.25 Solenoid trip block schematic. *(Elliott Company, Jeannette, Pa.)*

Even entire new disks may be added to an integral (solid) rotor with this technique. Some casing damage may be repaired in a similar manner. Better materials are often available to solve erosion problems. Blades can be protected by a stellite overlay in critical areas. Some coatings are available that may help with erosion, corrosion, or fouling problems. A bearing upgrade or something as simple as an at-speed balance could also extend the life of a turbine by eliminating a vibration problem.

16.4 Modification and Reapplication

Sometimes a situation arises when some type or level of modification of an existing turbine is required. An old turbine might be in use or a need might have arisen as a result of damage inflicted on a turbine during a wreck. A new turbine is certainly an option, or there might be used turbines on the market that could be refurbished to meet the requirements. There might even be another turbine available that could be reapplied.

If the turbine is damaged in a wreck, the discussion is entirely different. This is an emergency, and a quick solution is essential. A new turbine is usually out of the question because the delivery time might not be acceptable. Depending on the extent of the damage, most original equipment manufacturers (OEMs) can accomplish fairly major repairs in a short period of time by utilizing welding, plating, and other proven repair techniques. Blades and other specialized parts might be available in a reasonable amount of time.

Catastrophic damage requires a search for a suitable replacement. If desired, a suitable turbine that can be used with modifications might be available on the used equipment market. A suitable turbine may be overhauled with minor changes and be back in service in a fraction of the time required for a new unit. If no suitable turbine can be found on the used market, it might even be possible to reapply another of the user's turbines (possibly in less critical service) to replace the damaged turbine. It may then be feasible to replace the damaged turbine with either a new or a used unit.

The success of any project depends on a clearly defined plan consistent with the goal of the situation. It is critical in the case of modification to properly execute the plan. The lack of a defined goal and a plan increases the chance that the project will fail. In addition, a review of some key components must be conducted to achieve success.

An analysis of the driven machine is undertaken to determine the actual power requirement and speed, and to perform an evaluation of steam conditions at all expected operating points, including excess power margin. Pressure drops in supply piping, exhaust, and extraction piping need to be calculated accurately. On extraction units, the extraction pressure and flow must also be defined.

Many turbine standards may have been considered during original design, including corporate, API, NEMA, industrial, and local governmental standards. Decisions should be made to determine which specifications are required to be applied, including specifications that have been revised or enacted since the unit's original commissioning. Older turbines must be carefully analyzed, as many older units cannot meet current standards or may do so only at considerable added time and expense.

16.4.1 Casing

The pressure and temperature ratings of the casing should be checked when a change in inlet pressure, temperature, and/or exhaust pressure is required. Any such review should include casing and flange ratings to ensure that the casing is suitable for the new conditions. Multistage turbines with changes in pressure and flow require a verification of the first-stage maximum allowable pressure. In the case of a multistage extraction turbine, a change in flow or extraction pressure will also require review of the intermediate casing and extraction flange.

A change in inlet temperature requires a verification of the casing material. There are many other materials that have been used over the years, but the materials listed in Table 16.3 are the major casing materials currently in use. If new conditions exceed these limits, a thorough review of the metallurgy of the existing casing must be done. Typical temperature limits for these materials are given in Table 16.3.

Once the casing rating has been determined, the original hydrotest pressure must be determined. This is done to ensure that hydrotest pressure meets the new conditions based on the latest industry standards. The API guidelines refer to ASME Boiler and Pressure Vessel code section VIII. In some cases, the casing will need to be re-hydrotested to the new conditions.

TABLE 16.3 Casing Material Temperature Limits

Casing material	Maximum temperature (°F)
ASTM A278 Class 40	600
ASTM A216 Grade WCB	775
ASTM A217 Grade WC1	825
ASTM A217 Grade WC6	900
ASTM A217 Grade WC9	1000

If NEMA limits are applicable, then the casing should be verified for flange rating. For new construction, most OEMs verify that the casing design will meet the maximum design pressure and temperature plus 120 percent of the rated pressure combined with a temperature increase of 50°F. This verification is also included in the evaluation of the flange rating. Casing modification may be possible to achieve turbine upgrade in many cases. For example, an extraction section might be added or modified. It is also possible to convert a casing from condensing to noncondensing. Inspection of the interior boroscope parts can be added if desired. Inlays of materials have also been used to repair casing damaged by erosion.

Exhaust end improvement is also possible by converting old cast-iron casings with fabricated steel construction.

16.4.2 Flange sizing

If, during the modification of the turbine, the flow or the specific volume increases, the size of the inlet flange must be reviewed. The typical maximum value for the inlet velocity is 175 ft/s. Equation 16.1 determines the inlet velocity:

$$V = 0.0509Mv/d^2 \tag{16.1}$$

where V = velocity in flange, ft/s
M = mass flow, lb/h
v = specific volume of steam, ft^3/lb
d = diameter of the inlet, in

In smaller turbines, the inlet flange with the steam chest and control valve can easily be changed. On larger turbines, increasing the inlet size may be very difficult. In some turbines an additional inlet is added on the existing steam chest by welding on a flange followed by local stress relief. Figure 16.26 shows a flange modification from 8-in inlet to 10-in inlet.

Some turbines may have a blank flange that can be easily opened, which may be piped for additional inlet area. Figure 16.27 illustrates a typical steam end with dual-inlet capability.

In the worst-case scenario, the steam velocity for a trip and throttle (T&T) valve can be allowed to exceed the 175 ft/s limit as long as the correct pressure drop is taken into account and additional acoustic protection is provided to reduce the increased noise level.

The extraction and exhaust line sizes for a T&T valve also must be considered. The maximum value for an extraction line is 250 ft/s, while the maximum value for a noncondensing exhaust line is also 250 ft/s and for a condensing exhaust line it is 450 ft/s. Typically, there are not many options for upgrading an extraction or exhaust connection. The

Figure 16.26 Increasing the size of a flange. *(Elliott Company, Jeannette, Pa.)*

Figure 16.27 Inlet casing with dual-inlet flanges. *(Elliott Company, Jeannette, Pa.)*

most practical solution is to increase the exhaust header size as close to the turbine casing as possible. Again, the appropriate pressure drop calculations must be made to determine the pressure at the flange.

16.4.3 Nozzle ring capacity

Nozzle ring or nozzle block is the inlet to the control stage that ultimately controls the amount of steam the turbine will be able to pass. If the nozzle ring had additional area available during the original manufacture of the turbine, the turbine may be rerated by adding additional nozzles. Depending on the design, this can be as easy as installing a new nozzle ring. However, some small turbines have nozzles machined into the steam end, and modifying these turbines is difficult if not impossible. It may be cost effective to purchase a replacement turbine if this is the case.

If the change in flow is significant, the nozzle may have to be replaced with a nozzle ring that has an increased nozzle height. This normally requires increased blade height on the turbine rotor. The maximum nozzle height will be dictated by the height of the opening in the steam end of the turbine. The maximum number of nozzles in each bank or segment of the steam end is dictated by the original design parameters. In many cases, the maximum nozzle area in a steam end is matched to the volumetric area of the inlet of the turbine.

The same issues affect the extraction nozzle ring on an extraction turbine, although the extraction diaphragm in some instances can be changed or modified to allow for additional nozzle area. Some manufacturers offer designs with an advanced nozzle profile and improved structural integrity to achieve better performance and enhanced reliability. Hard facing of nozzles can also be done to protect against erosion.

16.4.4 Steam path analysis

The remainder of the steam path must be reviewed following the review of inlet nozzles. An increase in flow may require an increase in diaphragm nozzle height and rotor blade height in some or all stages. As stage flow increases, the heat drop across the stage will increase with the associated decrease in velocity ratio. The velocity ratio is defined as the ratio of the blade velocity at the pitch diameter to the steam jet velocity, as defined in Eq. 16.2:

$$V_o = DN\pi/(224\sqrt{\Delta h}) \tag{16.2}$$

where V_o = velocity ratio
D = pitch diameter of rotating blades
N = shaft rotational speed
Δh = enthalpy drop across the stage

In order to increase the efficiency, the diaphragm and blade path areas must be carefully matched. Each stage design has an optimum velocity ratio with the equivalent maximum efficiency. During the modification, some of the stages will require changes for mechanical or other reasons. Some of the stages will be acceptable with less than optimal efficiency. Decisions should be made after comparing the loss in efficiency with the cost of modifying those stages.

16.4.5 Rotor blade loading

With the change of power, flow, or speed of a turbine, each rotating row of blading requires review of its mechanical properties. This includes a review of the speed limits and analysis of both Goodman and Campbell diagrams. The blade mechanical speed limits are related to the disk and blade geometry, stress values for the material, and shroud design.

The Campbell diagram graphically compares the blade natural frequency to the turbine speed range. The natural frequencies of the blading can be obtained from the OEM or determined experimentally. If the speed range is not changed in the modification, the Campbell diagrams will not change except to adjust for any blading that has been changed. It has been found that a SAFE diagram provides more information for designers to make reliability decisions. This aspect of blading is discussed in detail in Chap. 11.

The Goodman diagram evaluates the combined effects of alternating stresses and steady-state stresses in the turbine root and base of vane locations. This diagram contains the Goodman line, which in theory represents the minimum combination of steady-state stress and alternating stress above which a blade fatigue failure could occur. The OEM will normally have minimum allowable values for a Goodman factor. There may be several minimum values depending on the location of the blade within the steam path; typically, partial arc admission and moisture transition stages have higher limits. A significant change in the stress levels of the blading may require replacement or redesign of the blading. In some rare cases, no completely suitable blading design can be found and low Goodman factors may need to be accepted.

Refer to Chap. 11 for a detailed discussion of Goodman, Campbell, and SAFE diagrams.

16.4.6 Thrust bearing loading

An increase in flow and speed may increase the thrust values in many cases. Depending on the age of the turbine, a replacement thrust bearing may be required. Bearings are available with higher thrust capacity that will fit in the existing cavity, such as chrome-copper-backed or directed lube bearings. In the most extreme cases, a new bearing housing with a larger thrust bearing may be required.

16.4.7 Governor valve capacity

To ensure correct flow area, the capacity of the governor valve must be checked during the design cycle. The valve area must be carefully matched to the area of the nozzle ring. In most cases the nozzle area should be the limiting point of the flow in a valve bank. Setting the valve as the limit will not provide the most efficient conversion of pressure into velocity.

If the valve and seat require an increase in size, most casings can accept a larger-size valve and seat. Some single-valve turbines allow the steam chest to be replaced by the next larger size.

16.4.8 Rotor

It is possible, and may be desirable, to rebuild the existing rotor. Proper factory inspection will determine the extent of repair. For example, in some cases weld repair may be possible for the dovetail, wheel, and shaft end. Shrunk-on wheels can be refurbished, and new rotor forging may enable the use of the latest staging and bucket design. Solid rotors will eliminate the need for shrunk-on wheels, thrust collars, and coupling hubs, thus achieving better reliability. Rotor rebuilding provides the advantage of using modern materials with current nondestructive standards.

16.4.9 Shaft end reliability assessment

Generally, the torque limit of a shaft end has been based on a shear stress calculation. The basic equation required to calculate shaft stress is given in Eq. 16.3:

$$\tau = 321{,}000\ T_0/(Nd^3) \tag{16.3}$$

where τ = shaft shear stress, psi
T_0 = power, hp
N = shaft speed, r/min
d = shaft end diameter, in

The limit of the shaft shear stress is a function of the material and the heat treatment as well as the shaft end design. Typical turbine shaft ends are single- or double-keyed, NEMA tapered, hydraulically fit, or integral.

With an increase in power or a decrease in speed, the capability of shaft ends must be reviewed. Almost all failures of shaft ends are due to fatigue. Variable load occurs during each rotation primarily due to misalignment of the coupling to the driven shaft. Many times, optimized coupling selection provides more reliability or more capability for the shaft end. At other times, shaft replacement can be avoided by replacing a gear-type coupling with a diaphragm-type coupling. This is because gear-type couplings have the potential of inducing both torsional and bending stresses, whereas diaphragm couplings tend to induce primarily torsional stresses and insignificant bending stresses. To ensure reliability, the design method must consider the influence of loads such as those arising from compressor surge, misalignment between the turbine shaft and the driven machine, for determining the size of the shaft end. The method should also include the effect of shrink fits, keyways, fillets, plating, and fretting on the fatigue life of the shaft material. A method based on the von Mises criteria has been used and a service factor has been used to estimate alternating stress and also to adjust mean torque. This includes the influence of the type of coupling on stresses. Following this method it has been shown that a diaphragm-type coupling provides better reliability to a shaft end than a gear-type coupling against bending fatigue.

$$1/n = (32/\pi d^3)([(K_{t1} \cdot M_f)/(2 \cdot K_d \cdot \sigma_f)]^2 + (3/4)\{1/(2\sigma_{yp}) + [(\mathrm{SF}-1)/(\mathrm{SF}+1)][K_{t2}/(K_d \cdot \sigma_f)]\}^2)^{1/2}(\mathrm{SF}+1) \cdot T_0 \tag{16.4}$$

where n = factor of safety
d = shaft diameter, in
M_f = moment factor due to coupling
K_d = factor due to fretting
σ_f = fatigue strength
σ_{yp} = yield strength
SF = service factor
K_{t1} = stress concentration factor
K_{t2} = stress concentration factors

For gear-type coupling:

$$M_f = [(x/D_p)^2 + (\mu + \sin\alpha)^2]^{1/2} \tag{16.5}$$

where x = face width of gear, in
D_p = pitch diameter, in
μ = coefficient of friction of gear teeth
α = maximum misalignment angle, degrees

For diaphragm-type coupling:

$$M_f = [(K_B \cdot \alpha/T_0)^2 + (\sin\alpha)^2]^{1/2} \tag{16.6}$$

where K_B = spring rate of coupling, in-lb/degree
α = maximum misalignment angle, degrees
T_0 = maximum continuous torque, in-lb

It is clear that the use of flexible coupling results in less alternating bending stress, thus providing more reliability against fatigue failure due to misalignment of shafts near coupling. Many manufacturers offer conversion from gear-type coupling at the shaft end to disc-type flexible coupling to add more reliability.

16.4.10 Speed range changes

Although this has been briefly discussed in other sections, care must be taken when making significant changes to the speed range. The most important is the speed limits for blades (see sec. 16.4.5, "Rotor Blade Loading"), but other areas require analysis, such as lateral critical speeds, torsional critical speeds, and coupling speed limits.

Lateral critical speeds are a function of the rotor design, bearing design, and bearing support design. Typically, the original critical speed is denoted on the nameplate for the turbine and also in the vendor literature. If no changes are being made to the rotor and the operational data agrees with the noted critical speed, this value normally will not change. Should changes be made to the rotor or bearing system, a new lateral analysis will need to be performed to confirm the lateral critical speeds.

Torsional critical speeds should be reviewed. Direct-drive units are not normally an issue, but a train with gears requires the torsional critical speed to be checked. Each body in the string must be modeled, as well as each coupling and gear set. The data to perform a torsional calculation may not be as readily available as that for the turbine lateral critical speed. Normally, this documentation is supplied by the vendor in the form of a report with engineering documentation.

The turbine governor will need to be modified or reprogrammed to meet the new speed range. The trip mechanism, which could be either mechanical or electronic, will also need to be adjusted. Other driven systems will require review, such as any gears, couplings, or shaft-driven oil pumps.

16.4.11 Auxiliary equipment review

Auxiliary equipment will require some review during any modification. This may include the T&T valve, other steam block valves, leak-off and sealing steam system, surface condenser system, relief valve sizing, lubrication and control oil systems, valve actuation system, and supervisory instrumentation.

The T&T valve may be acceptable as is for the change in flow, but it may be upgraded for reliability or the addition of a manual exerciser. OEMs are no longer actively pursuing this market, so an upgrade may be more expensive than a completely new valve of a current design. For extraction turbines, the nonreturn valve should also be evaluated to determine the acceptability of long-term continued operation. Other steam valves in the system should be reviewed for proper sizing and good mechanical operation.

The leak-off and sealing steam system must be reviewed to determine whether additional leak-off flow will be experienced under the new conditions. Both API 612 and 611 require 300 percent capacity in the leak-off system. If a change has been made to the first-stage pressure or the exhaust pressure, this additional capacity might be difficult to achieve without major rework. The gland condenser will require a review along with the ejector or vacuum pump to ensure that these are of adequate capacity. From a piping point of view, it may be a good decision to replace some or all of the leak-off and drain piping if significant corrosion or buildup is noted on the piping and instrument diagram (P&ID) for the piping from the turbine.

On a condensing turbine, the surface condenser system must be reviewed to ensure that the proper capacity is available with the current cooling-water temperature and available cooling-water flow. The hot well capacity and pump sizing must be verified to ensure that the correct hot well level can be maintained. Instrumentation connected with the condenser system should be reviewed and upgraded at this time.

When modifying a backpressure turbine, the relief valve sizing must be reviewed once the final flow capacity is determined to ensure that the exhaust casing is protected from overpressurization. With a condensing turbine, the relief valve or rupture disc sizing must be reviewed to ensure that the exhaust casing is prevented from going over the maximum pressure rating, which is normally 5 psig.

The lubrication and control oil system must be carefully checked to ensure that the required oil flow and the proper cooling capacity are available. If the control oil system is being modified, check to ensure that the proper oil pressure and flow are available.

16.4.12 Oil mist lubrication for general-purpose steam turbines

Plantwide oil mist systems have been in use at numerous reliability-focused refineries and petrochemical plants since the mid-1960s. In most of these installations, small and medium-size steam turbines have been provided with either pure oil mist (dry sump) lubrication or an oil mist purge. A cross-section view of a small steam turbine with oil mist purge (wet sump) is shown in Fig. 16.28.

In the United States, Canada, South America, the Middle East, and the Pacific Rim countries, oil mist applications have matured to the point where plantwide systems (see Fig. 16.29) are being extensively specified and installed by major design contractors. Oil mist certainly constitutes lubrication technology that cannot be omitted from an up-to-date steam turbine text.

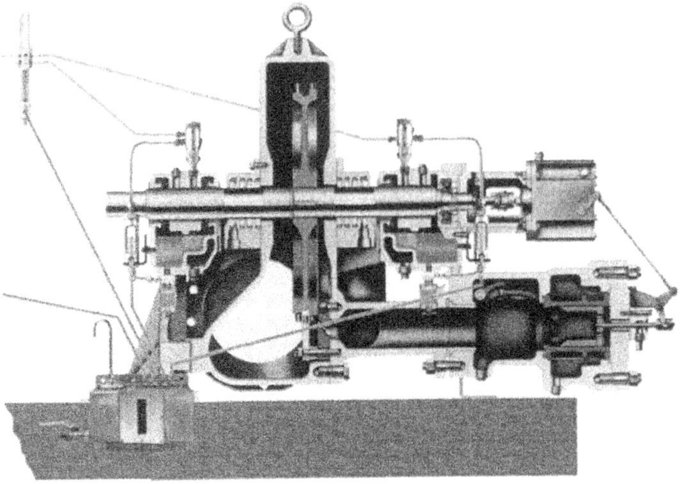

Figure 16.28 Cross section of a small steam turbine with sleeve-type radial bearings and an antifriction (rolling element) thrust bearing. Oil mist serves a purge in this wet sump application. *(Lubrication Systems Company, Houston, Tex.)*

Closed oil mist systems (see Fig. 16.29) have been available since they were first applied to high-speed bearings in the Swiss textile industry in the late 1950s. Today, closed systems are in use at several U.S. petrochemical plants. Although more expensive to install than the traditional open systems, closed systems allow an estimated 99 percent

Figure 16.29 A closed plantwide oil mist lubrication system with mist collection and coalescing tank shown to the left. *(Lubrication Systems Company, Houston, Tex.)*

of the lube oil to be recovered and reused. Unlike open systems, closed systems emit no oil mist into the atmosphere and are thus preferred by environmentally conscious users.

Wet sump (purge mist) vs. dry sump (pure mist) application. It is necessary to distinguish between the wet and dry sump methods of oil mist application. In the wet sump method, a liquid oil level is maintained and the mist fills the housing space above the liquid oil. Wet sump oil mist, also called *purge mist,* is essentially old technology and is used primarily on equipment with sleeve bearings. Dry sump oil mist describes an application method whereby no liquid oil level is maintained in the bearing housing. The principle is illustrated in Figs. 16.30 and 16.31. Lubrication is provided entirely by oil mist migrating through the bearing. The application of dry sump oil mist is advantageous for a number of reasons, including but not limited to lower bearing temperatures, the presence of nothing but uncontaminated oil mist, and the exclusion of external contaminants. However, one main reason often overlooked is that oil rings (see Fig. 16.32) are no longer needed with the dry sump oil mist application method.

Oil rings often represent outdated nineteenth-century technology, as they were initially developed for slow-speed machinery during the Industrial Revolution. Deletion of oil rings is one of the many keys to improved reliability for virtually *any* type or style of bearing. Oil rings

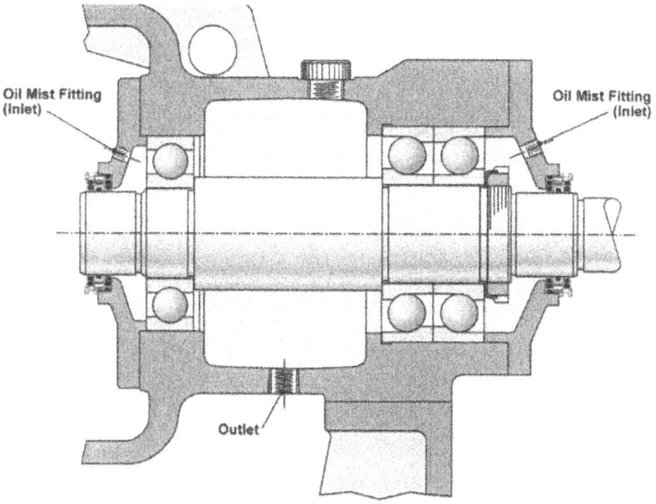

Figure 16.30 Optimized new-style application of pure oil mist in a typical bearing housing. Note that mist entries are located between a magnetic bearing housing seal and the bearing. Mist must flow through bearings on its way to the common drain location at the housing low point. *(AESSEAL plc, Rotherham, U.K., and Knoxville, Tenn.)*

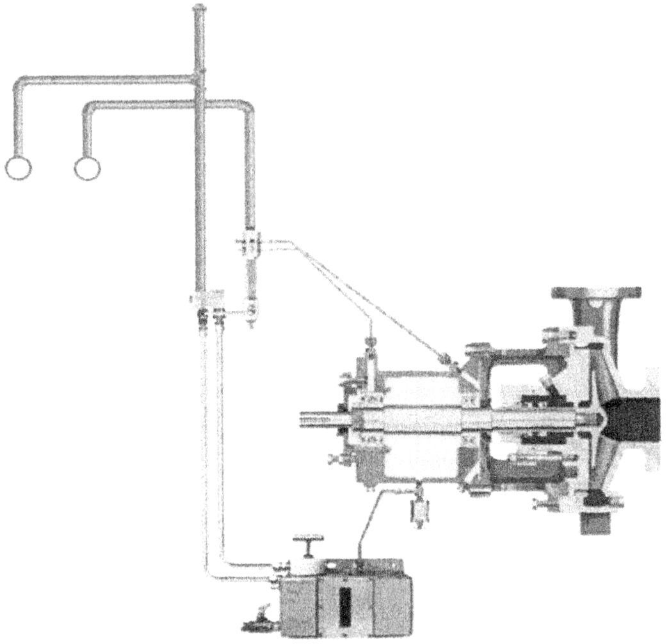

Figure 16.31 Oil mist supply and return headers in this closed system are routing oil mist to a dry sump pump bearing housing, per API-610. *(Lubrication Systems Company, Houston, Tex.)*

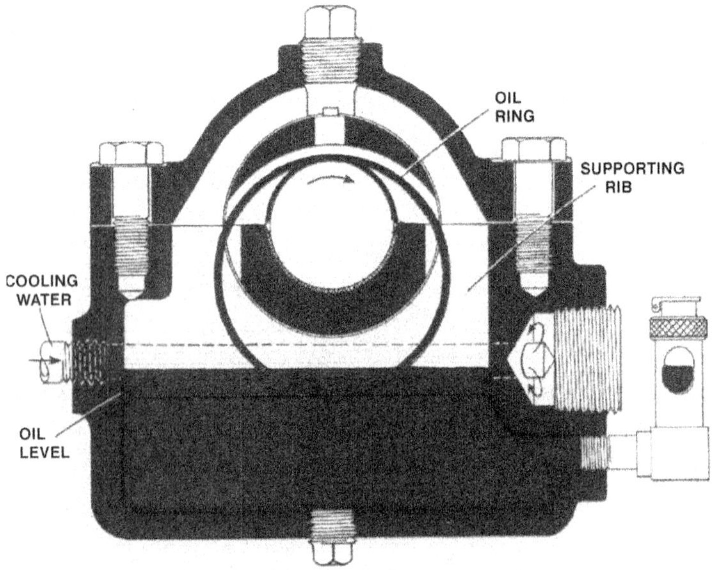

Figure 16.32 Bearing housing found on typical small steam turbines. Note the oil ring (slinger ring) needed for sleeve bearings in this (conventional) lubricant application.

are known to have journal surface velocity limitations, sometimes as low as 2000 ft/min, or 10 m/s. So as not to "run downhill," which might cause the rings to make frictional contact and slow down, ring-lubricated shaft systems would have to be installed with near-perfect horizontal orientation.

Moreover, frictional contact often results in abrasive wear, and such wear products certainly contaminate the oil. Oil rings will malfunction unless they are machined concentrically within close tolerances. These rings suffer from limitations in allowable depth of immersion, and to operate as intended they need narrowly defined and closely controlled oil viscosities. Because pure oil mist does not use oil rings, these are no longer issues. Note, however, that dry sump oil mist is recommended for rolling element bearings only. Wet sump or purge mist is used with sleeve-type bearings. For instance, it could be applied in the vapor space over the liquid oil in the turbine bearing housing of Fig. 16.32, where its only function would be to prevent atmospheric contaminants from entering the bearing housing.

Oil mist is easily produced, controlled, and applied. It should be noted that the eighth (2000) and subsequent editions of the API-610 Standard for centrifugal pumps have described sizing and configuration parameters for oil mist. Although not specifically addressed in the API pump standard, the same rules have been successfully applied to rolling element bearings in steam turbines for many years.

Modern plants use oil mist as the lube application of choice. Oil mist is readily produced in a suitably sized oil mist generator console (see Fig. 16.33). From there, the pipe headers shown previously in Figs. 16.29 and 16.31 distribute the mist to a wide variety of users throughout the plant. Mist flow to bearings is easy to control because it is a function of piping (header) pressure and final orifice (reclassifier) size. These orifices can be located at a convenient spot just upstream from the bearing housing, as was indicated in Fig. 16.30. Unless obstructed by an unsuitable (e.g., an elevated pour point) lubricant, reclassifiers have a fixed flow area that is selected based on bearing size and speed criteria.

Depending on make and systems provider, header pressures range from 20 to 35 in (500 to 890 mm) of H_2O. Modern units contain controls and instrumentation that will maintain these settings effortlessly. Mixing ratios are typically 160,000 to about 200,000 volumes of air per volume of oil. In older mist generators that incorporate gaskets and O-rings in the mixing head, the mixing ratio may drift out of this acceptable range unless these elastomers have periodically been replaced or properly serviced.

Figures 16.30 and 16.31 depict the most advantageous pure mist entry points, between a bearing housing seal and the adjacent bearing.

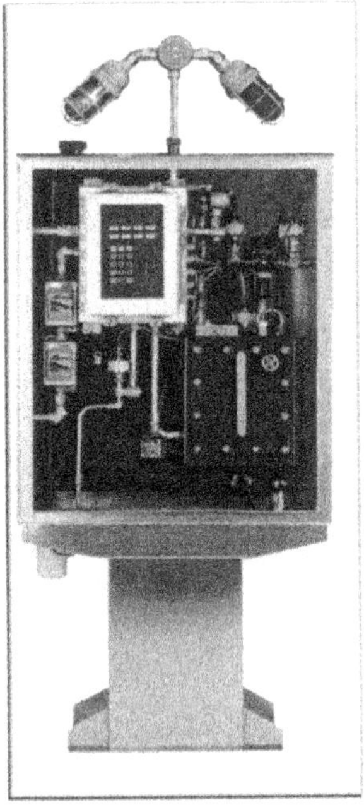

Figure 16.33 Oil mist generator console for (*left image*) plantwide multitrain and (*right image*) single-train lubricant applications. *(Lubrication Systems Company, Houston, Tex.)*

This routing ensures that oil mist leaving through the bottom vent opening of the bearing housing will have traveled through the bearings and can be collected for either reuse or approved disposal. The lip seal–equipped housing of Fig. 16.31 was further upgraded, in Fig. 16.30, by providing a "best available technology" dual-face magnetic bearing housing protector seal. Compared with plants using *non*optimized mist entry and escape locations (Fig. 16.34), the recommended configuration of Fig. 16.30 consumes about 40 percent less oil mist.

Forward-looking plants have used the latest, best-technology API method, that is, as shown in Fig. 16.31, since the mid-1970s. These plants have also recognized that mist entering at locations far from the bearings (as was done prior to 1975) might have difficulty overcoming bearing windage or fan effects. Windage is most often produced by the skewed (diagonally oriented) ball cages used in many angular contact bearings. If such windage were produced so as to push the mist away from the bearings, much of the mist would take the preferential path straight to the vent exit and only insufficient amounts of oil mist would reach the bearing rolling elements.

A larger quantity of oil mist or specially designed *directional* reclassifiers are needed with certain bearing types *unless* the superior API method of Figs. 16.30 and 16.31 was being used. This latter method easily overcomes windage, the flow-induced fan action caused by skewed ball cages.

Header temperature and header size. Ambient temperature has never been an issue for properly designed systems. Once a mist or aerosol of suitably small particle size has been produced, the oil mist will migrate to all points of application. Mist particle size is influenced mainly by the temperature constancy of both air and oil in a static mixing header located inside the console. The temperature of the environment has little, if any, influence on mist quality and lubrication effectiveness. Mist temperatures in headers have ranged from well below freezing in North America to over 122°F (50°C) in the Middle East.

However, regardless of geographic location, conscientiously engineered systems will incorporate both oil and air heaters in the mist-generating console; these heaters are needed to produce constant and optimized air/oil mixing ratios at the point where mixing occurs. The heaters must have low-watt density (low power input per square inch of surface area) in order to prevent overheating of the oil. Headers are sized for low mist velocity and there is no incentive to insulate these headers even at the low ambient temperatures of northern Canada.

Users who try to save money through omitting heaters in the console by employing undersized headers will not be able to achieve optimum

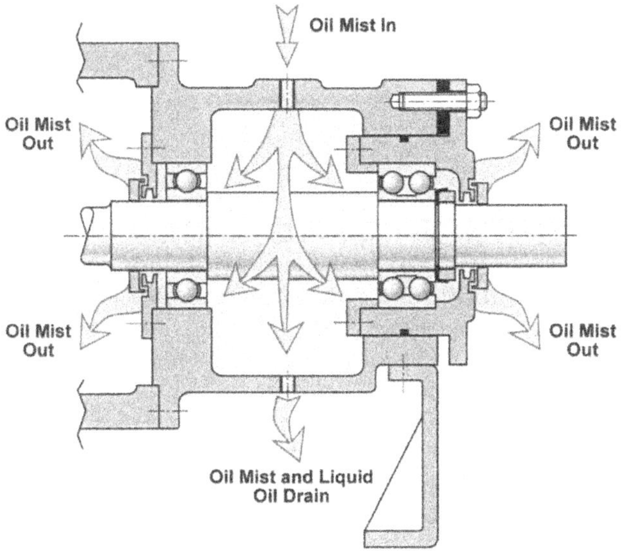

Figure 16.34 Nonoptimized old-style application of pure oil mist in a typical bearing housing. *(AESSEAL plc, Rotherham, U.K., and Knoxville, Tenn.)*

life cycle cost of their assets. Undersized headers may increase the flow velocity to the point where the small oil globules suspended in the carrier air experience too many collisions. The globules may then agglomerate into droplets large enough to fall out of suspension, causing excessively lean mist to arrive at the point to be lubricated.

Also, for decades, environmental and health concerns have been addressed by using oil formulations that are neither toxic nor carcinogenic. Such formulations are readily available to responsible users. Appropriate lubricants have also been formulated for minimum stray mist emissions.

Experience and concluding comments. Actual statistics from a facility convey an accurate picture of the value of properly applied oil mist technology. This petrochemical plant went on-stream in 1978 with 17 oil mist systems providing dry sump oil mist to virtually every one of the many hundreds of pumps and electric motors in the facility. As stated earlier, with the dry sump method per current API recommendation, the oil mist is introduced at a location that guarantees its flow through the bearings and to an appropriate vent location. There are neither oil rings nor any other provisions for the introduction of liquid oil on pumps and motors with rolling element bearings at the plant.

Over a period of 14 years, one qualified contract worker serviced these systems by visiting the plant one day each month. In this 14-year time period there was only one single malfunction, which involved a defective float switch in one of the 17 systems. The incident caused a string of motor-driven pumps to operate, without inducing even one bearing failure, for eight hours. In 1992, the combined availability and reliability of oil mist systems at this plant was calculated to be 99.99962 percent. As of 2004, over 60,000 pump sets were operating successfully on oil mist.

Being aware of the relative unreliability of conventional lubricant application methods involving oil rings and certain (not really very constant level) oil lubricator bulbs, knowledgeable reliability professionals can attest to the utility and overall advantages of properly engineered dry sump oil mist systems. The known advantages of such systems far outweigh their actual or perceived disadvantages. It is unfortunate that much information to the contrary is either anecdotal or pertains to systems that were not correctly designed, installed, maintained, or upgraded as new technology became available.

Only dry sump applications will lubricate, preserve, and protect both operating and standby rolling element bearings. To reemphasize, an oil mist system is an automated means of lube application for operating machines and also a protective system for standby equipment. Used in the dry sump mode, only clean, fresh oil will reach the bearings. In many instances, bearing operating temperatures achieved with dry sump oil mist lubrication are 10 to 20°F (6 to 12°C) lower than with wet sump lubrication. Industry experience with dry sump oil mist systems is well documented and the superiority of these systems over both conventionally applied liquid oil and wet sump oil mist lube applications has been solidly established.

Regrettably, there still are plants that operate on wet sump oil mist. Wet sump lubrication makes economic sense on sleeve bearings. The only function of wet sump lubrication is the exclusion of atmospheric contaminants by existing at a pressure slightly above that of the ambient air. Often, it is expected that the wet sump oil level will be maintained by an externally mounted constant-level lubricator. However, due to the slight pressurization and on-bearing housings equipped with traditional open-to-atmosphere constant-level lubricators, the oil level in the bearing housing will often be pushed down below the oil level in the lubricator. Arrangements of this type will always prove considerably less reliable than most of the alternative methods that are available today.

Also, unless appropriate bearing housing protector seals are used, free water may contaminate the bearing oil (See Fig. 16.35). Regardless

Figure 16.35 An excessive amount of free water is being drained from a turbine bearing housing.

of which method of lubricant application is ultimately chosen, it is best to avoid free water in the lubricant.

16.4.13 Problem solving

Many operating problems may be solved by some of the items that have already been discussed. Monitoring systems can uncover problems such as a bowed rotor or fouling. Better tools exist today to determine rotor dynamic characteristics than existed 30 years ago. The seals that were discussed can eliminate water contamination of the oil and excessive gland leakage. The electronic controls can cure some process control problems or speed fluctuations. By replacing improperly sized governor valves or worn valves and valve seats, speed fluctuation can be controlled. Performance problems must be analyzed on an individual basis, perhaps with sophisticated tools now available, such as computational fluid dynamics (CFD).

16.5 Summary

It may be difficult to determine whether an upgrade or a rerate is called for due to many required changes. Therefore, this chapter has concentrated on the details of the kinds of changes that have been made or should be made to achieve benefits without mentioning whether the modification in question is an upgrade or a rerate. Rerates, upgrades, and a combination of the two allow the design of the flow path components to be revised at every outage. This allows the turbine to be tuned to the current power requirements, improve reliability, and/or increase the overall level of turbine performance. Deciding whether to simply rerate a turbine, upgrade it, or some combination of the two should be accomplished with a typical cost/benefit analysis.

Generally, a simple rerate is the lowest-cost and shortest-lead-time option. Upgrades require more cost and time to complete, and a combination of rerating and upgrading is the most costly, longest-lead-time option. A turbine that is brought down unexpectedly due to failure or other problems may have time for only limited options on an expedited basis. On the other hand, a scheduled outage may allow for a detailed analysis to select the most appropriate options. This option provides time for the required engineering, manufacturing, and shipping to be accomplished so that the new components are available on-site when the outage occurs.

Appendix

A

Glossary

Turbomachinery and Related Equipment Terms

This group of definitions includes only major and general items, broadly applied.

ACTUAL CAPACITY is the quantity of gas actually compressed and delivered to the discharge system by the machine at rated speed, and under rated inlet and discharge conditions (and under rated interstage conditions in a multistage machine). Actual capacity is usually expressed in cubic feet per minute referred to first-stage inlet flange temperature and pressure.

AIR SPEED is an industry term denoting the average velocity of the gas flowing through the valve lift area. It assumes that the valves are wide open full stroke and that the piston moves at a constant velocity. It is thus a rating velocity expressed in ft/min. It is most accurately obtained by dividing the net piston area (in^2) by the inlet valve lift area of *one cylinder end* (in^2) and multiplying by piston speed (ft/min). Some evaluators divide the total cylinder piston displacement (ft^3/min) by the total cylinder inlet valve lift area (in^2) and multiply by 144. The results differ by a factor of 2.

CLEARANCE is the volume present in one end of the cylinder in excess of the net volume displaced by the piston during the stroke toward that end. It may not be the same for the two ends in a double-acting cylinder. An average is usually used.

COMPRESSION EFFICIENCY is the ratio of the theoretical work requirement (using a stated process) to the actual work required to be done on the gas for compression and delivery. Expressed as a percentage, compression efficiency accounts for leakage and fluid friction losses and thermodynamic variations from the theoretical process.

COMPRESSION RATIO is the ratio of the absolute discharge to the absolute intake pressure. It usually applies to a single stage of compression but may be applied to a complete multistage compressor as well.

CORROSIVE GAS is one that attacks normal materials of construction. Water vapor when mixed with most gases does *not* make them corrosive within the sense of the above definition. In other gases (CO_2, e.g.) it makes them corrosive.

DESIGN PRESSURE is a term also frequently used to denote maximum allowable working pressure.

DESIGN SPEED is the same as maximum allowable speed.

DISCHARGE PRESSURE is the total pressure (static plus velocity) at the discharge flange of the fluid machine.

Note: Pressures may be expressed as gauge or absolute pressures. psig plus atmospheric pressure equals psia. Note that psig does not define a pressure unless the barometric pressure (atmospheric) is also stated.

DISCHARGE TEMPERATURE is the gas temperature existing at the discharge flange of the compressor.

DRY OR WET GAS terminology has different meanings in different applications. One must be very careful when using these terms to be sure that the meaning intended is actually conveyed. For definitions see the Thermodynamic Terms.

INERT GAS, to a chemist, is one that does not enter into known chemical combination, either with itself or another element. There are four known gases of this type: helium, neon, argon, and krypton. To the engineer, however, the term usually means a gas that does not supply any of the needs of combustion.

INLET PRESSURE is the total gas pressure (static plus velocity) at the inlet flange of a dynamic fluid machine. Velocity pressure is usually too small to be considered at any point in a reciprocating fluid machine.

INLET TEMPERATURE is the gas temperature at the inlet flange of the compressor.

MAXIMUM ALLOWABLE SPEED (r/min) is the highest speed at which a manufacturer's design will permit continuous operation, assuming overspeed and governor mechanisms are installed and operated per manufacturer's recommendations.

MAXIMUM ALLOWABLE WORKING PRESSURE is the maximum continuous operating pressure for which the manufacturer has designed the machine (or any part to which the term is referred, such as an individual cylinder) when handling the specified gas at the specified temperature. (See Rated Discharge Pressure, Safety Valve Setting, and Design Pressure).

MECHANICAL EFFICIENCY is the ratio, expressed as a percentage, of the actual thermodynamic work required in a fluid machine to the actual shaft horsepower or kilowatt requirement observed or measured for the machine.

NONCORROSIVE GAS is one that does not attack normal materials of construction.

Note: The words *corrosive* and *noncorrosive* are of the relative type. They do not define with exactness, and there are differences of degree in the application of these terms. Specifications must clarify what is meant when these terms are listed.

OVERHEAD AND UNDERTOE are terms used to represent those parts of an actual indicator card that lie respectively above and below the discharge and inlet pressure as measured at reciprocating compressor cylinder flanges.

PERFECT INTERCOOLING is obtained when the gas is cooled to first-stage inlet temperature following each stage of compression.

PISTON DISPLACEMENT of a cylinder is the net volume displaced by the piston at rated machine speed, generally expressed in cubic feet per minute. For single-acting cylinders it is the displacement of the compressing end only. For double-acting cylinders it is the total displacement of both ends. For multistage compressors, the piston displacement of the first stage only is commonly stated as that of the entire machine.

PISTON SPEED is the distance, in feet, traveled by the piston in 1 min.

RATED DISCHARGE PRESSURE of any casing element is the highest continuous operating pressure to meet the conditions specified by the purchaser for the intended service. The rated discharge pressure is always less than the maximum allowable working pressure by at least 10 percent or 15 psi, whichever is greater, for the operation of safety valves.

RATED HORSEPOWER is the continuous power input required to drive equipment at rated speed and actual capacity under rated pressure and temperature conditions. For all machines *other than integral* *steam driven units,* it is the power required at the equipment input shaft. It does not include losses in the driver or in the transmission elements between driver and driven machine.

RATED SPEED (r/min) is the highest speed necessary to meet the specified service conditions. Rated speed and maximum allowable speed may be the same, but rated speed can never exceed the maximum allowable speed.

RATED STEAM INDICATED HORSEPOWER is the indicator card horsepower developed in the steam cylinders of an integral type steam driven unit operating under rated conditions.

SAFETY VALVE setting is to be no higher than the maximum allowable working pressure.

Note: Cylinders are not always installed to operate at their maximum rated discharge pressure. This is due to specific conditions involved in a specific compressor design. Limitations other than maximum cylinder design pressure are in control. The proper safety valve setting may be substantially below the maximum allowable working pressure.

SUPERCHARGING of a compressor cylinder occurs when the pressure in the cylinder at the end of the suction stroke is higher than the rated inlet pressure at the cylinder inlet flange. This occurs when a resonant condition exists in the inlet pipe, which brings a pressure peak at the end of the suction stroke. It results in greater than normal actual capacity and greater power requirement. It can seriously overload a driver.

TRIPPING SPEED is that speed at which the overspeed device is set to function. It is normally 110 percent of the rated speed.

VACUUM is a pressure below atmospheric.

VACUUM PUMP is a compressor operating with an intake pressure below atmospheric and usually discharging at atmospheric pressure or only slightly higher.

VALVE LIFT AREA is the minimum net flow area between the valve and seat when the valve is fully open. Usually, this is also the least area in a valve through which the gas must flow.

VOLUMETRIC EFFICIENCY is the ratio of the actual capacity to the piston displacement, expressed as a percentage.

Thermodynamic Terms

Some definitions given here are not as all inclusive as general thermodynamics might require, but they cover the ground necessary for fluid machinery applications. There are some items where authorities differ in definition and approach. In such cases a certain amount of judgment has been applied.

COMPRESSIBILITY is a volume ratio that indicates the deviation (as a multiplier) of the actual volume from that as determined by perfect gas laws.

CRITICAL PRESSURE is the saturation pressure at the critical temperature. It is the highest vapor pressure the liquid can exert.

Note: Critical conditions must be experimentally determined for each pure gas. When calculated for a mixture, they are called pseudo (pretended) critical conditions.

CRITICAL TEMPERATURE is the highest temperature at which a gas can be liquified.

DENSITY is the weight of a given volume of gas, usually expressed in lb/ft^3 at SPT conditions.

DEW POINT of a gas is the temperature at which the vapor (at a given pressure) will start to condense (or form dew). Dew point of a gas mixture is the temperature at which the highest boiling point constituent will start to condense.

DRY GAS is any gas or gas mixture that contains no water vapor and also in which all of the constituents are substantially above their respective saturated vapor pressures at the existing temperature.

Note: In commercial compressor work a gas may be considered dry (even though it contains water vapor) if its dew point is low at the inlet condition (say –50 to –60°F).

ENERGY of a substance is its capacity, either latent or apparent, to exert a force through a distance, i.e., to do work.

ENTHALPY (heat content) is the sum of the internal and external energies.

ENTROPY is a measure of the unavailability of energy in a substance.

EXTERNAL ENERGY is that energy represented by the product of pressure and volume. It may be regarded as the energy a substance possesses by virtue of the space it occupies.

GAS SATURATED WITH ANOTHER VAPOR—actually a gas is *never* saturated with a vapor. However the *space* jointly occupied by the gas and vapor may be saturated. This occurs when the *vapor* is at its dew point, the saturation temperature corresponding to its partial pressure.

HEAT is energy transferred because of a temperature difference. There is no transfer of mass.

INTERNAL ENERGY is that energy that a substance possesses because of the motion and configuration of its atoms, molecules, and subatomic particles.

ISENTROPIC (ADIABATIC) PROCESS is one during which there is no heat added to or removed from the system.

ISOTHERMAL PROCESS is one during which there is no change in the temperature.

KINETIC ENERGY is the energy a substance possesses by virtue of its motion or velocity. It occasionally enters into turbomachinery design calculations.

PARTIAL PRESSURE of a constituent in a mixture is the absolute pressure exerted by that portion of the mixture.

POLYTROPIC PROCESS is one in which changes in gas characteristics and properties are allowed for throughout the process.

POTENTIAL ENERGY is the energy a substance possesses because of its elevation above the earth (or above some chosen datum plane).

RATIO OF SPECIFIC HEATS (k) is the ratio of c_p over c_v. It may vary considerably with temperature and pressure.

REDUCED PRESSURE is the ratio in absolute units of the actual gas pressure to the critical pressure.

REDUCED TEMPERATURE is the ratio in absolute units of the actual gas temperature to the critical temperature.

SATURATED VAPOR PRESSURE is the pressure existing at a given temperature in a closed vessel containing a liquid and the vapor from that liquid after equilibrium conditions have been reached. It is dependent only on temperature and must be determined experimentally.

SATURATION PRESSURE is another term for saturated vapor pressure.

SATURATION TEMPERATURE is the temperature corresponding to a given saturated vapor pressure for a given vapor.

SPECIFIC GRAVITY is the ratio of the density of a given gas to the density of dry air, both measured at the same specified conditions of temperature and pressure usually, 14.696 psia and 60°F. It should also take into account any compressibility deviation from a perfect gas.

SPECIFIC HEAT (heat capacity) is the rate of change in enthalpy with temperature. It may be measured at constant pressure or at constant volume. The values are different and are known as c_p and c_v, respectively.

SPECIFIC VOLUME is the volume of a given weight of gas, usually expressed as ft^3/lb at SPT conditions.

SPT means standard pressure and temperature. As used herein it is 14.696 psia and 60°F.

TEMPERATURE is the property of a substance that gauges the potential or driving force for the flow of heat.

WET GAS is any gas or gas mixture in which one or more of the constituents is at its saturated vapor pressure. The constituent at saturation pressure may or may not be water vapor.

WORK is energy in transition and is defined as force times distance. Work cannot be done unless there is motion.

Appendix

B

Units of Measurement

Most users of major fluid machines are interested primarily in the quantity of gas handled between the initial and final pressure and temperature conditions and in the power required. The user may specify the conditions to be met in various terms and units. The designer must know positively what the buyer expects the equipment to do and what specific conditions are to be met. All too often the designer is not able, from the data given, to clarify the factors he or she must know.

It is the purpose of this section to outline and discuss the various units of measurement so that there may be a more perfect meeting of minds and better communication between the user and the manufacturer. Pages 326 through 332 contain conversion factors for the generally used units.

Pressure

Pressure is expressed as a force per unit of the area exposed to the pressure. Since weight is really the force of gravity on a mass of material, the weight necessary to balance the pressure force is used as a measure. Hence, as examples:

lb/in^2 (psi)

lb/ft^2

g/cm^2

kg/cm^2

Pressure is usually measured by a gauge which registers the *difference* between the pressure in a vessel and the current atmospheric pressure. Therefore a *gauge* pressure (psig) does not indicate the *true* total gas pressure. To obtain the true pressure, or pressure above zero, it is necessary to add the current atmospheric or barometric pressure, expressed in proper units. This sum is the *absolute* pressure (psia).

For all compressor calculations the *absolute* pressure is required. If gauge pressures only are given, the inquiry is not complete. The atmospheric or barometric pressure must *also* be specified.

Note: There is frequent confusion in transmission of pressures as to whether they are absolute or gauge. To reduce uncertainties of this type, the use of psig and psia is suggested.

Since the earth's atmosphere has weight, a measure known as the international atmosphere is sometimes used. It is the weight of the column of air existing above the earth's surface at 45° latitude and sea level. It is defined as being equivalent to 14.696 psia or 1.0333 kg/cm^2 and is usually expressed as *ATA* meaning atmospheres absolute. When specified as *ATM* it is often uncertain whether atmospheres *absolute* or atmospheres *gauge* is meant, and discretion must be used. Atmospheres absolute equals atmospheres gauge plus 1.

Occasionally, an inquiry may specify a *metric atmosphere* which is 1 kg/cm^2 or 14.233 psia. Normally, however, the international atmosphere is used throughout all engineering.

Also, since a column of a material of a specified height will have a weight proportional to its height, the height can be used as a force measure. It is reduced to a unit area basis automatically, since the total weight and the area are proportional. For example:

Feet of water (ft H_2O)

Inches of water (in H_2O)

Inches of mercury (in Hg)

Millimeters of mercury (mm Hg)

With the following exception, when pressures are expressed as height, they are *gauge* pressures unless specifically noted as absolute values.

Atmospheric pressure is measured by a barometer. It is designed to read the height of a column of mercury. The upper end of the tube containing the mercury is closed and is at zero absolute pressure. The lower end of the tube is submerged in a pot of mercury, the surface of which is open to the atmosphere. The weight of this column of mercury exactly balances the weight of a similar column of atmospheric air.

Although this gauge really measures a *differential* pressure, by design one of those pressures is zero, and the actual reading is the true *absolute* or total pressure of the atmosphere. 14.696 psia sea level pressure is equal to 29.92 in Hg.

Vacuum

Vacuum is a type of pressure. A gas is said to be under vacuum when its pressure is below atmospheric. These are two methods of stating this pressure, only one of which is accurate in itself.

Vacuum is usually *measured* by a differential gauge that shows the *difference* in pressure between that in the system and the atmosphere. This measurement is expressed, for example as:

Millimeters of Hg vacuum (mmHg vac)

Inches of Hg vacuum (in Hg vac)

Inches of water vacuum (in H_2O vac)

psi—vacuum (psi vac)

Unless the barometric or atmospheric pressure is also given, these expressions do not give an accurate specification of pressure.

Subtracting the vacuum reading from the atmospheric pressure will give an *absolute* pressure, which is accurate. This may be expressed as:

Inches of Hg absolute (in Hg abs)

Millimeters of Hg absolute (mmHg abs)

lb/in^2 absolute (psiA)

1 torr (torricelli) (1 mmHg abs)

Since the word *absolute* is often omitted, one must be very sure as to whether a vacuum is expressed in differential or absolute terms. It is equally important that the pressure (or vacuum) terms be clearly and unmistakably shown on all inquiries and proposals.

Use absolute rather than differential terminology and mention *absolute.*

Volume

The compressor designer must determine the actual volume of gas to be handled *at the inlet* of each compressor cylinder. In many inquiries, volumes are not so presented and translation is necessary. Parts of the theory sections of most compressor texts highlight this fact.

Since manufacturer's guarantees are based on volumes at conditions existing at the first stage inlet flange, it is highly advisable that the purchaser specify the *inlet volume* the compressor must handle, even though it is also specified in other terms. These specified volumes should also include actual moisture content.

Quantities handled by compressors are really *rates.* For example:

Actual ft^3/min (Aft^3/min)

Inlet ft^3/min (Ift^3/min)

ft^3/min
Standard ft^3/min (Sft^3/min)
Millions of ft^3/24 h (MMCFD)
m^3/s
lb/h
kg/h
Mol/h

A weight rate can readily be changed to a volume rate at specified or known inlet conditions if the properties of the gas or gas mixture are specified by the customer. It is again best to specify **volume at intake conditions.**

ft^3/min or ft^3/h, or ft^3/s as terms, frequently are inconclusive as to the temperature and pressure conditions at which the volume is measured. Unless specified, it is assumed to be at *intake* conditions.

Sft^3/min means ft^3/min at standard conditions. However, standards vary and some care is necessary. In the United States the *usual* standard is 14.696 psia and 60°F. Some chemical engineers will use one ATA and 0°C, but will usually be specific about the reference point. Europeans normally use one ATA and 0°C. When there is doubt, the manufacturer can only use judgment and endeavor to make it clear in quoting what basis has been used.

MMCFD, often called million cubic feet per day, is not the same to all people. It originated in the natural gas pipeline industry, where it usually refers to volume at 14.4 psia and inlet temperature. It is best that the purchaser specify the reference pressure and temperature.

Volume, wet or dry

Whether a gas is wet or dry refers (as far as compressor volume terminology is concerned) to its water vapor content. The one exception is the term *wet* natural gas. This indicates a natural gas containing hydrocarbons that are easily condensed rather than a gas saturated with water vapor. As applied to natural gases the term *wet* is entirely special.

Volume or weight rates are often given as dry (no water vapor). Specified conditions should include a flat statement that the relative humidity is zero (no vapor), 100 percent (saturated at inlet or some other specified temperature), or some in-between figure (partially saturated).

Water vapor in a gas that occupies space and the compressor must be sized to handle it. It is, therefore, a *must* that the amount of vapor be known. This is most important since many chemical process engineers make their calculations on dry gas, present the analysis of the dry gas, but neglect to pin down the water vapor content specifically. This can make a difference of several percent in capacity requirements. Precise specifications are clearly essential.

Temperature

Temperatures can be expressed in several definitely interrelated terms:

Degrees Fahrenheit (°F)

Degrees Rankine (°R)

Degrees Centigrade (°C)

Degrees Kelvin (K)

One degree Fahrenheit is equal to one degree Rankine, but the zero points are at different temperatures. The Rankine scale is zero at absolute zero which is 459.7° below zero in °F.

$$°F = °R - 459.7$$

Degrees Rankine (°R) is sometimes expressed as degrees Fahrenheit absolute.

$$°R = °F + 459.7 = °F \text{ abs.}$$

The Centigrade and Kelvin scales are similarly related. The Kelvin thermometer is zero at absolute zero or 273.2° below zero in °C. The Centigrade scale is zero at the temperature of a mixture of ice and water.

$$°C = °K - 273.2$$

$$\text{also } °K = °C + 273.2 = °C \text{ abs.}$$

One degree Centigrade is one-hundredth of the temperature difference between the freezing and boiling temperatures of water.

One degree Fahrenheit is 1/180 of the same temperature difference. Therefore:

$$\text{One } °F = 5/9 \text{ of one } °C.$$

$$\text{One } °C = 9/5 \text{ of one } °F.$$

Because of the different bases, it is convenient to use conversion tables and a good one will be found at the end of this Appendix B. Alternatively, use these formulas:

$$°F = 9/5\ °C + 32$$

$$= 9/5\ (°C + 40) - 40$$

$$°C = 5/9\ (°F - 32)$$

$$= 5/9\ (°F + 40) - 40$$

$$\text{Kelvin, } K = °C + 273.2$$

$$°\text{Rankin, } °R = °F + 459.7$$

Power

The mechanical engineer must determine the energy flow during compression, not only to properly size the driver needed, but to provide cooling media as required.

It is not always fully realized that the power or energy input to the compressor shaft is equal to the heat energy which must be removed from the system if it is to continue to operate. This heat energy includes that added thermodynamically to the gas being compressed and all frictional heat within the machine, both gas friction and the mechanical friction of moving parts.

This heat can all be accounted for in a complete heat balance as follows:

1. Heat contained in the gas as it leaves the compressor system
2. Heat radiated by frame, cylinders, coolers, and piping to the atmosphere
3. Heat conducted from frame to the foundation
4. Heat carried away by lubricating and cooling oil and radiated or transferred to cooling water for disposal
5. Heat carried away by the cooling medium in the cylinder jackets
6. Heat carried away by the cooling medium in the intercoolers and aftercooler

The total of these six items must equal in Btu/h the shaft horsepower input, also expressed in Btu/h. (One hp-h equals 2545 Btu/h).

Units used are well defined and as listed below are really rates:

Horsepower (hp)

Metric horsepower (mhp)

Kilowatt (kW)

Btu/hour (2545 Btu/hr = 1 hp h)

Ton of refrigeration (this is the heat required to melt 1 ton (2000 lb) of ice at 32°F to water at 32°F during 24 h).

Gas properties

Our companion volume, *Guide to Compressor Technology,* contains compressibility and temperature-entropy charts for a number of gases or mixtures of gases frequently compressed. These have been prepared by Dresser-Rand, formerly Ingersoll-Rand, in 1981 from the latest and best data available at the time. In some cases, several sources have been combined to obtain the range desired. It is believed these charts present the characteristics covered as accurately as existing knowledge permits.

Compressibility is shown as a multiplying factor to be applied to perfect gas volume at a specific condition to obtain actual volume.

Temperature entropy diagrams permit determination of the gas temperature at the completion of adiabatic (isentropic) compression for any compression stage. Entering the chart at a specific inlet temperature and pressure and moving horizontally to the discharge pressure, one reads discharge temperature. This frequently varies from values obtained by using perfect gas laws.

This discharge temperature is needed to determine compressibility from the other chart at *discharge* conditions, a factor that enters into many compressor calculations. Unless these calculations are reasonably precise, the driver—often a steam turbine—may not be properly specified or sized.

Temperature Conversion Chart

Note: The center column of numbers in boldface refers to the temperature in degrees, either Centigrade or Fahrenheit, which it is desired to convert into the other scale. If converting from Fahrenheit to Centigrade degrees, the equivalent temperature will be found in the left column, while if converting from degrees Centigrade to degrees Fahrenheit, the answer will be found in the column on the right.

Centigrade		Fahrenheit	Centigrade		Fahrenheit	Centigrade		Fahrenheit	Centigrade		Fahrenheit
–273.17	**–459.7**		–20.6	**–5**	23.0	11.1	**52**	125.6	54.4	**130**	266
–268	**–450**		–17.8	**0**	32.0	11.7	**53**	127.4	57.2	**135**	275
–262	**–440**					12.2	**54**	129.2	60.0	**140**	284
–257	**–430**		–17.2	**1**	33.8	12.8	**55**	131.0	62.8	**145**	293
–251	**–420**		–16.7	**2**	35.6	13.3	**56**	132.8	65.6	**150**	302
–246	**–410**		–16.1	**3**	37.4				68.3	**155**	311
–240	**–400**		–15.6	**4**	39.2	13.9	**57**	134.6	71.1	**160**	320
–234	**–390**		–15.0	**5**	41.0	14.4	**58**	136.4			
			–14.4	**6**	42.8	15.0	**59**	138.2	73.9	**165**	329
–229	**–380**		–13.9	**7**	44.6	15.6	**60**	140.0	76.7	**170**	338
–223	**–370**		–13.3	**8**	46.4	16.1	**61**	141.8	79.4	**175**	347
–218	**–360**					16.7	**62**	143.6	82.2	**180**	356
–212	**–350**		–12.8	**9**	48.2	17.2	**63**	145.4	85.0	**185**	365
–207	**–340**		–12.2	**10**	50.0	17.8	**64**	147.2	87.8	**190**	374
–201	**–330**		–11.7	**11**	51.8				90.6	**195**	383
–196	**–320**		–11.1	**12**	53.6	18.3	**65**	149.0	93.3	**200**	392
–190	**–310**		–10.6	**13**	55.4	18.9	**66**	150.8	96.1	**205**	401
			–10.0	**14**	57.2	19.4	**67**	152.6	98.9	**210**	410
–184	**–300**		–9.4	**15**	59.0	20.0	**68**	154.4	100.0	**212**	414
–179	**–290**		–8.9	**16**	60.8	20.6	**69**	156.2	102	**215**	419
–173	**–280**					21.1	**70**	158.0	104	**220**	428
–169	**–273**	–459.4	–8.3	**17**	62.6	21.7	**71**	159.8	107	**225**	437
–168	**–270**	–454	–7.8	**18**	64.4	22.2	**72**	161.6	110	**230**	446
–162	**–260**	–436	–7.2	**19**	66.2				113	**235**	455
–157	**–250**	–418	–6.7	**20**	68.0	22.8	**73**	163.4	116	**240**	464
–151	**–240**	–400	–6.1	**21**	69.8	23.3	**74**	165.2			
			–5.6	**22**	71.6	23.9	**75**	167.0	118	**245**	473
–146	**–230**	–382	–5.0	**23**	73.4	24.4	**76**	168.8	121	**250**	482
–140	**–220**	–364	–4.4	**24**	75.2	25.0	**77**	170.6	124	**255**	491

–134	–210	–346				25.6	78	172.4	127	260	500
–129	–200	–328	–3.9	25	77.0	26.1	79	174.2	129	265	509
–123	–190	–310	–3.3	26	78.8	26.7	80	176.0	132	270	518
–118	–180	–292	–2.8	27	80.6				135	275	527
–112	–170	–274	–2.2	28	82.4	27.2	81	177.8	138	280	536
–107	–160	–256	–1.7	29	84.2	27.8	82	179.6	141	285	545
			–1.1	30	86.0	28.3	83	181.4	143	290	554
–101	–150	–238	–0.6	31	87.8	28.9	84	183.2	146	295	563
–96	–140	–220	0.0	32	89.6	29.4	85	185.0	149	300	572
–90	–130	–202				30.0	86	186.8	154	310	590
–84	–120	–184	0.6	33	91.4	30.6	87	188.6	160	320	608
–79	–110	–166	1.1	34	93.2	31.1	88	190.4	166	330	626
–73.3	–100	–148.0	1.7	35	95.0				171	340	644
–67.8	–90	–130.0	2.2	36	96.8	31.7	89	192.2	177	350	662
–62.2	–80	–112.0	2.8	37	98.6	32.2	90	194.0			
			3.3	38	100.4	32.8	91	195.8	182	360	680
–59.4	–75	–103.0	3.9	39	102.2	33.3	92	197.6	188	370	698
–56.7	–70	–94.0	4.4	40	104.0	33.9	93	199.4	193	380	716
–53.9	–65	–85.0				34.4	94	201.2	199	390	734
–51.1	–60	–76.0	5.0	41	105.8	35.0	95	203.0	204	400	752
–48.3	–55	–67.0	5.6	42	107.6	35.6	96	204.8	210	410	770
–45.6	–50	–58.0	6.1	43	109.4				216	420	788
–42.8	–45	–49.0	6.7	44	111.2	36.1	97	206.6	221	430	806
–40.0	–40	–40.0	7.2	45	113.0	36.7	98	208.4			
			7.8	46	114.8	37.2	99	210.2	227	440	824
–37.2	–35	–31.0	8.3	47	116.6	37.8	100	212.0	232	450	842
–34.4	–30	–22.0	8.9	48	118.4	40.6	105	221	238	460	860
–31.7	–25	–13.0				43.3	110	230	243	470	878
–28.9	–20	–4.0	9.4	49	120.2	46.1	115	239	249	480	896
–26.1	–15	5.0	10.0	50	122.0	48.9	120	248	254	490	914
–23.3	–10	4.0	10.6	51	123.8	51.7	125	257	260	500	932

Units of Measure and Conversion Factors

Units	Symbol	Other units	
inch/inches	in		
foot/feet	ft		
meter	m		(SI)*
millimeter	mm		
square inch	in^2		
square foot	ft^2		
square meter	m^2		(SI)
square centimeter	cm^2		
square millimeter	mm^2		
cubic inch	in^3		
cubic foot	ft^3		
gallon (US liquid)	gal		
cubic meter	m^3		(SI)
liter	L		
pound mass (avoirdupois)	lbm		
kilogram	kg		(SI)
pound force (avoirdupois)	lbf		
kilogram force	kgf		
newton	N	$m \cdot kg/s^2$	(SI)
degree Fahrenheit	°F		
kelvin	K		(SI)
degree Celsius	°C		(SI)
British thermal unit (International Table)	Btu		
kilocalorie (International Table)	kcal		
joule	J	$N \cdot m$, $m^2 \cdot kg/s^2$	(SI)
kilojoule	kJ		
second (customary)	sec		
second	s		(SI)
minute	min		
hour (customary)	hr		
hour (metric)	h		
watt	W	J/s, $N \cdot m/s$, $m^2 \cdot kg/s^3$	(SI)
kilowatt	kW		
horsepower (electrical)	hp		
pound force/square inch	psi	lbf/in^2	
inches of mercury	inHg		
feet of water	ftH_2O		
pascal	Pa	N/m^2, $kg/m \cdot s^2$)	(SI)
kilopascal	kPa		
bar (metric)	bar		
millimeter of mercury	mmHg	torr	
torr	torr	mmHg	
centipoise	cp		

Density (Mass/Volume)			
Multiply	By	To obtain	
lbm/in^3	2.767990×10^4	kg/m^3	(SI)*
lbm/in^3	2.767990×10^1	kg/L	
lbm/ft^3	1.601846×10^1	kg/m^3	(SI)
lbm/ft^3	1.601846×10^{-2}	kg/L	
lbm/gal	1.198264×10^2	kg/m^3	(SI)
lbm/gal	1.198264×10^{-1}	kg/L	

Enthalpy (Energy/Mass)			
Multiply	By	To obtain	
Btu/lbm	2.326000×10^3	J/kg	(SI)
Btu/lbm	2.326000	kJ/kg	
Btu/lbm	5.555556×10^{-1}	kcal/kg	

Heat Capacity and Entropy (Energy/Mass-Temperature)			
Multiply	By	To obtain	
Btu/(lbm·°F)	4.186800×10^3	J/(kg·K)	(SI)
Btu/(lbm·°F)	4.186800	kJ/(kg·K)	
Btu/(lbm·°F)	1.000000	kcal/(kg·°C)	

Thermal Conductivity (Energy-Length/Time-Area-Temperature)			
Multiply	By	To obtain	
Btu·in/(hr·ft^2·°F)	1.442279×10^{-1}	W/(m·K)	(SI)
Btu·in/(hr·ft^2·°F)	1.240137×10^{-1}	kcal·m/(h·m^2·K)	
Btu·in/(hr·ft^2·°F)	1.730735	W/(m·K)	(SI)
Btu·in/(hr·ft^2·°F)	1.488164	kcal·m/(h·m^2·K)	

Dynamic Viscosity (Mass/Time-Length or Force-Time/Area)			
Multiply	By	To obtain	
cp	1.000000×10^{-3}	Pa·s	(SI)
lbm/(h·ft)	4.133789×10^{-4}	Pa·s	(SI)
lbm/(h·ft)	4.133789×10^{-1}	cp	
lbm/(s·ft)	1.488164	Pa·s	(SI)
lbm/(sec·ft)	1.488164×10^3	cp	
lbf·sec/ft^2	4.788026×10^1	Pa·s	(SI)
lbf·sec/ft^2	4.788026×10^4	cp	

Heat Flux Density (Energy/Time-Area)			
Multiply	By	To obtain	
Btu/(h·ft^2)	3.154591	W/m^2	(SI)
Btu/(h·ft^2)	2.712460	kcal/(h·m^2)	

Heat Transfer Coefficient (Energy/Time-Area-Temperature)			
Multiply	By	To obtain	
Btu/(h·ft^2·°F)	5.678263	W/(m^2·K)	(SI)
Btu/(h·ft^2·°F)	4.882428	kcal/(h·m^2·K)	

Fouling Resistance (Time-Area-Temperature/Energy)			
Multiply	By	To obtain	
$h \cdot ft^2 \cdot °F/Btu$	1.761102×10^{-1}	$m^2 \cdot K/W$	(SI)
$h \cdot ft^2 \cdot °F/Btu$	2.048161×10^{-1}	$h \cdot m^2 \cdot K/kcal$	

Power (Energy/Time)			
Multiply	By	To obtain	
Btu/hr	2.930711×10^{-1}	W	(SI)

Pressure or Stress (Force/Area)†			
Multiply	By	To obtain	
psi	6.894757×10^{3}	Pa	(SI)
psi	6.894757	kPa	
psi	6.894757×10^{-2}	bar	
psi	7.030696×10^{-2}	kgf/cm^2	
lbf/ft^2	4.788026×10^{1}	Pa	(SI)
lbf/ft^2	4.788026×10^{-2}	kPa	
lbf/ft^2	4.882428	kgf/m^2	
inHg(32°F)	3.38638×10^{3}	Pa	(SI)
inHg(32°F)	3.38638	kPa	
inHg(32°F)	3.38638×10^{-2}	bar	
inHg(32°F)	3.45315×10^{-2}	kgf/cm^2	
inHg(32°F)	2.540×10^{1}	mmHg	
torr (0°C)	1.33322×10^{2}	Pa	(SI)
torr (0°C)	1.0	mmHg	
ftH_2O (39.2°F)	2.98898×10^{3}	Pa	(SI)
ftH_2O (39.2°F)	2.98898	kPa	
ftH_2O (39.2°F)	3.047915×10^{2}	kgf/m^2	

Velocity (Length/Time)‡			
Multiply	By	To obtain	
ft/sec	3.048000×10^{-1}	m/s	(SI)
ft/min	5.080000×10^{-3}	m/s	(SI)

Mass Flow Rate (Mass/Time)			
Multiply	By	To obtain	
lbm/hr	1.259979×10^{-4}	kg/s	(SI)
lbm/hr	4.535924×10^{-1}	kg/h	

Volume Flow Rate (Volume/Time)§			
Multiply	By	To obtain	
ft^3/min	4.719474×10^{-4}	m^3/s	(SI)
ft^3/min	1.699011	m^3/h	
gal/min	6.309020×10^{-5}	m^3/s	(SI)
gal/min	2.271247×10^{-1}	m^3/h	
gal/min	3.785412	L/min	

Mass Velocity (Mass/Time-Area)			
Multiply	By	To obtain	
lbm/(h·ft^2)	1.35623×10^{-3}	kg/(s·m^2)	(SI)
lbm/(h·ft^2)	4.882428	kg/(h·m^2)	
lbm/(s·ft^2)	4.882428	kg/(s·m^2)	(SI)

Specific Volume (Volume/Mass)			
Multiply	By	To obtain	
ft^3/lbm	6.242797×10^{-2}	m^3/kg	(SI)
ft^3/lbm	6.242797×10^{1}	L/kg	
gal/lbm	8.345406×10^{-3}	m^3/kg	(SI)
gal/lbm	8.345406	L/kg	

Prefixes Denoting Decimal Multiples or Submultiples		
Prefix	Symbol	Multiplication factor
micro	μ	$0.000\,001 = 10^{-6}$
milli	m	$0.001 = 10^{-3}$
centi	c	$0.01 = 10^{-2}$
deci	d	$0.1 = 10^{-1}$
deca	da	$10 = 10^{1}$
hecto	h	$100 = 10^{2}$
kilo	k	$1\,000 = 10^{3}$
mega	M	$1\,000\,000 = 10^{6}$
giga	G	$1\,000\,000\,000 = 10^{9}$

Conversion Factors: Length			
Multiply	By	To obtain	
in	2.540×10^{-2}	m	(SI)
in	2.540×10^{1}	mm	
ft	3.048×10^{-1}	m	(SI)
ft	3.048×10^{2}	mm	

Area			
Multiply	By	To obtain	
in^2	6.451600×10^{-4}	m^2	(SI)
in^2	6.451600×10^{2}	mm^2	
ft^2	9.290304×10^{-2}	m^2	(SI)
ft^2	9.290304×10^{4}	mm^2	

Volume			
Multiply	By	To obtain	
in^3	1.638706×10^{-5}	m^3	(SI)
in^3	1.638706×10^{-2}	L	
ft^3	2.831685×10^{-2}	m^3	(SI)
ft^3	2.831685×10^{1}	L	
gal	3.785412×10^{-3}	m^3	(SI)
gal	3.785412	L	

Mass			
Multiply	By	To obtain	
lbm	4.535924×10^{-1}	kg	(SI)

Force			
Multiply	By	To obtain	
lbf	4.448222	N	(SI)
lbf	4.535924×10^{-1}	kgf	
kgf	9.806650 N (SI)		

Temperature	
$°K = (°F + 459.67)/1.8$	(SI)
$°C = (°F - 32)/1.8$	(SI)
$°F = 1.8°K - 459.67$	
$°F = 1.8°C + 32$	

Energy, Work or Quantity of Heat			
Multiply	By	To obtain	
Btu	1.055056×10^{3}	J	(SI)
Btu	2.519958×10^{-1}	kcal	
ft·lbf	1.355818	J	(SI)
ft·lbf	3.238316×10^{-4}	kcal	

* (SI) Denotes an International System of Units unit.
† Pressure should always be designated as gauge or absolute.
‡ The acceleration of gravity, *g*, is taken as $9.80665\ m/s^2$.
§ One gallon (U.S. liquid) equals $231\ in^3$.

Bibliography and List of Contributors

Asea Brown-Boveri Company, Baden, Switzerland

Burkhard, G. A.: "Steam Turbines For Industry," *BBC Review* 61, 1974 (1), pp. 4–8.

"The Coupling of Large Compressor and Steam Turbine Drivers," Publication No. CH-T 112032 E.

"Design and Operational Behavior of Steam Turbosets with Reduction Gears," Publication No. CH-T 110433 E.

Dien, W.: "Welded Turbine Shafts," *BBC Review* 60, 1973 (9), pp. 427–430.

"Erosion in the Low-Pressure Blading of Medium-Sized and Industrial Steam Turbines," Publication No. CH-T 110210 E.

Hohn, A.: "Large Steam Turbine Rotors," *BBC Review* 60, 1973, (9), pp. 404–416.

Hohn, A.: "Steam Turbines under Start-Up Conditions," *Brennstoff/Waerme/Kraft* (BWK), vol. 8, 1974.

Hohn, A.: "The Useful Life of Steam Turbine Parts," Publication No. CH-T 060143 E.

Wirz, K.: "Control Problems Associated with Industrial Steam Turbines," Publication No. CH-T 110303 E.

Luethy, A.: "Welding Techniques Used in the Manufacture of Gas Turbines," DVS-Tagung Berlin 1959, *Fachbuchreihe Schweisstechnik,* vol. 17, Frankfurt, 1960, pp. 102–106.

Muehlhaeuser, H.: "Modern Feedpump Turbines," *BBC Review* 58, 1971 (10), pp. 436–451.

Mohr, A.: "The Non-Destructive Testing of Welds and in Particular of Rotor Welds," *BBC Review* vol. 48, 1961, no. 8/9, pp. 475–484.

Reuter, H.: "Dampfturbinen-Bauarten," *BBC Nachrichten* 48, 1966 (7), pp. 438–445.

"Single-Cylinder Steam Turbines of Medium Output for Power Stations and Industry," Publication No. CH-T 110372 E.

"Solid Rotors for Steam Turbines," Publication No. 110232 E.

"Some Advantages of Welding Turbine Rotors," *Welding Journal* 47, 1968 (6), pp. 461–478.

Somm, E.: "A Means of Estimating the Erosion Hazard in Low-Pressure Steam Turbines," *BBC Review* 58, 1971 (10), pp. 458–472.

Spechtenhauser, A.: "High-Power Turbines for Compressor Drives," *BBC Review* 61, 1974 (1), pp. 17–27.

"Steam Turbines for Industry," Publication No. CH-T 110312 E.

Spechtenhauser, A.: "The Suitability of Reaction-Type Turbines for Driving Compressors and Pumps," Publication No. CH-T 112073 E.

"Welded Rotors for Steam Turbines," Publication No. CH-T 060072 E.

Cincinnati Gear Company, Cincinnati, Ohio

General Product Literature, 1990.

Coppus-Murray Turbomachinery Corporation, Burlington, Iowa

"Campbell And Goodman Diagrams For Murray Axial Steam Turbine Blades," Bulletin TT13-1M0391GI, 1991.
"Estimating Steam Turbine Condensing Pressures," Bulletin TT11-1M1290GI, 1990.
"Mechanical Drive Steam Turbines," Catalog 300A, 1986.
"Pressure Lubrication of Murray Multistage Turbines," Bulletin TT12-2M0291GI, 1991.
"Steam Turbine Generator Sets," Bulletin TT1-1M0290TG/2M0590TG, 1990.
"Thermodynamics Applied to Steam Turbines," Bulletin TT8-2M1290GI, 1990.

Dresser-Rand Company, Wellsville, N.Y. (formerly Worthington Turbine Division and Turbodyne Company)

Aldrich, C. K.: "Types of Extraction Turbines and Their Performance Curves," Publication No. ST-24.
Anderson, L. M.: "The Application of Electric Speed and Load Sensing Governors to Steam Turbine Synchronous Generator Units," Publication No. ST-2.
Anderson, Lester M.: "Checklist of Good Practices in Installation," Operation and Maintenance of Steam Turbines, Publication No. ST-9.
Ashline, E. C.: "Materials of Construction Single Stage Turbines," Publication No. ST-1.
Austin, N. M.: "Steam Turbine Diaphragms," Publication No. ST-5.
Beebee, D.: "Test Data Acquisition and Reduction System," Publication No. ST-26.
Lamberson, J., and R. Moll: "Performance Acceptance Test for a Non-Condensing Turbine Generator Drive," Publication No. ST-17.1.
Lamberson, J., and R. Moll: "Performance Acceptance Test for a Non-Condensing Turbine (Enthalpy Drop Method) Generator Drive," Publication No. ST-17.2.
Lamberson, J., and R. Moll: "Performance Acceptance Test for a Condensing Turbine Generator Drive," Publication No. ST-17.3.
Lamberson, J., and R. Moll: "Performance Acceptance Test for a Condensing Regenerative Cycle Turbine Generator Drive," Publication No. ST-17.4.
Lamberson, J., and R. Moll: "Performance Acceptance Test for a Non-Condensing Extraction Turbine Generator Drive," Publication No. ST-17.5.
Lamberson, J., and R. Moll: "Performance Acceptance Test for a Controlled Extraction Condensing Turbine Generator Drive," Publication No. ST-17.6.
Lamberson, J., and R. Moll: "Performance Acceptance Test for a Controlled Extraction Condensing Regenerative Cycle Turbine Generator Drive," Publication No. ST-17.7.
Lamberson, J., and R. Moll: "Performance Acceptance Test for a Double Controlled Extraction Condensing Turbine Generator Drive," Publication No. ST-17.8.
Lamberson, J., and R. Moll: "Performance Acceptance Test for a Non-Condensing Turbine Mechanical Drive," Publication No. ST-18.1.
Lamberson, J., and R. Moll: "Performance Acceptance Test for a Non-Condensing Turbine (Enthalpy Drop Method) Mechanical Drive," Publication No. St-18.2.
Lamberson, J., and R. Moll: "Performance Acceptance Test for a Condensing Turbine Mechanical Drive," Publication No. ST-18.3.
Lamberson, J., and R. Moll: "Performance Acceptance Test for a Non-Condensing Extraction Turbine Mechanical Drive," Publication No. ST-18.4.
Lamberson, J., and R. Moll: "Performance Acceptance Test for an Uncontrolled Extraction Condensing Turbine Mechanical Drive," Publication No. ST-18.5.
Lamberson, J., and R. Moll: "Performance Acceptance Test for a Controlled Extraction Condensing Turbine Mechanical Drive," Publication No. ST-18.6.
Loughlin, R. M.: "Steam Turbine Rotors," Publication No. ST-4.
Rubira, H.: "Turbo-Expanders," Publication No. ST-25.
Singh, M., and D. Schiffer: "Application of the Finite Element Method in Steam Turbine Design," Publication No. ST-15.

Singh, M.: "SAFE Diagram—A Turbodyne Evaluation Tool for Packeted Bladed Disc Assembly," Publication No. ST-18.
"Turbodyne SST-2 Standard Single Stage Turbines," Publication No. 4805-B6A.
Vargo, J. J.: "Blading," Publication No. ST-3.
Del Vecchio, R.: "Steam Turbine Water Washing," Publication No. ST-8.
"Worthington Large Mechanical Drive Steam Turbines," Bulletin 4850-B3706.

The Elliott Company, Jeannette, Pa.

"Bar-Operated Governor Valves," Form 2434-370.
"Blade Design," Reprint No. 136A.
"Elliott Multivalve Turbine," Bulletin H-37B.
"Governors And Control Systems," Reprint No. 135A.
"High Pressure Steam Turbine Design," Reprint No. 169.
"Motor-Driven Turning Gear," Form 1599A-576.
"Multivalve Turbine Bearings," Form 2441-3M74.
"Multivalve Turbine Casing," Form 2442-374.
"Multivalve Turbine Construction And Arrangement," Form 2443-374.
"Multivalve Turbine Materials of Construction," Form 2439 B 579.
"Multivalve Turbine Nozzles And Diaphragms," Form 2440-3M74.
"Multivalve Turbine Overspeed Trip System," Form 2438-374.
"Multivalve Turbine Rotor Assembly," Form 2437-374.
"Multivalve Turbine Shaft Seals," Form 2436-374.
"Pos-E-Trip Control," Form 1687A-376.
"Selection Guide for Multistage Steam Turbines," Bulletin H-48 590-FL.
"Single Valve Multistage Industrial Turbines," Bulletin H-36.
"Speed Control For Straight Condensing And Straight Non-condensing Turbines," Form 2447A-576.
"Steam Turbine Components," Reprint No. 132A.
"Thermodynamics," Reprint No. 134A.
"Turbine Blades," Form 2446-374.
"Turbine Selection," Reprint No. 133A.
"2000 PSIG (138 Bar) Steam Turbines," Reprint No. 138A.
"Type N Multivalve Turbines," Reprint No. 131.
"Type Q Multivalve Turbines," Reprint No. 130A.
"Type R Multivalve Turbines," Reprint No. 129A.
"YR Single-Stage Turbines," Bulletin H-31K590FL.

General Electric Company, Fitchburg, Mass.

Baily, F. G., R. J. Peterson, R. T. Genter: "Steam Turbines for Industrial and Cogeneration Applications," Bulletin GER-3614.
Beck, R. C., and L. H. Neidich: "Uprating Turbines for Increased Efficiency," Unnumbered bulletin.
Bienvenue, R. T.: "Steam Turbine-Generator Maintainability—A Means to Improve Unit Availability," Bulletin GER-3621.
Carlstrom, L. A.: "Steam Turbine Design Philosophy," Bulletin MS0A3-1.
Caruso, W. J.: "Rotor Dynamics Technology," Bulletin MSOA3-8.
Catlow, W. G.: "Modern Bucket Design," Bulletin MSOA3-4.
Finck, E. J.: "Turbine Efficiency—Key to Reduced Operating Costs," Bulletin MSOA3-6.
Industrial Steam Turbines, Advanced Technology for Improved Output and Reliability, GE Energy Bulletin GEA 13450A, 2006.
Leger, Donald R.: "Mechanical Drive Steam Turbines" Bulletin GER-3768.
Lopes, W. F.: "Geared Steam Turbine Generator Sets," Bulletin GER-3461.
Mahon, A. P.: "Rotor Blade Balance Procedures and Practices," Bulletin MSOA3-14.
"Mechanical Drive Turbines," Bulletin GEA6232.
"Mechanical Drive Turbines—the Heart of Your Operation" Bulletin GEA-12059.

Ruegger, W. A., and Donald R. Leger: "Recent Advances in Steam Turbines for Industrial and Cogeneration Applications."
Scoretz, M., and R. Williams: "Industrial Steam Turbine Value Packages," GER-4191A, 2005.
Vitone, E. T.: "Operating Experience with Large Industrial Steam Turbines," Bulletin MSOA3-2.
Vitone, E. T., W. G. Catlow, J. I. Cofer, B. E. Gans: "Recent Advances in Mechanical Drive Turbine Technology," Bulletin GER-3617.
Wellner, K. V.: "Casing Design," Bulletin MSOA3-9.

Maag Gear Company, Zurich, Switzerland

"Gearboxes for Generator Drives," Bulletin G-105E.
"The Maag Synchronous Clutch Coupling Program and Its Application," Bulletin G-104aE.

Mitsubishi Heavy Industries, Ltd., Hiroshima, Japan

"Mitsubishi Mechanical Drive Steam Turbines," Form HD70-ESTC01E1, 1993.

Philadelphia Gear Corporation, Philadelphia, Pa.

"Epicyclic Gear Drive Series," Bulletin OK-5000.
"Synchrotorque Hydroviscous Drives," Bulletin CEN-5M-10/90.

Salamone Turbo Engineering, Houston, Tex.

"Journal Bearing Design: Types and Applications," 1984.

Siemens AG, Power Generation Group, Erlangen, Germany (USA Branch: Siemens Power Corporation, Milwaukee, Wis.)

"Drum Rotor Blading for Industrial Steam Turbines," Bulletin No. A19100-U231-X-7600.
"Gearing for High-Speed Industrial Steam Turbines," Bulletin No. E25/2007-101.
"High Output, High Speed Industrial Steam Turbines, Type Range MG, EMG," Bulletin No. E25/1292-101.
"High Output, High Speed Industrial Steam Turbines, Type Range WK," Bulletin No. E25/2030-101.
"Industrial Steam Turbines," Bulletin No. 125259E25.
"Industrial Steam Turbines, LP Last-Stage Blades for Condensing Turbines," Bulletin No. A19100-U33-A119-V1-7600.
"Industrial Steam Turbines on Building Block Principle," Bulletin No. A19100-U331-A159-X-760.
"Industrial Turbines for Ethylene Plants," Bulletin No. E-25/2025-101.
"Industrial Turbines for High Initial Steam Conditions, Type Range HG," Bulletin No. E25/2036-101.
"Means of Improving Efficiency of Reaction Turbines," Bulletin No. 133009U33.
"Steam Turbosets for Economical Waste Heat Recovery," Bulletin No. A19100-E25-B038-X-760.

"Steam Turbosets for Pulp and Paper Mills," Bulletin No. E25/2023-101.
"Thrust Bearings for Industrial Steam Turbines," Bulletin No. A19100-U231-A202-X-7600.

Sulzer Brothers, Ltd., Winterthur, Switzerland, and New York, N.Y.

Publication e/25.01.10-Cgg 30.

VOITH TURBO G.m.b.H. & Co., K. G., Crailsheim, Germany (USA Branch: Voith Transmissions, Inc., York, Pa.

"Geared Variable-Speed Turbo Couplings and Multi-Stage Variable Speed Drives," Bulletin Cr 101 e/USA 11.12.2 Wa.
"MSVD Multi-Stage Variable-Speed Drive," Bulletin Cr 168 ea 11.7.4 Wa.
"Voith (Turbo) Fluid Couplings," Bulletin Cr 128 e 12.3.3 ho.

Others

Abdul-Wahed, M. N., J. Frene, and D. Nicolas: "Analysis of Fitted Partial Arc and Tilting-Pad Journal Bearings," ASLE Preprint No. 78-AM-2A-3, 1978.
Abramovitz, S.: "Fluid Film Bearings Fundamentals and Design Criteria and Pitfalls," *Proceedings of the Sixth Turbomachinery Symposium,* Gas Turbine Laboratories, Texas A&M University, 1977.
Adams, M. L.: "Axial-Groove Journal Bearings," *Machine Design,* April 1970, pp. 120–123.
Allaire, P. E., and R. D. Flack: "Design of Journal Bearings for Rotating Machinery," *Proceedings of the Tenth Turbomachinery Symposium,* Turbomachinery Laboratories, Texas A&M University, College Station, Tex., 1981, pp. 25–45.
Allaire, P. E., J. C. Nicholas, and L. E. Barrett: "Analysis of Step Journal Bearings—Infinite Length, Inertia Effects," *Trans. ASLE,* 22(4), 1979, pp. 333–341.
Bloch, H. P: "Less Costly Turboequipment Uprates Through Optimized Coupling Selection," *Proceedings of the Fourth Turbomachinery Symposium,* Turbomachinery Laboratory, Texas A&M University, College Station, Tex.
Chen, H. M.: "Active Magnetic Bearing Design Methodology—A Conventional Rotordynamics Approach," 15th Leeds-Lyon Symposium on Tribology, Sept 6–9, 1988, at Leeds, U.K. Published in workshop proceedings.
Chen, H. M., and Dill, J. F.: "A Conventional Point of View on Active Magnetic Bearings," NASA Langley Workshop on Magnetic Suspension Technology, Feb. 2–4, 1988.
Dietzel, F.: Dampfturbinen. 2nd ed, Carl Hanser, Munich 1970.
Eierman, R. G.: "Stability Analysis and Transient Motion of Axial Groove, Multilobe and Tilting Pad Bearings," M. S. thesis, University of Virginia, 1976.
Ettles, C. M. McC.: "The Analysis and Performance of Pivoted Pad Journal Bearings Considering Thermal and Elastic Effects," *J. Lubr. Technol.,* Trans. ASME, ASME Paper No. 79-LUB-31 (1979).
Falkenhagen, G. L.: "Stability and Transient Motion of a Hydrodynamic Horizontal Three-Lobe Bearing System," *The Shock and Vibration Dig.,* 7(5), 1975.
Falkenhagen, G. L., E. J. Gunter, and F. T. Schuller: "Stability and Transient Motion of a Vertical Three-Lobe Bearing System," *J. Eng. Ind.,* Trans. ASME, Series, B, 94(2), 1972, pp. 665–677.
Grammel, R.: "The explanation of the problem of the high bursting strength of rotating cylinders." *Ing. Archiv.* 16, 1947 (1).

Gyarmathy, G.: The Fundamentals of the Theory of Wet Steam Turbines. Review of the Thermal Turbomachinery Institute, No. 6, Zurich, 1962.

Hort, W.: "Schwingungen der Räder und Schaufeln in Dampfturbinen." VDI-z. 70 1962 (42) 1375–1381 and (43) 1419–1424.

Huppmann, H.: "Häufigkeit und Ursachen von Schäden an Bauteilen grosser Dampfturbinen." *Maschinenschaden* 43, 1970 (1) 1–6.

Jones, G. J., and F. A. Martin: "Geometry Effects in Tilting-Pad Journal Bearings," ASLE Preprint No. 78-AM-2A-2, 1978.

Kirillov, J. J., and A. S. Laskin: "Untersuchung der wechselnden aerodynamischen Kräfte in dem von einem nichtstationären Strom umflossenen Turbinengitter." (Russian.) *Energomashinostrojenie* 12, 1966 (12), pp. 29–32.

Kirk, R. G.: "The Influence of Manufacturing Tolerance on Multi-Lobe Bearing Performance in Turbomachinery," *Topics in Fluid Film Bearing and Rotor Bearing System Design and Optimization,* ASME Book No. 10018, 1978, pp. 108–129.

Löffler, K.: The Calculation of Rotating Disks and Cylinders, Springer, 1961.

Lotz, M.: "Erregung von Schaufelschwingungen in axialen Turbomaschinen durch die benachbarten Schaufelgitter." *Wärme* 72, 1965 (2) pp. 59–68.

Lund, J. W.: "Rotor-Bearing Design Technology; Part III: Design Handbook for Fluid-Film Bearings," Mechanical Technology Incorporated, Technical Report No. AFAPL-TR-65-45, 1965.

Lund, J. W.: "Rotor-Bearing Dynamics Design Technology, Part VII: The Three-Lobe Bearing and Floating Ring Bearing," Mechanical Technology Incorporated, Technical Report No. AFAPL-TR-64-45, 1968.

Lund, J. W.: "Spring and Damping Coefficients for the Tilting-Pad Journal Bearing," *Trans. ASLE,* 7(4), 1964, pp. 342–352.

Lund, J. W., and K. K. Thomsen: "A Calculation Method and Data for the Dynamic Coefficients of Oil-Lubricated Journal Bearings," *Topics in Fluid Film Bearings and Rotor Bearing System Design and Optimization,* ASME Book No. 100118, 1978, pp. 1–28.

Mansfield, D. H., D. W. Boyce, and W. G. Pacelli: "Guidelines for Specifying and Evaluating the Rerating and Reapplication of Steam Turbines," *Proceedings of the Thirty-Fifth Turbomachinery Symposium,* Turbomachinery Laboratory, Texas A&M University, College Station, Tex., 2006.

Moissejev, A. A.: "Einfluss des Axialspiels auf die Vibrationsfestigkeit der Überdruck-Laufschaufeln einer Axialturbine." (Russian.) *Energomashinostrojenie* 17, 1971 (5), pp. 21–23.

NEMA SM 23: "Steam Turbine for Mechanical Drive Service," National Electric Manufacturers Association, Washington, D.C., 1991.

Nicholas, J.C: "Hydrodynamic Journal Bearings—Types, Characteristics and Applications," *Mini Course Notes, 20th Annual Meeting, The Vibration Institute,* Willowbrook, Ill., 1996, pp. 79–100.

Nicholas, J. C: "Lund's Tilting Pad Journal Bearing Pad Assembly Method," *ASME Journal of Vibrations and Acoustics* 125, 2003 (4), pp. 448–454.

Nicholas, J. C: "Tilting Pad Bearing Design," *Proceedings of the Twenty-Third Turbomachinery Symposium,* Turbomachinery Laboratory, Texas A&M University, College Station, Tex., 1994, pp. 179–194.

Nicholas, J. C: "Tilting Pad Journal Bearings with Spray-Bar Blockers and By-Pass Cooling for High Speed, High Load Applications," *Proceedings of the Thirty-Second Turbomachinery Symposium,* Turbomachinery Laboratory, Texas A&M University, College Station, Tex., 2003.

Nicholas, J. C., and P. E. Allaire: "Analysis of Step Journal Bearings—Finite Length, Stability," *Trans. ASLE* 23 (2), 1980, pp. 197–207.

Nicholas, J. C., P. E. Allaire, and D. W. Lewis: "Stiffness and Damping Coefficients for Finite Length Step Journal Bearings," *Trans. ASLE,* 23(4), 1980, pp. 353–362.

Nicholas, J. C., E. J. Gunter, and P. E. Allaire: "Stiffness and Damping Coefficients for the Five Pad Tilting Pad Bearing," *Trans. ASLE,* 22(2), 1979, pp. 113–124.

Pinkus, O.: "Analysis and Characteristics of the Three-Lobe Bearing," *J. Basic Eng.,* Trans. ASME, 1959, pp. 49–55.

Raimondi, A. A., and J. Boyd: "An Analysis of the Pivoted-Pad Journal Bearing," *Mech. Eng.,* 1953, pp. 380–386.

Raimondi, A. A. and J. Boyd: "A Solution for the Finite Journal Bearing and Its Application to Analysis and Design: I, II, III," *ASLE,* 1957.
Rippel, H. C.: "Cast Bronze Bearing Design Manual," Cast Bronze Bearing Institute, Inc., 1971.
Salamone, D. J.: "Introduction to Hydrodynamic Journal Bearings," *Vibration Institute Minicourse Notes—Machinery Vibration Monitoring and Analysis,* The Vibration Institute, Clarendon Hills, Ill., 1985, pp. 41–56.
Senger, U.: "Die Dampfnässe in den letzten Stufen von Kondensationsturbinen." *Elektrizitätswirtschaft* 38, 1939 (14), pp. 354–360.
Shapiro, W., and R. Colsher: "Dynamic Characteristics of Fluid-Film Bearings," *Proceedings of the Sixth Turbomachinery Symposium,* Gas Turbine Laboratories, Texas A&M University, College Station, Tex., 1977.
Singh, M. P., and C. M. Ramsey: "Reliability Criteria for Turbine Shaft Based on Coupling Type—Geared to Have Flexibility," *Proceedings of the 19th Turbomachinery Symposium,* Turbomachinery Laboratory, Texas A&M University, College Station, Tex.
Tanaka, M.: "Thermohydrodynamic Performance of a Tilting Pad Journal Bearing with Spot Lubrication," *ASME Journal of Tribology* 113, 1991 (3), pp. 615–619.
Traupel, W.: *Thermische Turbomaschinen.* Vol. I. 2nd ed. Springer, Berlin/Heidelberg/New York, 1968.
Wygant, K. D: "The Influence of Negative Preload and Nonsynchronous Excitation on the Performance of Tilting Pad Journal Bearings," Ph.D. Dissertation, University of Virginia, Charlottesville, Va., 2001.
Young, L. A., J. N. Burroughs, and M. B. Huebner: "Use of Metal Bellows Noncontacting Seals and Guidelines for Steam Applications," Proceedings of the 28th Turbomachinery Symposium, Turbomachinery Laboratory, Texas A&M University, College Station, Tex., 1999, pp. 21–27.

SSS Clutch Company, Inc., New Castle, Del.

Various Technical Bulletins, 1986–1994.

Kingsbury, Inc., Magnetic Bearing Systems Division, Philadelphia, Pa.

"Kingsbury Active Magnetic Bearings," Technical Bulletin, 1991.

KMC, Inc., West Greenwich, R.I.

Various Technical Bulletins, 1994.

MTI, Latham, New York, N.Y.

Publicity material on Active Magnetic Bearings, 1990–1994.

Waukesha Bearings Corporation, Waukesha, Wis.

"Active Magnetic Bearings," Technical Bulletin, 1991.

Glacier Metal Co., Ltd., London, UK, and Mystic, Conn.

"Magnetic News Bulletin," GMB 11/91.

Index

Active magnetic bearing (AMB), 75–79
  application range, 75–76
  comparison vs. hydrodynamic bearings, 77
  control current, 78–79
  design, 77, 78
  operating principles, 77
  power loss, 78
  stability, 77
Actual (approx.) steam rate (ASR), 271, 276–278, 303–313
Actuator, hydraulic, 148–151
Adjustment devices, 31
Admission:
  sections, 33–36
  valves, 144, 145
Aerodynamic characteristics, of blading, 124
Aerodynamic forces, 216
Aerodynamic seal forces, 186
Air gap, relationship in active magnetic bearings, 79
Air stripping, 136
Airfoils, 113
Alignment, of shafts, 157, 160, 162
Alignment changes, how resisted, 30
Alignment devices, 31
Alternating stresses, 117
Amplifier, 138
Antiresonance, 173
API (American Petroleum Institute):
  Spec 612, 81, 82
  Spec 671, 90
Application ranges for turbines, 1, 12, 13
Approximate steam rate:
  of multistage turbines, 309–313
  at part-load conditions, 312
Aspect ratio, 116
ASR (actual approximate steam rate), 271, 276–278, 303–313
Automatic extraction, 12, 15
Automatic extraction/admission, 16
Auxiliaries, 125–136
Axial clearance, 237
Axial displacement, of couplings, 157, 159, 162
Axial groove bearing(s), 51
Axial thrust, 232

Babbitt, 71–73
  load limit, 71
  temperature limit, 72, 73
  temperature profile, 72, 73
  temperature sensing, 74
Balance piston, 3, 222, 232
Balance tolerance, 94
Balancing, 85, 90, 92–94
  arbors, 90
  rotor, 92–94
Balancing holes, 89
Bar lift mechanism, 17, 32, 34, 145
Barring gear (turning gear), 18, 127–129
Base steam rates:
  of multistage turbines, 291–299
  of single-stage turbines, 276–278, 288, 289–291
Base-mounted oil system(s), 22, 23
Bearing characteristics, 171–174
Bearing collar, 69
Bearing set bore, 68
Bearings, 51–80
  active magnetic type, 75–79
  axial groove, 53, 56
  geometric parameters, 51
  lemon bore (elliptical), 51
  multilobe, 51
  offset split type, 51

Bearings (*Cont.*):
plain journal, 52
pressure dam, 51, 56
rocker pivot, 68
summary of types, 51
tilting-pad, 51, 56, 61
Bending oscillations, 117
Bending stresses, 117, 122, 123
Bibliography, 399–405
Blade efficiency, 124
Blade foil profile, 120
Blade natural frequency, 193
Blade stiffness, 120
Blade(s) (buckets), 18, 109–114
comparisons, impulse vs. reaction, 224–232
double-covered, 18
drawn, 109, 113
integral, 18
low-pressure, 5, 109, 111
materials, 111
milled, 109–112
packets, 112
root attachment, 111
Boiler, 10
Bore holes, in rotors, 180
Bottoming turbine, 13
Bucket lock(s), 84, 85
Bucket shroud, 85
Bucketing procedure, 84, 85, 89, 90
Buckets (blades), 18, 109–114, 348–351
double-covered, 18
drawn, 109, 113
integral, 18
low-pressure, 5, 109, 111
milled, 109–112
root attachment, 111
Building-block principle, 24–27
Built-up rotors, 81, 82, 84
Bulk oil temperature, 74

CAD/CAM (computer-aided design/manufacturing), 25
Cam drive, 32, 33
Cam lift design, 145
Cam lift valves, 16, 18, 34
Cam valves, 14, 33
Campbell diagram (interference diagram), 122, 123, 191, 192, 202, 208, 213
Campbell plane, 201, 204, 205
Carbon-ring packing, 47
Casing construction, 14, 29
steels employed, 8
Center of curvature, in preloaded bearings, 60
Centrifugal separator (Centrifuge), 134
Charge gas service, 11
Chord length, 115, 116
Chord ratio, 120
Circumferential swirl, 186
Classes of construction, 8
Clearance ratio, 69
Clearances, internal, 5, 237
Clutches, 246, 252, 260
Coefficient of friction, in couplings, 162
Cogenerated power, 2, 15, 16, 21
Combination bearing, 74
Combination rotors, 81, 83
Compliance plot, 192, 195
Computer-aided design (CAD), 25
Condenser, 48
Condensing designs, 9–13
Conductivity measurements, 240, 241
Construction, classes/categories, 8
Construction materials, 87, 88, 99, 105, 111
Contoured valve geometry, 33
Control, of steam turbines, 9–14, 18–21, 147, 148
Control air, 147, 148
Control considerations, 18–21
Control current, in active magnetic bearings, 80
Control parameters, 18–21, 147, 148
Control range, 17
Controlled steam pressure/flow, 29
Conversion charts, 392–398
Cooling steam, 321
Correcting element, 138
Coupling(s), 157–163
eddy current, 253, 255, 256
selection criteria for, 159
solid, 15
Critical speed(s), 168–170
comparisons, 223, 224
influence of couplings, 162
maps, 166
(*See also* Resonance)
Cross-coupling, 66, 170
stiffness, 66
Curtis staging, 5

Dampers, 174
Damping characteristics, 68, 72, 118, 123, 167–184, 197
Damping wire, 121–123, 228

Dead band, 141
Deflection pad™ tilt pad, 73–75
Deposits:
on blading, 235
removal, 235–241
Deposits (steam inclusions), effect on strainer design, 35
Destabilization:
due to cover and labyrinth forces, 185–187
due to shaft deflection, 182
Desuperheater, 237, 238
Diagram(s), steam balance, 9–11
Diaphragm coupling, 157–159, 161
axial forces generated, 68
Diaphragm(s), 36–45
Differential expansion, 162
Diffusion coating, 231
Direct-drive generator sets, 21
shipping limitations, 21
Disks, 82–87, 101
Disk-type rotor, 219
Dissolved salts, 119, 120
Double automatic extraction, 12, 16
Double-covered bucket design, 18
Double-seated valve, 138, 139, 144
Double-shell design, 17, 29
Dovetail cutter, 112
Dovetail design, 18
Dovetail fit, 84
Dowel, 42
Droop, 141
Drum-type rotor, 219
Dry gas seals, 48
Dual trip valve, 33, 34
Dynamic response, 198
Dynamic stiffness:
concept, 166
of oil film, 167
of supports, 167, 170, 173–184

Eccentric pin, 31
Eccentricity, in shaft systems, 53
Eddy current coupling, 253, 255, 256
Efficiency:
of Curtis stage, 313
of Rateau stage, 313
Efficiency, improved through multivalve arrangement, 8
comparisons, 223, 259, 263
of gear-driven machines, 243–247
Elastomeric coupling, 157
Electric power generation, 16
Electrohydraulic transducer, 149
Electronic governor, 148–150
Electronic valve position feedback, 2
Elliptical (lemon bore) bearing, 51
Emergency stop valve(s), 35, 127–130
Epicyclic gears, 245, 249–251
Equilibrium position, in bearings, 53
Erosion, 230–232
Ethylene service, 11
Exciting frequencies, 195–197
Exhaust size requirement:
for multistage turbines, 301–303
for single-stage turbines, 287, 288
Extraction, automatic, 12
Extraction control, 9, 150, 152–155
Extraction diagram, 150, 153
Extraction diaphragm, 17
Extraction turbine:
performance, 320–328
stage requirements, 324–328

Factor of safety, of blades, 190, 191
Failure line, on Goodman diagram, 189
Fan losses, 115
Feedback, electronic, 2
Finite element calculations, 97–100, 106, 193, 195
Flanged shaft end, 90, 91
coupling, 158
Flexible couplings, 157–163
Flexible support, 14
Flex-plate design, 7, 75, 147
Flexural rigidity, 106
Flexure pivot™ tilt pad, 74
Floating carbon rings, 47, 48
Flow distortion, 203, 227
Flow separation, 117
Flow velocity, 121
Fluctuating forces (periodic forces), 200–203, 227
Flugelized construction, 40, 89
Fluid inertia, in bearings, 172
Flyball governor, 19
Flyball weight, 137, 138
Force vector, 178–180
Forces, on rotor blading, 1, 117, 122, 123
Foundation design, 174
Fourier decomposition, 209
Fracture investigations, 99
Freestanding blades, 118

Gap configurations, 118
Gas stripping, 136

Gear coupling, 157, 159, 162, 163
Geared power, 2, 243
Geared turbines, 21, 22, 243
Geared variable-speed couplings, 257–264
Gears:
double helical, 248
single helical, 161, 248
spur gear units, 243–245
Gland condensers, 48, 133, 135
Gland seal systems, 133, 135
Goodman diagram, 189, 191
Governing valves, 33
Governor drive, 145, 146
Governor(s), 137–155
back-pressure, 20
direct-acting, 143
electronic, 19, 148–150, 154
flyball, 19
PG-type, 145
PH-type, 148
relay-type, 143
speed, 20
speed and extraction, 20
system terminology, 140–143
Grid valve(s), 18, 145
Grid-type valve, 145
Guide blades:
for impulse turbines, 5, 37
for reaction turbines, 114, 115, 116
Gyroscopic effect, 166

Hand valve(s), 6, 35, 144
Heat flow, 98, 107
Heat sink, 9
Heat stability test, 87, 88
High-speed turbines, 244
Hunting, of governors, 141
Hydraulic actuator, 148–151
Hydrodynamic brake, 260
Hydrodynamic converters, 257–264
Hydrodynamic journal bearing, 70
thrust bearing loads, 74, 75
Hydroviscous drive(s), 253–258
Hysteretic damping, 180

Impedance, 166, 173–177
Impulse blading, 3
Impulse principle, 4, 14
Impulse vs. reaction:
comparison, 219–241
efficiencies, 220
Inlet size requirement:
for multistage turbine, 300, 301
for single-stage turbines, 285, 286
Innovations, accounting for progress, 1
Inspection openings, 31
Instability, 67, 180–187
of bearings, 67
Installation, 23, 25
Integral bucket design, 18
Integral wheel, 83
Interference diagram (Campbell diagram), 191, 192
Intermediate shaft, 158
Interstage packing, 14, 41
sealing, 14, 41
Isochronous governors, 142, 143
Isochronous speed, 141
Isothermals, 98, 106

Jacking arrangements, 124–128
Jet gas expander blades, 109, 110
Jets, high-speed, 3
Journal bearing(s), 52–67
J-strips, 46–48, 116, 117, 221, 230

Keyways, 85
Kingsbury™ thrust bearing, 76, 77
KMC flexure pivot™ bearings, 75, 76

Labyrinth packing, 41, 44–46
Lacing bars (lacing wire), 121–123, 228
Lacing wire (lacing bars), 121–123, 228
L/D ratio (slenderness ratio), 50, 69, 115
Leakage minimization, 3
Leakoff, sealing steam, 48, 133
valve steam, 140
Leakoff steam, 48
Lemon bore bearing, 51
Life extension, of lube oil, 136
Lift points, 23
Lifting beam, 145
Load changes, 7
Load orientation, on bearings, 178
Locating dowel (key), 42
Locked rotor, force on blading while locked, 4
Locking buckets, 84, 85
Locking pin(s), 85
Log decrement, 187
Low-pressure staging, 120–123
Lube oil console(s), 125–128
Lube oil containment, 265, 266

Lube oil purifier(s), 134–136
Lube oil reclaimer(s), 134–136
Lube/hydraulic system:
  for turbines, 23
  for turbocouplings, 264
Lubricating oil system, 125–128

Maag synchronous clutch, 246, 252
Mach number effects, 124
Magnetic flux, 77–80
Magnetic force, 80
Magnetic pickup, 148
Magnification factor, 190
Main shutoff valve(s), 35
Materials of construction, 87, 88, 99, 105, 111
MCR Service, 11
Mechanical drive turbines, application range of, 1
Mechanical losses:
  of single-stage turbines, 279–282, 288, 289–291
Mechanical shaker (exciter), 192
Mechanical-hydraulic governor, 138, 146
Misalignment, of shafts, 67, 68, 74, 157, 160
Mixing valve, water/steam, 239
Mode curve (shape), 170, 175, 193, 194, 198–200
Mode shape (curve), 170, 175, 193, 194, 198–200
Modifications, 329–376
  auxiliary equipment, 366
  casing, 359
  flange, 360–362
  governor, 364
  nozzle ring, 362
  reason for, 331
  rotor, 364
  shaft end, 364–366
  speed range, 366
  steam path, 362
  thrust bearing, 363
Moisture drainage provisions, 31, 36, 44, 45
Mollier chart, 267–270
Mollier diagram, 270
Multistage steam turbine(s):
  approximate steam rates, 309–313
  basic efficiency, 305, 307
  correction factors, 306, 307
  dimensions, 304
  exhaust size, 301–303
  inlet nozzle size, 300, 301
  part load steam rate, 312
  selection, 291–308
  theoretical steam rate, 291–299
  weights, 304
Multivalve construction, 5, 7
Multivalve turbines, 145, 303–313

Natural frequencies of vibration, 109, 116, 117
Negative damping, 182
NEMA classifications, 143, 144
NEMA governors, 137
Nonautomatic extraction, 9
Noncondensing designs, 2, 9, 14
Nonsynchronous vibration, 180
Nozzle box, 17, 18, 235
Nozzle finish, 38
Nozzle passing frequencies, 196
Nozzle ring, 5, 35, 37
Nozzles, stationary, 5, 37, 114, 115

Oil film stiffness, 167–170, 173
Oil life extension, 136
Oil mist, 362–375
Oil pumps, 125
Oil purification, 134–136
Oil supply, 125–128
Oil systems, 22
Oil temperature, 74
Oil thickness, 70–73
Oil wedge, 68–70
Oil whirl, 55, 67, 168, 186
On-stream cleaning, 237–241
Operating experience, of different rotors, 81, 82
Operating limits, combination of, 9
Optimal pitch ratio, 115
Optimum components, 24, 25
Orbits, 179
Overload, 6
Overspeed control (overspeed trip device), 132
Overspeed trip, 137, 139

Packaged units, 21–23
Packeted blading, 112
Packing, labyrinth type, 41, 45, 46
Packing leakage, 14
Pad inertia, effect on bearing behavior, 172

Pad machined bore, 68
Part load:
  influenced by hand valves, 6, 35
  steam rate at, 312
Partial admission excitation, 196
Partial arc forces, 178–180
Pedestal, 75
Peening (riveting), 85
Periodic forces, 200–204
PG governor, 145–148, 151
PH governor, 148
Pilot valve, 138, 146
Pins, 120
Pitch, 115
Pitch diameter, and speed relationship, 82, 83
Pitch ratio, 115
Pivoting bearings, 67–68
Planetary gears, 245–251, 260
Poppet valves, 14, 17
Power cylinder, 146
Power loss, in active magnetic bearings, 80
Preload, 172
  in bearings, 53–55, 59, 60
Prepilot, 138
Pressure dam bearing, 68
Pressure-reducing valves, 21
Process steam, 9
Prohl, M. A., 165, 214, 215
Propane service, 11
Propylene service, 11
Provenzale, G. E., and Skok, M. W., 214

Reaction, reason for addition of, 5
Reaction blading, 3
Reaction turbine(s), 30
Relative velocity, 4
Relay governor, 143
Reliability assurance, 35
Relief scallop (groove), in pressure dam bearings, 68
Remote trip valve, 128–130
Rerates, 329–376
Residual unbalance, 92
Residual velocity pickup, 40
Resonance, 174, 175, 191, 197, 198
  of blades, 109, 116, 119, 121–123, 197, 198
  possible, 202, 206
  of rotors (*see* Critical speed), 170
  true, 202, 206
Response calculation, 170
Reversing blades, 5
Rigid coupling, 158, 159, 163
Ring, axial locating, 87
Riveted shrouds, 112
Riveting (peening), 85
Rocker pivot bearing, 68, 74
Roots, of blading (bucketing), 111–116
  straddle type, 120
Rotor disks:
  for impulse turbines, 82–87
  for welded reaction turbine rotors, 101
Rotor dynamics, 165–188
Running speed excitation, 195–197

SAFE diagram, 197–218
SAFE plane, 202, 204, 205
Salts, dissolved in steam, 119
  (*See also* Water washing)
Scoop tube, 263
Sealing steam, 48, 133, 135
Sealing strips, 46, 47, 116, 117, 221, 230
Self-equalizing tilt pad bearing, 74–76
Self-exciting vibration, 186
Sensing element(s), 137, 150
Separate base, 24
Sequential opening, 32
Servomotor, 33, 35, 138, 146, 150
Set point, 143
Shaft eccentricity, 53
Shaft ends, 90, 91
Shaft misalignment, 67, 68, 365, 366
Shaft orbits, 179
Shaft whirl, 168
Shaker test, 192
Shape factor, 216
Shell cracking, 30
Shielded electric arc welding, 40
Shipping limitations, of direct-drive generator sets, 21
Shock, 203
Shrink ring, 87
Shroudless construction, 90
Shroud(s), 85, 112, 114, 116, 118, 119, 220, 222, 226
Shrunk-on wheels, 83
Signature analysis, of vibration, 181–185
Single automatic extraction, 16
Single-casing turbine, 2, 29
Single-stage steam turbine(s):
  base steam rates, 276–278, 288, 289
  exhaust size requirement, 287, 288
  general specifications, 272, 273
  inlet size requirement, 285
  mechanical losses, 279–282

Single-valve designs, 5, 146, 276–302
Skid-mounting, 23
Slenderness (L/D) ratio, 69
Sliding support, of bearings, 173
Soderberg diagram, 189–191
Solar gears (sun gears), 246, 251
Solid coupling, 15, 158, 161, 163
Solid rotors:
  for impulse turbines, 81, 82, 89, 90
  for reaction turbines, 95–100
  stresses in, 95–97
Specific loading, 228
Speed compensation, 143
Speed droop, 141
Speed range, 18
Speed reference, 148
Speed regulation, 140
Speed rise, 141
Speeder spring, 148
Sphere-shaped nut, 33
Spherical bearings, 68
Spherical seats, of bearings, 173
Spoke type diaphragm, 40, 41
Spool valve design, 18
Spring-mass damper, 174
SSS clutch, 246–253
Stability, 141
Stability test, 87, 90
Stage:
  composition of, 35
  temperature limits, 40
Stage performance, 313–320, 324–328
Stage pressure increase, due to deposit formation, 235
Staging, velocity-compounded, 5
Standard components, 24–27
Standardization, 24, 25, 120
Star gears, 246, 247
Stationary blading, 5, 37, 114, 115
Stationary nozzles, 5, 37, 114, 115
Steady state stresses, in blading, 190, 191
Steam balance considerations, 8
Steam balance diagrams, 9–11
Steam chest, 33
Steam flow, through turbine stages, 6
Steam inclusions (deposits), 35
Steam jet force, 3
Steam strainer, 35
Steam swirl, 186
Steam temperature, 83, 84
Steels, used in casing construction, 8
Stiffness, of bearings, 67, 68, 79
Stimulus, 228, 229
Straddle roots, 120
Strain gauge measurement, 122
Strainer(s), 35
Stress concentration, 190, 191
Stress corrosion cracking, 120
Stresses:
  in solid rotors, 97
  in welded rotors, 106
Stripping technology (water removal from lubricants), 136
Structure concept, 25
Submerged arc welding, 38, 39
Subsynchronous vibration, 180–187
Sun gears, 246, 251
Superheat correction factors, 284, 285
Support, flexible, 14
Swirl, of steam in labyrinth area, 186
Syn Gas Service, 11
Synchronous clutches, 246
Synchrotorque drive, 254–258

Taper pins, 120
Temperature:
  of dry and saturated steam, 283, 284
  superheat correction, 284, 285, 288, 289–291
Temperature differentials, 83
Temperature limits, of stages, 40
Theoretical steam rate (TSR), 272, 274, 275, 292–299
Thermal coefficient(s) of expansion, 120
Thermal distortion, of bearings, 172
Thermal expansion, 14
Thermal stresses, transient, 14, 30, 107
Thermodynamic terms, 380–383
Third-party cogeneration, 15
Thrust bearing(s), 69–79
  materials, 71–73
Thrust collar(s), 159, 161
Thrust transmission, 159, 161, 232
Tilting-pad bearing(s), 56–67, 74, 75
  with subsynchronous vibration, 183, 184
Tip speed (velocity limitation), 83, 113
Titanium nitride, erosion protection, 231
Topping turbine, 10, 12
Torque, 162
Torque converters, 257–264
Torsional vibrations, 117
Transfer function analysis (TFA), 192, 194
Transient mechanical loads, 189
Transient thermal stresses, 14, 30
Transition zone, 116, 119, 120
Transmitting element, 138

Trip cylinder, 132
Trip device(s), 129–133
Trip oil circuit, 133
Trip-throttle valve(s), 35, 129–131
Trip valve, 33
Trunnion support, 173
TSR (theoretical steam rate), 272, 274, 275, 292–299
Turbomachinery terms, 377–383
Turning gear, 18, 127–129

Ultrasonic inspection, 90, 100
Unbalance tendency, 100
Uncontrolled extraction, 9, 29
Unit loading, of bearings, 68, 69
Units of measurement, 385–391
Upgrades, 329–376
  brush seal, 48–49, 332–336
  buckets, 348–352
  efficiency, 329, 331, 332
  electronic control, 352–356
  labyrinth, 332–336
  monitoring system, 356–358
  performance, 331, 332
  seals, 332–348
  wavy face dry seal, 336–348

Vacuum dehydrator, 135, 136
Valve, double-seated, 138, 139, 144
  governing, 144
  Venturi-type, 139
Valve opening sequence, effect of, 178
Valve position feedback, 2
Valve sequencing, 7
Valve steam leakoff, 140
Valve stem, 18
Velocity limitation (tip speed), 83
Velocity pickup, 5, 40
Velocity ratio, 7
Velocity-compounded (Curtis) staging, 5
Venturi-type valve, 139, 145
Vibration behavior, of freestanding blades, 119, 224
Vibration limits, 92, 93
Vibration resonance, 109
Vibratory stress, 189–196
Volume coefficient, 220
VORECON converter, 257, 259–264

Wake(s), 203
Wander, of governors, 141
Water contamination, of lube oil, 135, 136, 265–266
Water erosion, 230–232
Water injection, 237–241
Water removal, 31, 36, 44, 45, 231, 233
Water washing, 235–241
Water-steam mixing, 239
Weight:
  of rotors, 102
  of turbines, 247, 304
Welded construction:
  blades, 225
  diaphragms, 39
  nozzle rings, 36
  rotors, 96, 97, 100, 101–107
Wet steam, effect of, 119
Whirl, 168, 186
Wide speed range, 18
Windage shield, 44, 45
Wobble plates, 7, 74, 147

Z-lock covers, 18, 112
Z-type shroud, 112

CPSIA information can be obtained at www.ICGtesting.com
Printed in the USA
BVOW06*0041030216

435196BV00001B/1/P

9 780071 508216